Mapping Scientific Method is a superb addition to studies of the scientific method. Challenging the idea of any singular scientific method, the authors of this volume narrate the richness of disciplinary methods, and the innovations and imagination of the sciences. Taking on the 'method ladenness' of knowledge, Chadha and Thomas have assembled a path breaking volume that adds to our understanding of the Eurocentrism of science, and more importantly offering us alternate genealogies and methods from the histories and sociologies of the sciences of South Asia. Eschewing claims of an idealized and false unity of science, the authors call for a multiplicity and diversity of method. They present on-the-ground complexities of how science is done in India in a variety of the natural and social sciences, and the humanities. Deeply committed to a project of reclaiming 'science' as a critically important site for dealing with the complexities of the world, the volume reckons with science's deep and wide global roots. With our growing interest in decolonization, this anthology will prove to be an indispensable collection of how we might diversify not only our methods and methodologies, but also our history, sociology, and anthropology of the sciences.

Banu Subramaniam, University of Massachusetts, Amherst, USA.

How has the career of the 'scientific method' shaped our ways of knowing the world? This innovative and important collection suggests that decolonizing knowledge requires a head-on engagement with this question. And, that it is something more than cutting and pasting 'other' people's histories into dominant historical and cultural narratives. It necessitates nuanced and localised immersion in the history of methods across disciplines at sites beyond the Euro-American academia. This, the volume argues, carries the potential for renewing the possibilities of critical thinking itself. As contributors to the volume lucidly demonstrate, such reflections also allow for an understanding of the post-colonial condition as well as alternatives to the hegemonies of both western scientific method and its caricatures in the non-western world.

Sanjay Srivastava, University College London, UK.

MAPPING SCIENTIFIC METHOD

This volume explores how the scientific method enters and determines the dominant methodologies of various modern academic disciplines. It highlights the ways in which practitioners from different disciplinary backgrounds – the humanities, the natural sciences, and the social sciences – engage with the scientific method in their own disciplines.

The book maps the discourse (within each of the disciplines) that critiques the scientific method, from different social locations, in order to argue for more complex and nuanced approaches in methodology. It also investigates the connections between the method and the structures of power and domination which exist within these disciplines. In the process, it offers a new way of thinking about the philosophy of the scientific method.

Part of the Science and Technology Studies series, this volume is the first of its kind in the South Asian context to debate scientific methods and address questions by scholars based in the Global South. It will be useful to students and practitioners of science, humanities, social sciences, philosophy of science, and philosophy of social science. Research scholars from these disciplines, especially those engaging in interdisciplinary research, will also benefit from this volume.

Gita Chadha is a faculty member at the Department of Sociology, University of Mumbai, India. Her areas of academic interests are sociological theory, feminist epistemologies, feminist science studies, and visual cultures. Her publications include *Feminists and Science: Critiques and Perspectives in India*, Vol. 1 and Vol. 2 (2015, 2017) and *Reimaging Sociology in India: Feminist Perspectives* (2018).

Renny Thomas is Assistant Professor of Sociology and Social Anthropology at the Department of Humanities and Social Sciences, Indian Institute of Science Education and Research (IISER) Bhopal, Madhya Pradesh, India. He is the author of *Science and Religion in India: Beyond Disenchantment* (2022).

SCIENCE AND TECHNOLOGY STUDIES

Series Editor: **Sundar Sarukkai,** *former Professor of Philosophy, National Institute of Advanced Studies, Bengaluru, India, and Founder-Director, Manipal Centre for Philosophy and Humanities*

There is little doubt that science and technology are the most influential agents of global circulation of cultures. Science & Technology Studies (STS) is a well-established discipline that has for some time challenged simplistic understanding of science and technology (S&T) by drawing on perspectives from history, philosophy and sociology. However, an asymmetry between 'western' and 'eastern' cultures continues, not only in the production of new S&T but also in their analysis. At the same time, these cultures which have little contribution to the understanding of S&T are also becoming their dominant consumers. More importantly, S&T are themselves getting modified through the interaction with the historical, cultural and philosophical worldviews of the non-western cultures and this is creating new spaces for the interpretation and application of S&T. This series takes into account these perspectives and sets right this global imbalance by promoting monographs and edited volumes which analyse S&T from multicultural and comparative perspectives.

THE SOCIAL CONTEXT OF TECHNOLOGICAL EXPERIENCES
Three Studies from India
Anant Kamath

ON SCIENCE
Concepts, Cultures and Limits
Tuhina Ray and Urmie Ray

MAPPING SCIENTIFIC METHOD
Disciplinary Narrations
Edited by Gita Chadha and Renny Thomas

For more information about this series, please visit: www.routledge.com/Science-and-Technology-Studies/book-series/STS

MAPPING SCIENTIFIC METHOD

Disciplinary Narrations

Edited by
Gita Chadha and Renny Thomas

Routledge
Taylor & Francis Group
LONDON AND NEW YORK

First published 2023
by Routledge
4 Park Square, Milton Park, Abingdon, Oxon OX14 4RN

and by Routledge
605 Third Avenue, New York, NY 10158

Routledge is an imprint of the Taylor & Francis Group, an informa business

British Library Cataloguing-in-Publication Data
A catalogue record for this book is available from the British Library

Library of Congress Cataloging-in-Publication Data
A catalog record has been requested for this book

ISBN: 978-1-032-27352-5 (hbk)
ISBN: 978-1-032-28874-1 (pbk)
ISBN: 978-1-003-29890-8 (ebk)

DOI: 10.4324/9781003298908

Typeset in Sabon
by Deanta Global Publishing Services, Chennai, India

DEDICATED TO ALL THOSE WHO BELIEVE IN
THE POSSIBILITY OF LESS VIOLENT, MORE JUST
EPISTEMIC UNIVERSES

CONTENTS

BIOGRAPHICAL NOTES

Editors

Gita Chadha teaches at the Department of Sociology, University of Mumbai, India. She has developed frameworks for feminist archiving at the Research Centre for Women's Studies, SNDT Women's University, designed and taught courses in Feminist Science Studies and integrating science and social science teaching at the Tata Institute of Social Sciences, Mumbai and HEIRA, Bangalore, respectively. Her major works include two co-edited volumes on *Feminists and Science*, *Re-Imagining Sociology in India: Feminist Perspectives*, a special issue of the Contributions to Indian Sociology titled *Feminisms and Sociologies: Insertions, Intersections and Integrations*, a special issue of the Review of Women's Studies of the Economic and Political Weekly titled *Feminist Science Studies: Intersectional Narratives of Persons in Gender-Marginal Locations in Science*, and two special issues of EPW Engage called *Feminist Dilemmas: Moving beyond the List-Statement Binary* and *Gender and COVID 19: Perspectives from the Margins.*

Renny Thomas is Assistant Professor of Sociology and Social Anthropology at the Department of Humanities and Social Sciences, Indian Institute of Science Education and Research (IISER) Bhopal, Madhya Pradesh, India. Before joining IISER, he was a faculty member at the Department of Sociology, Jesus and Mary College, University of Delhi, India. He was the Charles Wallace Fellow in Social Anthropology at Queen's University Belfast, Northern Ireland, UK (2017–2018). He has published journal articles and book chapters on diverse themes including science and religion, anthropology of atheism, machines and rituals, and science and social justice. He is the author of *Science and Religion in India: Beyond Disenchantment* (London: Routledge, 2022).

CONTRIBUTORS

Yasmeen Arif teaches Sociology in the School of Humanities and Social Sciences, Shiv Nadar University, Delhi NCR, India. Her book *Life, Emergent: The Social in the Afterlives of Violence* (2016) explores a politics of life across multiple global conditions of mass violence. A new book project is tentatively titled *Life, Per se: The Government of Identity* and explores neo-identities in contemporary democracies, labour work, and race–caste–faith orientations. She currently engages with Italian thought as a medium with which to consider these anthropological and sociological queries on life. A constant parallel genre of research and writing has been a commitment to the geo-politics of knowledge production and equitable epistemologies in the contemporary. A large part of her research questions considers the urban and the city as a dominant foil. Her field experiences have been in Beirut, Lebanon, and Delhi, India.

Gita Chadha teaches at the Department of Sociology, University of Mumbai, India. She has developed frameworks for feminist archiving at the Research Centre for Women's Studies, SNDT Women's University, designed and taught courses in Feminist Science Studies and integrating science and social science teaching at the Tata Institute of Social Sciences, Mumbai and HEIRA, Bangalore, respectively. Her major works include two co-edited volumes on *Feminists and Science*, *Re-Imagining Sociology in India: Feminist Perspectives*, a special issue of the Contributions to Indian Sociology titled *Feminisms and Sociologies: Insertions, Intersections and Integrations*, a special issue of the Review of Women's Studies of the Economic and Political Weekly titled *Feminist Science Studies: Intersectional Narratives of Persons in Gender-Marginal Locations in Science*, and two special issues of EPW Engage called *Feminist Dilemmas: Moving beyond the List-Statement Binary* and *Gender and COVID 19: Perspectives from the Margins*.

Probal Dasgupta has a PhD in Linguistics from New York University. He has taught in the USA, Australia, and India. Probal retired in 2018 from the linguistic research unit of the Indian Statistical Institute, Kolkata.

He is an Honorary Member, the Linguistic Society of America, 2004, and was awarded the Vidyasagar-Dinamoyee Prize (Government of West Bengal) in 2021. Probal has over 500 publications on linguistics, literature, philosophy, and translation studies. His best-known book is *The Otherness of English: India's Auntie Tongue Syndrome* (1993).

Kamala Ganesh is a retired Professor from the Department of Sociology, University of Mumbai, India. Her main areas of academic and research interest include gender and its interface with kinship, ageing, culture and identity, Indian diaspora, and feminist methodology. Her research interests have spanned gender and kinship, archiving for women, feminist methodology, culture and identity, and Indian diaspora studies. Recently she has written on disciplinary history, especially of the Department of Sociology, Mumbai University, where she was formerly Professor and Head. Currently she is Prof. M.N. Srinivas Chair Professor at the Institute for Social and Economic Change, Bangalore. Her book *Boundary Walls: Caste and Women in a Tamil Community* won the Silver medal of the Asiatic Society of Mumbai. The book co-edited by her was *Zero Point Bombay: In and Around Horniman Circle*, which was listed by *The Guardian* in 2013 as among the 10 best books set in Mumbai.

Rahul Govind teaches at the Department of History, University of Delhi, India. He studied at St. Stephen's College, University of Delhi, and Columbia University, and has held fellowships at the Indian Institute of Advanced Study, the Centre for Study of Developing Societies, and the University of Cambridge. His research on metaphysics, jurisprudence, political economy, and imperialism has been published in *Indian Economic and Social History Review, Telos, Studies in History, Journal of the Indian Council of Philosophical Research, Jesus and Mary College Review, National Law School of India Review, Social Scientist, Law Culture and the Humanities*, and the *Economic and Political Weekly*. He has edited a volume of *Studies in the Humanities and Social Sciences* titled 'Ascertaining Certainty: Self and Cosmos' and published a book, *Infinite Double* (IIAS Shimla 2015).

Amber Habib is Professor of Mathematics at Shiv Nadar University, India. His research interests are in representation theory and mathematical finance. For about 20 years, he has been involved in efforts to make mathematics education at the school and college level more fulfilling through special topics, projects, and an appreciation of the myriad links of mathematics with other disciplines. He has contributed to curriculum development at various universities and regularly teaches in the Mathematical Training & Talent Search Programme. He has written one textbook *The Calculus of Finance* (2011) and co-authored another two:

A Bridge to Mathematics (2017) and *Exploring Mathematics through Technology* (2021).

Sasheej Hegde teaches sociology at the University of Hyderabad, India. His research and teaching have concerned a subject area intermediate between 'philosophy', social and political theory, and culture critique: the question, specifically, of the enabling histories with which one works and the conceptual basis of human inquiry and socio-political activism. His current work actively negotiates the design of inquiry across disciplinary domains, while also opening up to new formulations in the intellectual history of Indian nationalist thought. He is the author of *Recontextualizing Disciplines: Three Lectures on Method* (2014)

Chitra Kannabiran is a scientist at L.V. Prasad Eye Institute, Hyderabad, India. She has a doctorate in biochemistry from the Indian Institute of Science, Bengaluru, and post-doctoral research experience in laboratories in the USA. Her professional experience covers different areas of biology including biochemistry, molecular biology, and genetics. She has been working in human genetics for over two decades with a specific focus on understanding the genes involved in the development of several blinding eye diseases. Apart from these, she is interested in the social dimensions of science and in its history and epistemology.

Neetha N. is Professor at the Centre for Women's Development Studies (CWDS), New Delhi, India. She holds a PhD in Economics from the Centre for Development Studies, Jawaharlal Nehru University. Her research interests include labour and employment issues of women, care work, paid domestic work, gender statistics, and female labour migration. She was a member of the International Panel on Social Progress in 2018, and she has several publications in reputed journals and books.

Aditya Nigam is a Professor at the Centre for the Study of Developing Societies, Delhi. His recent work has been concerned with the decolonization of social and political theory. In particular, he is interested in theorizing politics, populism, democracy, and capitalism from the vantage point of the non-West. He also comments on politics and current affairs.

He is the author of *The Insurrection of Little Selves: The Crisis of Secular Nationalism in India* (2006), *Power and Contestation: India Since 1989* with Nivedita Menon (2007), *After Utopia: Modernity and Socialism and the Postcolony* (2010), *Desire Named Development* (2011), and *Decolonizing Theory: Thinking Across Traditions* (2020).

Saraswati Raju is a former Professor of social geography at the Jawaharlal Nehru University, New Delhi, India. She has published in the areas of gendered marginalities in labour market, access to literacy/education/skills, demographic concerns and gender and space. Her books include

Colonial and Post-colonial Geographies of India (2006), *Doing Gender, Doing Geography: Emerging Research in India* (2011), and *Gendered Geographies: Space and Place in South Asia* (2011). She is a recipient of the Janet Monk Service Award (2010) given by the Women Specialty Group of Association of American Geographers for exemplary contribution to the study of gender/feminist concerns in geography. She has also been awarded the 2012's Distinguished Service Award for Asian Geography, instituted by the Asian Geography Specialty Group, Association of American Geographers, USA.

Ram Ramaswamy is presently Visiting Professor of the Department of Chemistry at the Indian Institute of Technology, Delhi, India. He taught at the Jawaharlal Nehru University from 1986 till his retirement in 2018 in the School of Physical Sciences. Educated at Madras University, IIT Kanpur, and Princeton University, his research interests have meandered over the years, through aspects of theoretical chemical dynamics, classical and quantum chaos, statistical physics, and computational biology. Ram is an elected Fellow of the Indian Academy of Sciences, the Indian National Science Academy, and the World Academy of Sciences, and he served as President of the Indian Academy of Sciences from 2016 to 2018.

Sundar Sarukkai works primarily in the philosophy of the natural and the social sciences. He is the founder of Barefoot Philosophers and is currently a Visiting Faculty at the Centre for Society and Policy, Indian Institute of Science, Bangalore, India. He is the author of the following books: *Translating the World: Science and Language*, *Philosophy of Symmetry*, *Indian Philosophy and Philosophy of Science*, *What Is Science?*, *JRD Tata and the Ethics of Philanthropy*, *Philosophy for Children*, and two books co-authored with Gopal Guru – *The Cracked Mirror: An Indian Debate on Experience and Theory* and *Experience, Caste and the Everyday Social*. He is the Co-Chief Editor of the Springer *Handbook of Logical Thought in India*, the Series Editor for the Science and Technology Studies Series, Routledge.

Sabah Siddiqui is Assistant Professor of Psychology at Krea University and Honorary Research Fellow at the University of Manchester. She has published and presented papers on diverse topics concerning mental health, gendered violence, cultural and postcolonial studies, the sociology of religion, and the history and philosophy of psychology. Since 2010, she has investigated how medical science and traditional/alternative medicine intersect in mental health service provision in India. She has also explored the place of fiction in social science methodologies through the trope of ghost stories. Her published books include *Religion & Psychoanalysis in India: Critical Clinical Practice* (2016) and *Islamic Psychoanalysis and Psychoanalytic Islam* (Ed., 2019).

Sharmila Sreekumar teaches literature in the Department of Humanities and Social Sciences, Indian Institute of Technology, Bombay, India. Curiously, this circumstance appears to symptomatise her drive to re-inhabit the disciplinary (literary studies) within/amidst lavish, multifold urges to interdisciplinarity. Her research enquiries tend to be wayward and vagrant. Much of her published work, however, tugs at the twin impulses of gender and genre. She is the author of the book *Scripting Lives: Narratives of 'Dominant' Women in Kerala*, 2009.

K. Sridhar is a theoretical particle physicist and works at the Azim Premji University in Bangalore. Prior to this, he was Professor of theoretical physics at TIFR, Mumbai, India. He has published extensively in physics including a book, *Particle Physics of Brane Worlds and Extra Dimensions*. He has also edited a volume on Integrated Science Education, *Breaking the Silo: Integrated Science Education in India*. Sridhar is also a writer of literary fiction, has published a work of fiction called *Twice Written*, and has recently completed his second novel, provisionally entitled *Ajita*. He has interests in philosophy, especially of science, and is something of a flâneur in the visual art scene.

Renny Thomas is Assistant Professor of Sociology and Social Anthropology at the Department of Humanities and Social Sciences, Indian Institute of Science Education and Research (IISER) Bhopal, Madhya Pradesh, India. Before joining IISER, he was a faculty member at the Department of Sociology, Jesus and Mary College, University of Delhi, India. He was the Charles Wallace Fellow in Social Anthropology at Queen's University Belfast, Northern Ireland, UK (2017–2018). He has published journal articles and book chapters on diverse themes including science and religion, anthropology of atheism, machines and rituals, and science and social justice. He is the author of *Science and Religion in India: Beyond Disenchantment* (2022).

ACKNOWLEDGEMENTS

This volume was not an easy one to work on. There were moments of pause, doubts, frustration, and anxiety. A volume like ours is a collaborative and collective effort. This volume would not have been possible without the help, friendship, support, and generosity of many.

We thank Sundar Sarukkai for inviting us to edit this volume. As the series editor, he gave us all the needed support. It was his timely help and intervention at different stages that motivated us to complete the volume.

We thank all our contributors for agreeing to write for us. It was because of a shared commitment to the theme that this volume could be conceived, curated, and actualised. A special thanks to Sasheej Hegde, Sanjay Srivastava, Yasmeen Arif, and Savyasaachi for walking some of this path with us.

Rhea D'Silva was with us throughout the process. She extended meticulous and rigorous support to edit the chapters at all stages. We thank her for her efforts. We thank Arnav Sethi for writing excellent chapter summaries which we have included as aids for teaching and learning. The volume has gained enormously due to the contributions of Rhea and Arnav.

An interdisciplinary volume is difficult to review, considering that most of us are trained within the silos. We thank the anonymous reviewers for their important and insightful suggestions that have helped us craft the volume in a better fashion. Further, we thank our contributors for agreeing to do an internal peer review. Amber, Chitra, Neetha, Probal, Sabah, Saraswati, Sharmila, Sridhar, Sundar, Yasmeen, thank you for your intellectual camaraderie.

We thank Sasheej Hegde for doing the Afterword to the volume.

We would like to thank Banu Subramaniam and Sanjay Srivastava for agreeing to endorse our volume.

At Routledge, we would like to thank Shoma Choudhury, our commissioning editor for her support and immense patience, especially during the lockdown. We would also like to thank Anvitaa Bajaj for her timely help and support.

ACKNOWLEDGEMENTS

We are grateful to our students. In many ways, we are responding to students who have asked us questions of methods and disciplines over these years. We thank students from the Department of Sociology, University of Mumbai; the Department of Sociology, Jesus and Mary College, University of Delhi; and IISER Bhopal for engaging with us.

We would like to thank the Department of Sociology, University of Mumbai, and the Centre for the Study of Social Systems, Jawaharlal Nehru University (JNU), for being our academic homes at different stages of our academic lives.

Renny would like to thank his parents, Gracy and Thomas, for their constant and continuous support. Gita would like to thank K. Sridhar and Ira Chadha Sridhar for their intellectual, emotional, and practical support at every stage of producing this volume, the brainstorming over cups of tea – and over video calls – upheld a shared commitment to an academic way of life.

Gita Chadha and Renny Thomas

Introduction

METHOD-O-LOGICAL DIVERSITY: SEEKING DISCIPLINARY NARRATIONS

Gita Chadha and Renny Thomas

Neither our scientific knowledge, nor the constitution of our society, nor traditional statements about the connections between our society and our knowledge are taken for granted any longer. As we come to recognize the conventional and arti-factual status of our forms of knowing, we put ourselves in a position to realize that it is ourselves and not reality that is responsible for what we know. Knowledge, as much as the state, is the product of human actions.

Steven Shapin and Simon Schaffer (1985)

India as a culture area will be nowhere, I think, in the world of knowledge, the sciences and the arts if it does not first defy the European monopoly of the scientific method, established in modern times.

J.P.S. Uberoi (1984)

The relationship between gender and science is a pressing issue not simply because women have been historically excluded from science, but because of the deep interpenetration between our cultural construction of gender, and our naming of science. The same cultural tradition that names rational, objective, and transcendent as male, and irrational, subjective, and immanent as female, also, and simultaneously, names the scientific mind as male, and material nature as female ... Modern science is constituted around a set of exclusionary oppositions, in which that which is named feminine is excluded, and that which is excluded – be it feeling, subjectivity, or nature – is named female. Actual human beings are of course never fully bound by stereotypes, and some men and some women – and some scientists – will always go beyond them. But at the same time, stereotypes are never idle.

Evelyn Fox Keller (1992: 43)

DOI: 10.4324/9781003298908-1

Introduction: Setting the Frame

We use method to make sense of the world in our daily lives. Even if it might appear that we do not do so, refrains like 'there is a method in our madness' are commonly used to gain logical credibility for our sense-making activities. Method – and its logic – becomes a justificatory strategy for our ideas and beliefs.

We use academic methods, more specifically as ways to organise, enquire, analyse, and explain the world at different levels of abstraction and of scale. These then lead to the making of different domains and separate disciplines in academia. In modernity, what has united academic disciplines is the common assumption that they follow a common method which is distinct from what is commonly used in our everyday worlds. As in our everyday worlds, method – and its logic – becomes the justificatory strategy for knowledge claims even in academia. There are both convergence and divergence between everyday sense-making activities and academic methods.

From the early days of academic life, we hear the word 'method' everywhere – from class rooms to informal conversations with colleagues and students. In academic disciplines, the question of method is crucial in developing models of research and theories of explanation. Method becomes the definite marker of the identity of disciplines and how they distinguish themselves not only from everyday thinking but also from each other. Given the fact that most academic disciplines of today emerged in the context of Western modernity, the scientific method has become central to the disciplinary discourse on method. In fact, the scientific method of thinking and reasoning has gone beyond the academic disciplines, finding complex convergences and divergences with everyday thought on this journey. Scientific method, in many *avatars*, has become a wider tool for shaping modern societies on the path of progress. It has become a part of the larger societal imagination and has led to the insistence on 'scientific management' of resources, the need to adopt a 'scientific attitude' to life, and the building of what we in India call the 'scientific temper'.

Across the Silos

Writing about methods in multiple and diverse academic disciplines with their distinct stories of origin and distinct histories of development is a challenge. An attempt to understand what method means across the silos of the natural/physical sciences, the social/human sciences, and the humanities is ridden with difficulties of registering – and translating – conceptual languages specifically developed by and for each discipline.

Given the paradigmatic status of the scientific method in modern academics, our volume has practitioners engaged with domain-specific claims of 'becoming scientific' – of scientificity – in their origin stories, elucidating the historical necessity and context of these claims. It is from these stories that we can tell the many strengths of the scientific method as it emerged as a definite tool of rationality and of critical thinking. The essays in the volume attempt to find, and tell, the stories of method in their own disciplines, stepping in and out of the disciplinary gates.[1]

We further invite the practitioners to engage with how these claims might have been contested in their discipline domains and what have been the contours and outcomes of these contestations. Through the telling of these contestations, we hope to highlight the limitations of the method of science and how its hegemonic presence might have erased or suppressed alternative epistemic understandings, equally reasonable and critical that are available within the disciplines.

Most importantly, we invited the contributors to indicate how despite the contestations to the scientific method per se, every discipline reaffirms its claim over truth, objectivity, authenticity, and validity. We ask our contributors to tell us how disciplines at times expand the notion of what is science and at other times abandon some of that project. Our aim is to see how different ontological assumptions might shape different epistemic claims, depending on the subject matter of the research on hand in these disciplines. In a sense then, our attempt is to seek engagements with methodology. We use method to mean methodology of research and not just as a tool for research.

In times of specialisations and super-specialisations, it is rare for practitioners to dwell upon common questions, questions that are relevant within disciplinary boundaries and those that transcend disciplinary boundaries. Method is one such question. Also, the authors have different styles of writing, different ways of approaching a question, and different ways of answering the question of method. These multiple narrations are important ways of thinking and reflection. Though we have many philosophical debates on methods: what is method, and what should be the nature of method, we do not have material where practitioners of multiple disciplines ranging from mathematics to sociology are assembled together to reflect and unpack the hidden histories and politics of methods. Our assemblage of essays attempts to open the *black box* of method. We hope this assemblage will loosen the hold of a narrow scientific method in knowledge production and will make space for complex, richer, and plural methodological cultures within disciplinary silos, in the larger academic domain and eventually in the world at large. More importantly, we hope to demonstrate that the assumed unity of methods is probably just that: an assumption, and that academic methods probably need to be brought together in their diversity rather than an ideal unity.

The required disclaimer we make is: the idea of this exercise is not to take the discourse of method towards any form of anti-science positions or the polemics of the science wars (Sokal 1996a, Sokal 1996b; Chadha 1998, Nanda 1998, Chadha 1999). And our academic pursuits in science

3

criticism are definitely distinct, in intent and terrain, from the science bashing of right-wing political cultures of dominant groups.

Our Disciplinary Background(s)

As scholars in the social studies of science and technology in India, both of us have been addressing questions in science criticism. More specifically, the question of knowledge production in science and the nature of scientific knowledge (Chadha 2017a; Thomas 2018, 2022) has been an area of deep interest to both of us. As sociologists, our own discipline's origin story has fascinated us and we have also been interested in the question of critical and integrated science studies (Shah and Chadha 2015; Chadha 2017b). Fashioned around physics, sociological positivism – the dream to craft a social physics – was the earliest paradigm that set the disciplinary grammar for sociology. Interestingly, this paradigm met with its discontents and critique within the origin story. Known as the positivist–interpretivist debate between Emile Durkheim and Max Weber, it led to the articulation of the distinction between natural and social sciences –in terms of both ontological assumptions and epistemological directions. This complexity of our parent discipline is an important genealogy to trace our practice of science criticism. Though in contemporary sociology we have moved considerably away from this debate, questions of what constitutes an appropriate method for valid knowledge production necessarily refer back to the early grammar of this debate. In the first half of the 20th century, the sub-fields of sociology of science and sociology of knowledge largely privileged science and exempted it from any sociological scrutiny (Chadha 2015). It was only from the second half of the 20th century onwards that we see the emergence of critical approaches to science in sociology. These approaches were propelled by social movements like the pacifist movement that began to look closely at science and its role in the violence of war. These critical approaches opened up ways of examining the inherent violence within the scientific method. In the last quarter of the 20th century, we begin to see – again through the urgency that comes from social movements – how critical perspectives of gender and race challenge the method of science and point to the role it plays in reproducing social organisation of power, all the while being performative of objectivity, value neutrality, and openness. And yet, as sociologists we find ourselves unable to abandon the project of science. Instead we find ourselves trying to extend, expand, and reframe questions of science from within sociology, from within its origin story. As David Bloor, when asked about how sociologists can claim validity for their own disciplinary knowledge, which is also 'scientific', if they critique knowledge of the natural sciences, said,

I am more than happy to see sociology resting on the same foundations and assumptions as other sciences. This applies whatever their

4

status and origins. That the sociology of knowledge stands or falls with the other sciences seems to me both eminently desirable as a fate, and highly probable as a prediction.

(Bloor 1976: 144)

Within this framework, we curated this volume as a way to understand journeys of method in disciplines other than, and including, ours.[2]

Backdrop of the Contemporary Moment

Even as we write this book, the question of scientific method surfaces with an interesting urgency for those who aim to evaluate modernity and its institutions in a critical and reflexive manner. For those of us in the field of science criticism, it becomes an important juncture for evaluating the metanarrative(s) – produced by nation states, industry, and civic society – around science, while looking closely at the narrative(s) of/from the practice of science. We set out two contemporary moments as the backdrop for our essays on the question of, critically and reflexively, evaluating science – but primarily the kernel of scientific method – in academic disciplines.

A campaign like *The March for Science* mounted in the last decade by progressive ideologies across the world has emerged as an arduous attempt to counter the rise of cultural nationalism, global capitalism, and a right-wing political economy. Campaigns such as these take the method of science outside of the academic disciplines and make it a larger project of spreading a scientific temper and of promoting scientific rationality. This rationalist movement, like many before it, place a faith in science – its method, theories, knowledge, and institutions – to lead us out of the surrounding political darkness. Like all proponents of the rationalist movement, *The March for Science* propagates the superiority and desirable pervasiveness of scientific rationality in all aspects of life. The ideology of scientism defines such movements, and campaigns. For its proponents, *The March for Science* is not only a protest against political conservatism and religious orthodoxy[3] but is also a protest against national and global cuts on scientific funding from national budgets. While individual scientists and the scientific community at large continue to stay apolitical in India these campaigns have surprisingly seen many of them take to the streets. Campaigns such as these are undoubtedly significant for a vibrant civic society but they come with a bottleneck for science criticism. Sadly, they put forth science as the only rational and critical tool for social transformation without tapping into the potential of other ethical and moral tools available in civilisations and cultures – like religion, art, and humanism. In effect, they produce science as an isolated tool of social transformation rather than one of many in a combination of moral and critical methods available to us for resistance and transformation. More dangerously, they erase the role of science in producing and reproducing non-emancipatory ideologies of gender, race, sexuality, and caste to name a few. They demand a non-reflexive and monogamous commitment to science in all its forms. At their best, these efforts promote a rational way of thinking of traditions – that may or may not work – and at their worst they promote a cardboard version of scientific rationality amounting to a scientism that works only in an echo chamber. Further, campaigns like these polarise most academics to take sides. For scholars of science criticism in India, this becomes an intellectual *cul-de-sac* that does not allow for the critical evaluation of science and scientific knowledge. Though these campaigns bring scientists out of their insulated world, ironically they curb the critical development of science criticism in academia. And yet given the larger political field, most scholars like us have to defer our voice in the public sphere. We are left with no other choice but to resort to a 'strategic essentialism' in alliance with these campaigns, hoping that an eventual 'internal' critique will lead to dialogues in the future.

In another contemporary moment, that of COVID-19, our worlds have become engulfed in the whirlwind of a global pandemic indicating an increased interdependence and connectedness between communities and nations. While the 'master narrative' of nations and communities has been to promote modern science and modern medicine as a way out of the pandemic disease condition, the discourse has been complex. In fact, the response of the scientists and the scientific community towards their knowledge has seen some nuance (Kang 2021). First, we have seen that the nature of research in science is ridden with difficulties of method. This has been a time for scientists to recognise the contingency of their knowledge claims; they have faced the fact that their knowledge too is uncertain, contingent, and precarious. While many a scientist and science enthusiast placed their trust only in science, we have seen that even scientists have articulated the limitations of what science can do in understanding the virus, in predicting its spread, and in finding cures and prevention mechanisms. They recognise that there are no quick fixes that the scientific method can produce, it needs time and will produce results only in due course. In that interim period, a lot will happen which is beyond the control of science and scientific management, something that science criticism has pointed out in the last six decades. Second, the fact that modern science operates, often as an ally, within pressures of capitalist political economies, where not only the nation states but also the pharmaceutical companies control the course that research and medicine will or will not take, making it amply clear that science operates within the un-freedoms of modernity. This moment has once again brought home the fact that we have to put measure on what science can and cannot do (Jasanoff 2019, Horgan 2019, Maani and Galea 2021). The narrative that the method of science is both neat and free stands challenged. For us in science criticism, this has once again demonstrated what we have said all along: that science in practice is more complex and messy than the neat

6

notion of a scientific method of knowledge production presented to us in the metanarrative of modernity. In fact, COVID-19 clearly brought forth the promise and precarity of science and the scientific method in an unprecedented fashion. Most importantly, it pushed us to face the fact that science, like many other institutions, produces both risk and trust simultaneously. The loyalty to science, and modernity, therefore needs to become more and more a matter of strategic preference rather than one of moral superiority.

At this dramatic juncture of our contemporary moment, where we see both a surge in the self-assured and naïve scientistic assertion of a campaign like *The March for Science*, on the one hand, and the sheer power of scientific research, however precarious and however controlled, on the other. Against this backdrop, we attempt to explore one of the central – and perennial – questions of knowledge production in science and technology studies: what is the scientific method, how does it travel across disciplines, how does it meet its discontents, and how is it reclaimed. We mount, through each essay, bits and pieces of the answer.

Why Study Method

Questions of method are latent, relegated, and taken for granted in most academic disciplines. These questions are the *sotto voce* of disciplinary discourses, never sufficiently articulated either in pedagogy or in research. Method and the questions that surround it are often neatened and presented in a linear and non-discursive manner to the students of the discipline, to the rest of academia and to the public at large. If we cast a look back into history, we might find that every modern academic discipline contains debates on method within its historical and philosophical foundations. Often, in the inner courtyards of disciplinary practice, these debates are excavated only in times of epistemological distress caused by questions on the nature of truth, knowledge, and practice. Generally, during periods of 'normalcy', where the rush is to generate and transmit content that is useful to the worlds we live in, questions of method are mostly 'assumed' and rarely unpacked. If and when imparted to students of the discipline, these debates focus more on how to apply method rather than how to think of it. For instance, questions about what is method, is there a method at all in what we do, how is it similar or different from methods of other disciplines, what are the connections with, and departures from forms of sense-making used by people in everyday worlds are questions left to the margins of teaching and learning practices of most disciplines. One observes that method as means is imparted with a 'taken for granted' certitude that blunts a critical engagement with the question of method itself. The method question thus becomes silent in most disciplinary discourses. Debating method is made to seem, by several practitioners, like an exercise in fruitless semantics. When we look at the everyday practices of many disciplines, we see that method is not often discussed in a detailed manner during the period of training. It is assumed that method is part of every discipline. In certain disciplines such as sociology and history, studying a separate course in 'research methodology' is part of their training, but in other disciplines such as physics and literature, a course on research methodology is absent in most of the universities and colleges. If one looks at the course structure of universities and colleges in India, one can see that they start studying the fundamentals of their disciplines, such as Fundamentals of Physics, Fundamentals of Mathematics, without having any discussion on method. A student who enters a PhD program in most of the disciplines is unfamiliar with what method means in the research they do. The mantra is 'we know how to do research and we learn how to do research'. Writing about feminist science criticism and the question of how method is reified, Sandra Harding says, 'It is not just particular research methods that are the target of feminist criticism, but the fetishisation of method itself' (Harding 1989: 18).

Studying Method for Itself

This being the problem, the term method acquires several narrow registers of meaning in the general academia. It often means tools for doing research. The tools are generally finite, validated, and prescribed by what is considered acceptable in particular academic disciplines.

What is forgotten, or put aside, is the fact that the process of research necessarily entails making significant and attentive choices about method. Decisions about how to acquire material and what techniques to use for analysing it are crucial to determining the nature of research. Underlying the discussions on method are methodological propositions of knowledge-making that frame the discourse. In a sense then, the process of knowledge production is method laden. These propositions about the nature of reality and the nature of knowledge set methodological[4] paths. In a sense then, the ontological and epistemological assumptions made are the basic matter of method. This method ladenness of knowledge makes the study of method important. A study of the methods is a study of these assumptions. We submit that the study of method is integral to knowledge-making practices within disciplines. Given that disciplines are windows only, the study of method must necessarily acquaint the practitioner with methods across and beyond disciplines so that an informed and mindful disciplinary choice can be made. The need to examine method for its historical contexts, its philosophical foundations, and its formal characteristics is compelling for an intellectual understanding of the disciplinary practice. Various scholars have studied method and specifically scientific method and analysed the contexts, histories, myths, and politics of scientific method (Gower 1996, Thurs 2015, Hegde 2014, Cowles 2020). Historical scholarship on objectivity in science also demonstrates the complex pasts and negotiations of objectivity

8

and its relationship with science and method (Daston and Galison 2007). Through the essays in this volume, we hope to open doors, mark thresholds, and understand the boundaries of methodological discourses in disciplinary practices. We also hope that critical attempts at subversions and transgressions of disciplinary discourses will help us map and chart methodological trajectories as we move in and out of disciplines. We believe that studying the life of method in each discipline is in many ways studying the very history of these disciplines. It is important to note that a discussion on method was crucial in all forms of radical thinking of disciplines and disciplinary formations as we know from the philosophical discussion on methods from thinkers like Paul Feyerabend (Feyerabend 1975, 1981). Through this volume, we attempt to understand how the practitioners of disciplines deal with the method question in their everyday academic activity and in what ways they think about the past, present, and future of methods in their disciplines.

Theory and Method

Apart from the need to engage with the question of method as a question in itself, it is important to engage with method because of its obvious relationship with theory. In many content-driven disciplines in the physical or natural sciences, the distinction between theory and method often gets blurred in practice. In others like the social sciences and humanities, the distinction is marked by separating the study of method and theory.[5] In both instances, the connection between theory and method is lost. Questions of how theory is generated do not form an active part of the pedagogy of theory. In a sense then, a double injustice gains legitimacy. Not only does method get relegated into the background but theory is also truncated from its methodological foundations. Theory thus gets reduced to its substantive claims about the nature of the world that a discipline studies rather than seeing these claims as being generated and shaped by the complex mechanisms of method. Very often research scholars of social sciences come with the request for theoretical packaging for their dissertations simply because they have to 'add on' theory. 'What theory should I use' is a question that must come organically and through adequate engagement with what method is to be followed – instead it comes as an afterthought, as a necessary evil. Does theory inform method or method inform theory is a perennial epistemological question. The umbilical connection between the two is undeniable. Method is the way of doing research, a path that leads us to theory. Hence seeing method in theory becomes as important as seeing the theory itself. To develop the ability to witness the presence of method – or its absence – is to develop the ability to testify for – and choose – a particular theory. This will make us recognise that our truth claims, multiple and often contradictory to each other, are both methodologically governed and theory laden. Taking a radical position on how the incommensurability of theories is resolved, Feyerabend argues that, in this resolution there is no scientific method. He says, 'What remains after we have compared the possibility of logically comparing theories by comparing sets of deductive consequences are aesthetic judgments, judgments of taste, metaphysical prejudices, religious desires, in short, what remains are our subjective wishes' (Feyerabend, 1975: 285).

Scientific Method and the Scientification of Disciplines

Any discussion on method in modern academic disciplines has to be on the scientific method: on the emancipatory and liberation potential it contains, on how it gains a paradigmatic status within disciplines and in academia at large, and how it has become hegemonic. This paradox of the scientific method is indeed a story that needs to be told time and again.

One of the hallmarks of Western modernity and the European enlightenment was the development of the scientific method. The origin stories of many academic disciplines lie in the birth, and the dreams therein, of the scientific method. Scientific method emerged in history by severing links with magical, theological, and metaphysical methods. It promised a better, more progressive, more rational world view. It held within it the promise not only of a better method, but almost by extension, also of a better world, a better society. It also held within it the promise of criticality and democracy. It freed knowledge from the stronghold of unreason. The entire historiography of modern Western science till mid-20th century tells us that story. Our volume too might draw upon that story

The Scientific Method

If we look carefully, at what constitutes this scientific method, we realise that the textbook understanding still dominates most discourses around it. The classic understanding of scientific method equates it to the hypothetico-deductive method, a method that relies extensively on causal analysis, using deductive and inductive logic. It further relies heavily on experimental verifications. Till the 1960s, 'the dominant view of science was that scientific knowledge is a product of logical reason applied to observational and experimental data acquired through value-neutral and context-independent methods, leading to a single unified account of an objective and determined world' (Keller and Longino 1996: 1). At its simplest, it is a method of verification based on an epistemology of empirical testing of truth claims. Knowledge produced by this method lays claim to universality, standardisation, and predictability. The methodological assumptions of the scientific method assume an ontological knowable universe that can be approximated and mirrored. The standard view of scientific method is that there is a universe 'out there' that can be known through the application of the scientific method of controlled experimentation and observation that can be reined in for theory.

Due to its 'successes' in the understanding of the physical natural world, it was gradually adopted as the method for the understanding of the human natural world. Its normative status made it paradigmatic for almost all the disciplines that emerged as 'sciences' in that period. This includes the social sciences that modelled themselves on the natural and physical sciences. Its shining success brightened many domains of understanding and control, of discovery and invention, and of explanation and engineering. We hope to speak of some of that shine. This scientification of academia became part of all the disciplines and began to define the very essence of academic knowledge. Since scientification also came to be used as identity markers for disciplines, the use of scientific techniques becomes central to its prac- tice, creating its own set of problems discussed in the earlier sections. In a derivative mode, disciplines such as economics and psychology in the social sciences got taught in many places as 'sciences'.

Simultaneously, through the imperialist and colonial project, science and the scientific method spread across the world. Science was an important instrument, less visible than others, in casting the world in the mould of Western modernity. Shiv Visvanathan divides the development of modern science and technology in India into three broad phases: the phases of initia- tion, education, and institution-building (Visvanathan 1985: 8–14). Like everything else in colonialism, it was both liberating and violent. The scien- tific method gradually gained a hegemonic status for producing, validating, and legitimising all knowledge across nations and civilisations, freeing us from the 'old'. Through it all, in several new nations, like India, science became an important tool of a 'renaissance', accepted by the colonised. It gave new aspirational and critical tools to imagine the colonial national modern in opposition to the traditional. In fact, social reformers like Raja Ram Mohan Roy who said

> The Sangscrit system of education would be the best calculated to keep this country in darkness, if such had been the policy of the British legislature. But as the improvement of the native popula- tion is the object of the Government, it will consequently promote a more liberal and enlightened system of instruction; embracing mathematics, natural philosophy, chemistry, anatomy, and other useful sciences.
>
> (Baber 1998: 197)

The hegemony of modern Western science was complete because the dominated accepted the virtues of the dominating. It became impossible for any form of knowledge that did not follow the principles of scientific method to survive in the larger world of knowledge production. In fact,

11

the enlightenment modern planted its roots across the world through the complex mechanisms of orientalism. As the historian of science Daniel P. Thurs argued,

> If we return to a simplistic view, one in which the scientific method really is a recipe for producing scientific knowledge, we lose sight of a huge swath of history and the development of a pivotal touch-stone on cultural maps. We deprive ourselves of a richer perspective in favour of one both narrow and contrary to the way things actually are.
>
> (Thurs 2015: 218)

Historian of ideas Jason A. Josephson-Storm in his path-breaking work *The Myth of Disenchantment: Magic, Modernity, and the Birth of Human Sciences* (2017) demonstrates that the world of science, human sciences, and scientific method had always been a messy realm. He shows that, paradoxically, even the most important practitioners of scientific method from sociologist Max Weber to physicist and chemist Marie Curie did not entirely inhabit the world of scientific method. He shows that the grand narrative of disenchantment – the idea that scientific rationality will uncover the mysteries of the world replacing magic and religion and art had always been a myth.[6] This clearly shows that it is difficult to articulate a neat success story of scientific method even in the *most scientific* of the scientific disciplines. There had always been questions, scepticisms, and challenges. Contributors in this volume write about many such questions, scepticisms, and challenges their respective disciplines experienced and explore the possibilities of a plural way of doing disciplines without thinking about the burden of following *the* scientific method. We believe that such an exercise will make the disciplines more accommodative, democratic, and intellectually diverse.

Scientification

The overreliance on science and scientific method has concerned scholars in the social studies of science and technology for the last six decades. This has been debated by philosophers, historians, and sociologists of science. To some extent, it has also been a matter of debate among scientists. In the name of scientific method and scientificity, the ideology of scientism also gets validation and acceptance in many disciplines. This happens when we proffer science as a totalising system that has the superpower to transform all ignorance, all evil, and all regression. In this volume, we argue that scientism as an ideology has to be necessarily challenged when we think of method in disciplines. Thinking critically about method is to also challenge the hidden scientism in modern disciplines. The idea of scientism as Tzvetan Todorov argued is more dangerous, 'For people are not usually proud of being ethnocentric, whereas one can take pride in professing a "scientific" philosophy' (Todorov 1993: 12). This form of scientism still guides our thinking and debates on methods in many disciplines.[7]

We are interested in this volume to talk about the connection between scientific method and power. Historically as well, the demarcation of natural philosophy and theology was made using 'scientific' method. Over a period of time, disciplines have undergone many transformations, and practitioners have asked important questions about the scientific nature of their own disciplines. Social Anthropology for example has started rethinking about the way they write and think about ethnography as a method.[8] Scholars now even argue for using prose and poem as narrative forms, especially through Literary Anthropology, an emerging field that challenges the mainstream method in the discipline.[9] Anthropology as a discipline has become more nuanced, accommodative, and pluralist by re-examining and rethinking its own scientificity and method. This volume examines the ways in which different disciplinary practitioners engage with scientific method in their own disciplines, and they show how productive it is to ask questions of method in their disciplines. Asking questions about method in a discipline is also to ask questions about the structures of power and domination in disciplines.

We discuss in detail through various disciplines how modern sciences and knowledge became 'valid', 'useful', and 'acceptable' because of its close link with power. In that process, we also discuss how various forms of knowledge practices became 'invalid', 'useless', and 'unacceptable', 'unscientific' because they lacked the 'scientific method'. This volume is an enquiry into the possibilities of thinking about the past, present, and future of scientific method in disciplines, ranging from physics to sociology.

Scientification is part of all the disciplines and every discipline will have questions asked about its scientific nature and scientificity. Since scientification is used as an identity maker, techniques became an important category in defining the very nature of disciplines.[10] Techniques then define the validity of a discipline, and therefore all disciplines have a compulsory paper on techniques and research methods. It is an interesting space to think about the politics of knowledge production. It is through techniques and research methods that we classify what is scientific and non-scientific in a discipline. What is scientific has to be objective, and things that are not objective are non-scientific. Both science and technology studies (STS) and feminist scholarship inform us how the powerful defined their identity as objective to erase the experiences of outsiders as subjective and unscientific (see Haraway 1988, Subramaniam 2014, 2016).

Scientification is dangerous especially when we study the worlds of people; the so-called natives, where anthropologists, for instance, have the power to write about the community that they are studying or live with, because they use scientific method. This creates the difference between anthropologists, who use scientific methods, and the people whom they study and write about. How do we deal with the power relation that exists between anthropologists/practitioners of scientific method, and the people/subjects of scientific method? Can we think of an alternative or better way to deal with *the people*? Can we think through conversations and experiences, and not treating the people that we study as mere 'data'?[11] The very idea of treating human beings as 'data' comes from the principles of scientific method, and we need to rethink about the very usage 'data' in all disciplines. One of the major concerns in the volume is to look for an alternative view of methods in different disciplines and see if a rethinking of methods leads to a new understanding of disciplines. And the chapters show that there are possibilities of addressing scientification and scientism in disciplines without necessarily being anti-methods.

We also ask this question to rethink the dominance of one form of methods that almost exist as unquestionable in various disciplines, for instance, disciplines such as economics and psychology. They are taught in many places as 'sciences', and the practitioners of these disciplines do take pride in being the most 'scientific' among social sciences. They also don't mind being labelled as 'least scientific' among the 'pure' sciences. Many universities therefore offer courses such as Master of Science (MSc) in Economics and MSc in Psychology as it gives more credibility to the discipline and it attracts students. They will also be happy to say that they are doing a degree in Science. Here, we need to think about the authority and power of science, and that is precisely why disciplines such as economics and psychology try to get a 'science tag' to be accepted and to be respected. For example, the Indian Institute of Science Education and Research (IISER) Bhopal, India, offers a BS programme in Economic Sciences. The usage *Economic Sciences* is not free from the power of science and scientific authority. What gave the Western modern science authority and power was the scientific method, and we can see how various disciplines try to embrace scientific method to be part of the world of power and authority. The chapters in the volume ask questions of power and authority in academic disciplines and attempt to show how that power came into being in the disciplines and how that power continues to shape the identity of many of the academic disciplines.

What does scientification do to these disciplines? What are the challenges therefore practitioners have if they want to come out of the *many scientism* that exist in their respective disciplines? We enquire through various disciplines how there can be a possible way of talking about it and what alternatives can they suggest? In what ways can we differentiate between methods and various forms of scientism that exist in all disciplines? Can we think of a different way of doing physics, chemistry, biology, psychology, history, economics, and linguistics without being pressurised to follow *a/ the* scientific method? By asking questions about the possibilities of plural epistemologies, we also enquire into the various histories through which the process of *scientification* emerged in many disciplines. We also enquire into various changes that took place in these disciplines to deal with the question of method. We enquire into the various forms of resistance that many of these disciplines witnessed, and the ways in which these forms of resistance changed the identity of these disciplines. The authors have addressed these questions in engaging ways in their chapters.

This volume asks if we can think of disciplines beyond the fixed nature of *the* scientific method, and we believe that such rethinking of disciplines will invite more debates in class rooms when we teach courses on methods. This volume is an invitation for more such future work, as we believe that these reflections play a major role in reshaping and redesigning our disciplines. In this volume, distinguished scholars and practitioners of diverse disciplines work in Indian universities and institutions ask this question of scientific method also as a way of democratisation of scholarship and as a part of decolonising scholarship on methods. This volume is one of its first kind emerging from south Asian academia. We invite the readers, be they students, practitioners, or anyone who is interested in the question of method in disciplines to read the chapters of their choice and come with more questions. We hope that this volume will help all of us to engage with the question of method critically and will help us understand its varied histories, practices, contexts, and politics.

Conclusion: Towards Science Criticism

The role of science and technology in producing risk societies has been marked in the mid-20th century. According to J.P.S. Uberoi 'The ruling scientific theories of nature are even more dangerous than the ruling western theories of man' (Uberoi, 1978: 14). Science criticism as it developed through social and people's movements challenged the violence inflicted by science on our bodies and worlds. As a consequence, social scientists began looking at science critically. Naturally, scientific method too came under scrutiny. Shiv Visvanathan, for example, in his deeply incisive account of the vivisectionist nature of science, discusses the works of scientists Claude Bernard and François Magendie. According to Visvanathan:

> Bernard's work reflected the intrinsic violence of science as vivisection. Vivisection is the infliction of pain for experimental purposes of understanding and control, where pain and suffering are justified in the pursuit of scientific knowledge as an absolute value. François Magendie 'sacrificed' 4,000 dogs in making a distinction between sensory and motor nerves. Some of the early vivisectors might have been sadists, but Bernard exemplifies the schizophrenic attitude of 'normal science' to vivisectionist violence. Bernard remarked that 'the physiologist is not an ordinary man: he is the scientist possessed and absorbed by the scientific idea he pursues. He does not

15

hear the cry of animals, he does not see the flowing of blood; he sees nothing but the idea and is aware of nothing but the organism that conceals from him the problem he is seeking to resolve'.

(Visvanathan 1988: 263)

The decades from the early 1960s to the late 1990s saw rich intellectual debates that challenged the epistemic and ethical supremacy of science, and scientific method. Further, praxis-based epistemologies driven by the need for social transformation from the perspectives and standpoints of people from marginal social locations have driven the academia to re-examine its unchallenged dedication to the scientific method as the paradigm of progress and truth set out by modern Western science. Be it through the environmental movements or the people's movement, the relationship between scientific method and structural power of dominant groups and the nation states gets firmly established. When translated into disciplinary registers, these movements reveal the situated nature of scientific knowledge and method. Feminist epistemology, especially feminist science studies for instance, has demonstrated the way scientific method was used to support patriarchal structures of modern sciences. They pointed out the need to look at science, and gender, using critical perspectives that science had forgotten. As Sandra Harding says,

Many of the most powerful examples of feminist research direct us to gaze critically at all gender, to take women's experiences as an important new generator of scientific problematics and evidence, and to swing around the powerful lenses of scientific inquiry so that they enable us to peer at our own complex subjectivities as well as at what we observe.

(Harding 1989: 26)

Further, we see that critical race studies, postcolonial critiques, the pacifist, and the environmental movements all indict modern Western science – and its method – for reproducing epistemic injustice and violence to epistemes of marginal people and cultures (Thomas 2020, 2022). For instance, critics show how in the discourse around modern biomedicine, systems of health and medicine like Ayurveda, Siddha, tribal medicine, Chinese traditional medicine, and Unani medicine are 'othered' as 'unscientific'. In times of late modernity, the promise of science has turned around, and on its head, in multiple ways. From being the institution that would lead us out of hegemonic structures, it has become a part of these structures. What went wrong? Or was it always wrong? Or is nothing really wrong? And wrong for what? For whom? Questions around the twofold problems we have identified – the scientification of the disciplines in the academia and the scientism prevalent in sections of our society – are important to ask. Who will ask these questions? Are people in the academic silo of the natural sciences equipped to do this? Can they be trained to do this? While we might think that they are the best constituency to do this, we also know that an 'outsider' view helps in critique. Who could be the best outsiders? Shouldn't those in the other silos and those in the humanities and the social sciences qualify? If yes, should they not start with their own disciplines because these are also 'scientificated'? In many of these, albeit the 'soft' ones like sociology, efforts towards asking these questions already exist. In others, the 'hard' ones like economics need to develop more rigorously. It is important for us, as academics, to begin reflecting on the need to strengthen the field of science criticism in academia. This field, as yet, has no home and must find one. Unlike literary criticism, which is now housed and practised in departments of literature, science criticism is developing on the fringes of social sciences.

Like literary criticism, science criticism attempts to contextualise its 'text'. Both are academic exercises and might or might not impact the world in which science and literature are consumed. Yet, the importance of literary or science criticism to the development of literature or science, respectively, cannot be minimised. We envisage the future of science criticism to proceed on similar lines, though we sense the many challenges too. However, as Jonas Salk wrote in his introduction to Latour and Woolgar's *Laboratory Life* (1979), 'Scientists often have an aversion to what non-scientists say about science. Scientific criticism by non-scientists is not practiced in the same way as literary criticism by those who are not novelists or poets' (Salk 1979: 11).

Scientific knowledge is constructed, by its practitioners and enthusiasts, as being produced by a supposedly unbiased method. It is a text perceived as being 'above context'. Hence, any attempt in science criticism to place science within its social-historical and political context is mostly seen as unnecessary work done by those who do not know or understand science. Much like C.P. Snow indicated in his classic text, *The Two Cultures* (Snow 1959), attitudes from the scientific communities are hostile and generally go like this:

> These radical critics of science seem to be having little or no effect on the scientists themselves. I do not know of any working scientist who takes them seriously. The danger they present to science comes from their possible influence on those who have not shared in the work of science but on whom we depend, especially on those in charge of funding science, and on new generations of potential scientists.
>
> (Weinberg, 1993, p. 151)

The adage 'those who do not do science, do science criticism' is popular in cultures of science, very similar to what we see in literary cultures: 'those who cannot write, critique'. It is therefore necessary and recommended that informed science criticism, as an academic field, be developed within the academia and housed within universities and research institutes. However, the unfortunate science wars that happen, almost like a backlash, with an interesting regularity push these debates into problematic boxes of pro- and anti-science positions. These attitudes in the academic domains foreclose the development of science criticism, particularly if and when there is no academic anchor for these. These attitudes of suspicion and a naturalised arrogance of scientific communities in academia resist the opening of the natural sciences, on the lines of the opening of the social sciences that began in the middle of the 20th century.[12] Problems that riddle the social study of science, according to Barry Barnes, is that science 'is hopelessly conflated with the ideas of what it ought to be, or must be' rather than what it is. He hopes that science criticism, as an academic field, will continue its 'attempts to take science as it finds it' (Barnes 1974: ix) and would not be deterred by the fact that 'in the current intellectual milieu where simply to talk of science in other than reverential terms may be seen as criticism' (Barnes 1974: ix). According to Sandra Harding, science alone is the kind of activity which, unfortunately, demands that it

> must be understood only in terms of its enthusiasts' understanding of its own activities – in terms of the unselfconscious, uncritical interpretations 'the natives' provide of their beliefs and activities. That is, scientists report their activities, and philosophers and historians of science interpret these reports so that we can 'rationally' account for the growth of scientific knowledge in the very same moral, political, and epistemological terms scientists use to explain their activities to funding sources or science critics.
>
> (Harding 1986: 35).

These attitudes foreclose the development of the field of science criticism.

An important dimension of the hostility to the development of science criticism is with reference to language. It is quite obvious that the split between the world of science, on the one hand, and the humanities and the arts, on the other, is largely due to the fact that these two worlds can hardly comprehend each other's languages and that there is a crying need to develop a common vocabulary where there can be an exchange of ideas, at least at a non-technical level. There is a widespread attitude within the scientific community that the language and concepts of science (couched as they often are in mathematics) are beyond the understanding of those in the humanities. Scientists also have to debunk the use of technical language and concepts in the humanities as being a humbug. This is part of the process by which science tries to project its own method as being supreme. Nowhere is the above attitude of scientists better illustrated than in the example discussed below. This is taken from a book called *Conceptual Foundations of Quantum Field Theory* (Cao 1999) which is an edited volume of the proceedings of a conference that was organised at Boston University in 1996 by a group of philosophers working in the area of philosophy of science and was aimed at providing, as the Chair of the conference Tian Yu Cao put it, 'a forum for the desirable, mutually beneficial but difficult exchange of views and ideas between physicists and mathematicians, on the one side, and philosophers and historians, on the other'. At the conference, Cao himself gave a talk called 'Renormalization Group: A Puzzling Idea' (Cao 1999: 268–286). As a philosopher of science, Cao discussed a technique in theoretical physics called renormalisation group and attempted to discuss some of the conceptual difficulties with this particular theoretical idea. The Chairperson for this session was a well-known theoretical physicist from the University of Maryland, Michael Fisher. When Cao finished his talk after several interruptions and vehement protests from Fisher, an interesting discussion followed. Fisher invited another theoretical physicist, Francis Low, to join the discussion and the two together proceeded to shred Cao's arguments to pieces. So far the discussion concentrated on Cao's misunderstanding of physics and how he was using concepts of physics like 'renormalisation group' wrongly. Interestingly, Fisher and Low objected to the use of words like 'ontology' and 'instrumentalism' by the philosophers and challenged Cao to define them (Cao 1999: 284). Cao offered definitions of these terms, but the physicists were unwilling to listen and became aggressive and impatient with the speaker. At some point in the discussion another physicist, Carlo Rovelli, interjected, saying

> I was invited to this conference by philosophers and understood this invitation as a request for dialogue. I may be wrong, but what I have heard mainly from the physicists' side in this conference is an attitude of lecturing: 'you do not understand this or we'll explain you that'. Several requests of discussions have been dismissed with an attitude that sounded to me almost closed-minded arrogance ... Today some discussion has emerged after Cao's lecture. But an immediate reaction against terms such as 'instrumentalism' and 'ontology' has followed. I think that in order to have a dialogue, we have to simply learn the meaning of words used by the other side. Do my fellow physicists think that only their difficult words have meaning?
>
> (Cao 1999: 285–286)

To this Fisher retorted, 'These are terms we don't normally use and so we find it not useful'. Rovelli's reply was,

> Exactly. There may be terms which we don't normally use but they may still be useful for communicating with people outside our community. There are two different languages here, and we need an

19

effort to understand the language and subtleties of the other camp. Philosophy is subtle and if we do not understand it is not necessarily because they are confused. We need an attitude of reciprocal respect ...

at which point he was abruptly dismissed by Fisher who said 'No! You cannot ask respect for people who are not prepared to define their terms! Any physics term that comes up here, there's not a physicist who would not attempt to define it, and philosophers have the same responsibility' (Cao 1999: 285–286). Most attempts at dialogue, which we think must be the toolbox for science criticism, meet this fate. It is this that has to change. Our attempt in this volume is a step in that direction.

ADDENDUM: CHAPTER SUMMARIES

The essays in this volume are difficult to introduce. They do not sit together with ease, and yet they speak to each other. It is important that our readers pick and choose what they would like to read and use. Since we are especially aiming at pedagogues aiming to teach method in interdisciplinary spaces, we thought it best to provide detailed chapter summaries. We present these as part of our Introduction. These summaries are written by Arnav Sethi, a doctoral candidate at the Department of Sociology, Delhi School of Economics, University of Delhi.

Probal Dasgupta brings out the interminable contestations within linguistics – a discipline that strikes the general public as largely opaque and abstract. Dasgupta relates this unusual opacity to the fact that linguistics is yet not accepted as a science with incontestable authority in the domain of language. He argues for the relevance and significance of substantivism – a school of thought within linguistics with which he identifies and continues to be associated – that treats the unfinished journey of linguistics towards social acceptance as a core concern that the discipline ought to address. As a crucial point of departure, Dasgupta takes up the common sense underlying scientific discourse that classified happenings into actions and mere events. In this scheme, science studies the patterns of necessity and chance; not actions, reflecting free will that agents have chosen to perform. Concrete choices that fall outside science count as discourse and studies of discourse cannot be systematised into a science. In other words, patterns of 'mere' events that seemingly nobody is responsible for fall under the purview of science. But Dasgupta asks, where does such an explanation leave language? He clarifies that saying anything

– whether spoken or written – amounts to a freely chosen action and so is technically outside the purview of science. On the other hand, language is a super-pattern of event-patterns, within which speakers and writers make their discursive choices. With this discussion of the watertight binary between language and discourse, Dasgupta introduces the reader to the subject matter of linguistics – a science dedicated to languages that flourished since the 18th century as a systematic and global enterprise. Throughout the chapter, the author invites the reader to consider nuanced conceptual formulations. An important one is the joint enterprise of Traditional Lexicography-and- Grammar (TLG) that codifies written language standards and its notion of language limits possibilities to the validly writeable, as acceptable in the normative codification. In the latter part of the essay, Dasgupta introduces the reader to substantivism, wherein morphology is separated from syntax and other components. Substantivists refrain from chopping up words into morphemes, as they recognise not only the connections and disconnections between these sensitive phenomena but also that word-parts may encode rudimentary semi-meanings. With the examples of Hindi, Indonesian, and Swahili, Dasgupta calls for a focus upon the postcolonial moment wherein linguists are forced to take all languages and consider contemporary problems, one of which is the spectrum of noun categorisation in these languages. He argues that substantivism is better equipped to deal with noun categorisation as it leaves words intact. Dasgupta argues that linguistics pursues a 'precarious exactitude' and once this begins to register, the discipline's self-understanding will change as well. Progress of the discipline in this way is a prerequisite, not only for more specific realignments of North–South equations but also for the health of democratic society more generally.

Sharmila Sreekumar takes the watertight separation of literature from literary studies as a fundamental entry point into the larger argument. While the former is frequently considered a creative pursuit and/or an art, the latter, though not precisely a science, is seen to be associated with knowledge or learning. Sreekumar elucidates the provocative relationships that literary studies establishes with science – as it tries to consolidate itself, while seeking methodological and definitional possibilities. Hers is not a prescriptive rendering of what such a 'legitimate' invocation of science and scientific method should look like. Instead, she delineates the historical impulses that have characterised literary studies since its inception. Sreekumar's anecdotal writing style allows a brief description of certain events that culminated in partitioning intellectual life into 'arts' and 'sciences'. These watershed

events marked the formalisation of the academic split between literature and 'scientific' literary criticism. A group of young scholars – the
New Critics –saw themselves as pioneering the latter. They sought to
make criticism more scientific, precise, and systematic. New Criticism
constituted its object of enquiry differently and vociferously claimed
that literary work was an autonomous object, disengaged from the
writer, reader, and other socio-biographical histories. In these ways,
the New Critics invested their intellectual energies solely upon compositional specificities and the facticity of words in the literary work, and
empiricism undergirded this newfound enterprise. The literary work
was thus seen as amenable to positivist observation and methodical
study. Sreekumar argues that the new critical methods systematised by
New Criticism drew their authority from scientism and were crucial
for consolidating and legitimising the claims and knowledge-regimes
of literary studies. An important method was the practice of 'close
reading' that put the literary work to scrutiny for microscopic observation. This 'specialised seeing' was reinforced by New Criticism's
stance of dispassionate objectivity, or disinterestedness about ulterior,
political, and practical considerations. As Sreekumar puts it, the language of New Criticism was transparent and planar, without unseen
depths and poetic intensities. Sreekumar considers Indian responses to
contemporary debates in literary studies, which are constitutive and
symptomatic of the larger crises in the discipline. With this enrich-
ing discussion of the crisis-laden nature of literary studies since the
1930s, Sreekumar concludes that neither does the discipline possess a
consensual paradigm that unites the disciplinary community; nor does
it strive for triumphal journeys from error to truth. Instead, any proposed 'findings' are interpretations and arguments that become preludes for new objects of scrutiny. Such a nuanced perspective allows
Sreekumar to point towards recent shifts that spark refurnished provocations with science.

Sundar Sarukkai tells us that the idea of method has a long history that can be traced back to the Indian and Greek philosophical
traditions. He characterises method as a series of steps and/or a set of
rules to be followed for specific reasons and for reaching some desired
goal. But the notion of method itself is an object of philosophical
enquiry and hence, philosophy's enquiry on method is a self-reflexive
one. With this significant caveat, Sarukkai invites the reader to his
own approach that focuses on the idea of method and offers a more
inclusive understanding of philosophy by drawing on examples from
Indian philosophy. He contends that while there is no paucity of literature on philosophical traditions in Asia and Africa, any discussion on

method gets restricted to the small group of 'Western' philosophers. The author delineates the important theme of the distinction between appearance and reality that leads to methodological suspicion about what is perceived. Sarukkai says that philosophy deals with this in a special way, since metaphysics as a discipline offers answers and methods to address the problem of the nature of the real. He also elucidates that debate is an important methodological principle for philosophy. The detailed classifications of the various types of debates, fallacies, and so on are common methodological principles in Indian and Greek philosophies. Sarukkai pertinently argues that it is important to recognise the commentarial tradition as a method in philosophy, even though it is not explicitly discussed as one. Sarukkai argues that methods are necessary for any meaningful knowledge-acquisition activity to adequately deal with the presence of the human subject or knower. Different methods differently take into account the role of the human subject in any description of the external world, the perception of which may range from being completely independent of human cognition to being completely a product of human consciousness. In this context, Sarukkai's succinct discussion of the implications of the 19th-century project that proposed to model philosophy on science – called 'scientific philosophy' – and the subsequent development of phenomenology is intriguing. Sarukkai argues that while these methods are seemingly polar opposite, they are in fact paradoxically performing similar philosophical tasks. While science has to find methods to systematically erase the presence of the human in any description of the external world, phenomenology, on the other hand, has to find methods to bring this external world into subjective human experiences. For Sarukkai, one of the most profound breaks between philosophy and science occurs in the removal of the human subject's presence from processes of scientific knowledge production. There is nothing in the activity of contemporary science that is concerned with righteous and good living. These practices of everyday life were very much part of knowledge production enterprises before the radical break with modern science but are now relegated to the world of religious institutions. Sarukkai concludes with the qualitative distinction between 'doing' philosophy and using philosophy that manifests itself as a difference between understanding methods in philosophy, as against looking at philosophy itself as a method.

Amber Habib describes the hierarchical structure of mathematical knowledge and comments upon its reliability and validity. He invites the reader to a consideration of mathematics' fundamental components: axioms, proofs, and theorems. Habib asserts that all mathematicians work with a common choice of axioms and rules of logic, and the degree of agreement amongst them is not found in any other discipline. He scrutinises the common view of mathematical activity and knowledge-making as offering certainty and increased rigour. The chapter is then centred around three fundamental issues: problems of ahistoricity, the subject matter proper of mathematics, and its relevance or importance. Habib discusses Indian mathematics as a significant tradition that has not subscribed to the common view that the essence of mathematics is a logical deduction from axioms. There was no search for universal axioms in the Indian tradition and it still produced a rich body of mathematics over 2 millennia covering number theory, algebra, geometry, and trigonometry, as well as the initial steps of infinite series and calculus. Habib focuses on the axiomatic approach that is considered to represent the final or ideal state of modern mathematics and traces its roots in the works of the Ancient Greeks. There were problems in using numbers to measure geometric quantities like length and area. Habib notes the 'demotion' of geometric axioms from self-evident truths to convenient principles to be adopted as per mathematicians' needs that had a liberating effect on the discipline: mathematicians sought basic principles within purely mathematical concepts and viewed the discipline as primarily concerned with relations rather than objects. As axioms were no longer seen as self-evident truths, the certainty of mathematics became questionable. A new intuitionist approach, according to which, existence and presence of a mathematical object required actual construction and demonstration gained ground. While most continued to be convinced of the discipline's solid ground, there were considerable moves away from attempts to show that everything could be proved towards the pragmatic task of actually proving theorems. Habib also argues that proof depends on social context: the mere correctness of the proof by itself is not a sufficient condition to make it valuable to the community of mathematicians. Mathematicians also attribute value to human understandings associated with the proof. For this reason, the issue of computer-generated proofs is contentious. Habib also brings out the paradox that while mathematics has assumed a central position in the acquisition of knowledge, it has withdrawn inwards and away from direct contact with its applications. Habib shows us that the general perception of mathematics as sterile manipulation of symbols may have its roots in school and university education. He concludes with a call for balance between deductive logic and full-blown axiomatism, while matching the needs of school mathematics to the discipline – so that its horizons are broadened.

Chitra Kannabiran's chapter focuses on the historical shifts in employing not one but many scientific 'methods' in the field of genetics. Kannabiran argues that the development of biology, within which genetics is subsumed, has been historically shaped by both reductionist and anti-reductionist perspectives. She argues that the idea of the scientific method as the logical process of arriving at connections between theory and data doesn't always bear testimony to actual practice. She discusses both inductivist and deductivist approaches, while highlighting examples of each, and also points out that the distinction between the two may often be blurred. With the example of Darwin's work, she posits that no data *is* neutral: it is always an entity produced by a scientist. Data *is* made, not given, and the attribution of meanings depends upon observers' interests and frames of reference. An important contribution of the chapter is its close scrutiny of contemporary attempts to create large databases in human genetics, as exemplified by the Human Genome Project (HGP), founded on the assumption that the human genome is key to health and disease. Kannabiran articulates her position against the genetic determinism and other underlying assumptions of the HGP. She points out that genome sequencing didn't provide practical solutions to tackle common diseases that are major public health concerns, since these are caused by several factors, including several genes and environmental influences. In this way, Kannabiran argues that the HGP is reductionist insofar as it reduces human health and disease, with all its socio-economic complexities, to a set of genes so that the properties of an organism are supposedly contingent predominantly upon its genes. But Kannabiran concedes that the HGP did provide an impetus for creating new methods and technologies in the field of genetics that are capable of acquiring and analysing big data. And it is this transition to big data that has established a new 'paradigm' in science. By employing the prisms of gender and race, Kannabiran puts to scrutiny the supposed claim of 'value neutrality' of science. She concludes the essay with a call to transcend mono-causality and genetic determinism, so that the complex etiologies of most diseases may be recognised and appreciated.

K. Sridhar's paper argues that critically scrutinising the question of method in physics is important and relevant not only for the discipline but for all the sciences. Sridhar elucidates the several revolutionary 'paradigm shifts' in modern science, beginning with the pioneering work of Newton that displaced several long-standing ideas of the Aristotelian system, thus establishing the world view of modern science. Sridhar argues that in many ways, Newton's 'new physics' actually brought questions of method in science to the fore. As science

25

began to constitute its subject matter, it took up several questions that had hitherto constituted the domain of theology and philosophy. According to Sridhar, theoretical research in physics is unique because there is a very clear separation of experimentation and theory, owing to physics' amenability to mathematical reasoning. He succinctly describes the guiding principles that constitute the skeletal framework of classical physics and form the basis of any scientific theory. With the paradigm shift from the Newtonian worldview towards the newer theoretical frameworks of relativity and quantum mechanics, certain fundamental challenges were posed to the classical view. In fact, as Sridhar puts it, quantum mechanics challenges every guiding principle of the Newtonian system, since the separation of the physical system from its environment/observer is no longer valid. Invoking the correspondence principle, he argues that the theory that is correct over a larger range of phenomenal experience (in this case relativity and/ or quantum mechanics) reduces in some situations to match with the theory that works in more limited domains (in this case Newton's mechanics). Sridhar argues that fundamental physics suffers from 'final-theory anxiety': searching for a large, macro, and generalisable theory that will explain every empirical reality. As a possible 'solution' to this pervasive 'problem', he suggests focusing on physicists' subjective interests, redefining and rethinking existing lines of enquiry, and fundamentally altering ahistorical research practices.

Aditya Nigam discusses the conflictual relationship between the empirical discipline of 'political science' and the field of 'political theory'. He brings out the debate between proponents of scientificity, who favoured quantitative methods and empirical evidence geared towards discovering the seemingly invariable laws of political behaviour; and political theorists who moved towards normative philosophy and critical theory. With a description of these debates that have characterised the discipline since its point of inception, the chapter focuses as much upon contemporary debates, in the present moment of 'decolonisation'. Nigam argues that it is now widely accepted that political theory – Western by definition – is highly parochial and can only be rescued by opening out to other intellectual traditions. So, he tells us that recent critiques of the discipline have forced it to look beyond its scientific and universalistic pretensions, in order to try to hear 'other voices'. Throughout the chapter, Nigam focuses on 'context-specific mode of theorising' as an important concern. According to him, the theory/empirics divide has a special resonance in political studies and the power of the disciplinary-institutional apparatus manifests itself in the fact that these battles continue to be replayed in political science departments in India. Nigam undertakes a brief examination of the practice of political reflection and analysis in India to argue that the sharp divide between the empirical and the theoretical did not actually have the same resonance that it had in the USA and elsewhere in the West. This is exemplified in the work of Rajni Kothari – an empirical political scientist engaged in theorising about caste, non-party political processes, and Dalit discourse that do not rely on quantifiable data in the way that the erstwhile election and democracy studies did. Furthermore, politically engaged theorising received an impetus since the 1980s with the powerful current of feminist and Dalit scholarship. The existing disciplinary methodologies of a scientistic nature made no sense here, since the endeavour was meant to produce histories where none existed. The author asserts that the theory/empirics divide acquires a very different and troubling resonance when the field is the non-West, but all the theory has been produced in Western academies. Nigam concludes with an instantiation of his own work to elucidate his methodological moves. He finds it important to engage with actual empirical history and texts from earlier times, while engaging with political thought from across the world. As he puts it, the move towards decolonising method is not about 'returning' to some pristine source of a precolonial past but about broadening the base of a shared disciplinary understanding, by going beyond the narrow confines of European experience.

Kamala Ganesh discusses the tensions between two major factions of the discipline of anthropology as manifest in the reactions to the American Anthropological Association's move to dropping the word 'science' in its own self-definition. These included those who espoused the discipline as a science and those who saw it as a humanistic discipline. Over the years, this move was followed by a powerful argument for bridging the sciences and humanities. Ganesh's central argument in this chapter is that anthropology in the coloniser countries of the West tackled the divide between the sciences and humanities in a distinctive way, due to the way in which it constituted its subject matter of studying other cultures, along with the methodology of ethnography with participant observation, which evolved for this. While in the West, anthropology became self-aware quickly and its fieldwork methods provided a unique vantage point for a critique of positivism, sociology, and anthropology in India developed as separate disciplines, with Indian scholars studying their own society and culture. This struck at the heart of anthropology's definition as the study of 'other cultures' and so rendered the cardinal distinction between the two disciplines as invalid. It was untenable to demarcate sociology as the study of complex, large scale, industrial society, with quantitative methods; and anthropology as that of small scale, 'tribal', 'primitive' cultures, with qualitative methods. Ganesh points out that sociology and social anthropology were thus institutionalised and professionalised as a unified field and social anthropology was distinguished from archaeology, linguistics, and physical anthropology. Ganesh tells us that anthropology's early affinity to scientific rigour had to do with subfields like archaeology and physical anthropology that directly used methods of the natural sciences and were amenable to structured observations, measurements, and calculations. Ganesh points out that the positivist empirical science of the time was exemplified in several additions of the Notes and Queries volumes meant for travellers and non-anthropologists in the colonies, to generate data for anthropologists to analyse in Europe. As part of this brief description of the history of anthropology, she also discusses the polemics around anthropology's affinity to science over several decades. This internal churning within anthropology in the West with respect to issues of identity and methodology continued in the 1980s with the advent of 'New Ethnography' that critiqued anthropology's colonial history as well as hierarchies between researcher and the researched, along with other aspects of the subjective nature and hidden biases of knowledge emerging from the ethnographic method. Taking into account the role of social location was also manifest in the feminist critiques that highlighted gender biases within anthropology, as they showed how a supposedly scientific activity was deeply gendered. Ganesh argues that the end of colonialism brought about a new genre of fieldwork and ethnographies by anthropologists from the former colonies working in their own societies and cultures. They produced critical analyses of the mode of studying 'other cultures' by Western anthropologists – work which ultimately became part of anthropological self-reflection that offered another powerful critique of scientific methodology.

Neetha N. brings out the proximity of economics to scientific and positivist methodologies as well as its use of quantitative data that seemingly justifies its claim to a superior status. This desire to stay close to the 'hard sciences' led to the development of neoclassical economics in the 19th century and continues to dominate economic thinking. She tells us that the central methodological principle in economics is the formulation of new hypotheses based on existing theories that can be tested using scientifically collected data following statistical principles. Neetha argues that certain underlying assumptions and ceteris paribus clauses are central to the falsification process, which also gives economics its scientific label. She discusses the methodological individualism and some reductionist ontologies of mainstream economics, including that of the rational economic man – the discipline's unit of analysis. Neetha alludes to several assemblages that are concerned about the theoretical and practical limitations of neoclassical economics. She posits these as heterodox approaches that ask different sets of questions related to the objective analysis of contemporary socio-economic phenomena. She brings to light feminist economics that takes gender inequality as its primary concern. Feminist economists have exposed the claim of gender neutrality and shown the value-embedded nature of the mainstream economic analysis. Neetha argues that implicit gender meanings often involve value judgments, wherein masculinity tends to be valued more than femininity.

One of the most important interventions of feminist economics is questioning the fundamental assumption of the rational economic man, supposedly driven by self-interest, which ignores gender stratification and women's roles. A central feminist criticism is that of the setting aside of unproductive activity – seen as a reflection of the dominance of neoclassical economics. The implicit assumption that unpaid work and care work are separate and independent from the monetised economy continues to be central in mainstream teaching and research in economics. Neetha brings out the troubling fact that feminist intervention has mostly been parallel as mainstream economics has conveniently neglected any serious engagement with these questions. While feminist economists may not have been as successful in altering the fundamental foundations of the discipline, they have nevertheless been successful in exposing several myths about the discipline's 'objectivity', rigour, and scientificity. The author calls for an open approach to methodological possibilities and argues that quantitative methods could in fact not only enrich feminist analyses but critical feminist insights too could strengthen quantitative methods.

In his chapter, **Rahul Govind** attempts to bring together Marxism and two traditions of the history of science to invite the reader to a critique of the popular understanding and imagination of science: that scientific advancement is a progressive force that perfects itself in a linear and cumulative chronology. Govind thinks of this as a subordination of time and its contingencies to 'logical structure', sought to be replicated in the social sciences and history, which consider science a paradigmatic method. Govind comments on the Anglo-American tradition of the history of science, embodied in the works of Popper, Kuhn, and Feyerabend; and the French tradition of Canguilhem and Foucault. While Marx critiques the science of classical political economy for attempting to subordinate history to the laws of the economy, the historians of science critique conceptualisations of science where the laws of nature are supposedly revealed in a cumulative history of advancement. Both Marxism and the history of science then critique the axiomatic separation of method (or theory) and nature (or history). Govind describes Marx's articulation of the possibility and need to change what appear to be 'objective conditions'. As is widely known, Marx critiques bourgeois political economists for their treatment of the contemporary state of the political economy as consisting of 'natural laws' that are eternal and independent of time and action. The author then points out that for philosophers of science Bachelard and Popper, scientific practice and progress could not be explained in terms of logic or cumulative history of results seemingly approximating a defined truth. Both would agree with the Marxist argument that scientific practice is inherently social and that there is no a priori logic or nature to be described and explained. On the other hand, Kuhn explicitly uses a history of science to make a distinction between 'normal science', which proceeds 'cumulatively' through resembling or modelling the existing paradigm; and 'revolutionary' science that sees a crisis in terms of revaluation of fundamentals and paradigm shift, which aren't cumulative.

As part of the French tradition, Govind describes the work of Canguilhem, who argues that history serves as 'epistemology's laboratory' and that the object of a history of science is essentially incomplete. According to Canguilhem, biological sciences have their own specific objects and one cannot transplant mechanical laws from other sciences into the sphere of biology. Canguilhem critiques such an attempt that tries to unify the normal and the pathological, by basing their distinction merely on quantitative variations. Foucault pushes this problem of biological normativity to its extreme by refuting the standard history of the origins of rational psychiatry. Congruent to the Marxist arguments, Govind concludes by arguing that scientific research in any form has to be conceived of as a social category and/or activity; and no nature or law awaits a description of itself as a logical structure.

Sabah Siddiqui focuses on the underlying epistemic assumptions, theoretical apparatus, and methodological as well as ethical dilemmas of the discipline of psychology, which is located at the interface between the subjectivity and objectivity debate. She takes a bird's eye view of the institutionalisation of psychology globally and in India, and finds that psychology is broadly characterised by a push towards positivist methodology and a subsequent pull of critical theories. Siddiqui tells us that the first module of the undergraduate course 'Introduction to Psychology' itself is an introduction to the field with a focus on method; and within the first few hours, psychology is introduced as a science of behaviour and cognitive processes. The emphasis on science manifests itself as the insistence on producing measurable descriptions of observable individual differences. Right at the outset, psychology is placed within the fold of the positivist scientific method that insists on experimentation and statistical analysis. The history of the discipline gets narrated as the move towards an objective, reliable, and universal science of human behaviour and cognition, which privileges quantitative and statistical methods over qualitative and constructivist ones. The conundrum, as Siddiqui presents it, lies in the fact that while psychologists are aware that they deal with human phenomena that is largely variable, they are required to capture the wide range of this variability according to the scientific principles emerging from the strict rationality of the scientific method. So, Siddiqui traces the history of the discipline – since its inception in Wilhem Wundt's first experimental lab in Germany in 1879 – to delineate how psychology has specifically developed its notions of method, validity, and rigour. Siddiqui shows that Wundt's scientific method involved conducting experimental introspection on sensory data that humans acquire through subjective activities of feelings, emotions, and volition. This method of experimental introspection was later criticised by behaviourists who felt that it was not scientific enough. Siddiqui points out that newer forms of empiricism continue to be dominant. Siddiqui revisits the polarised debate between those proposing hypothetico-deductive methods and those critiquing them, instead proposing qualitative methods. But she warns that qualitative methods are not necessarily less reductive and more ethical. She also poses provocative questions about an 'inferiority complex' of Indian psychologists and chooses to make a case for a postcolonial analysis wherein science and scientific method are implicated as Western hegemonic practices that strangle traditional/cultural systems of knowledge.

Saraswati Raju's chapter begins with an important operational definition of geography as a discipline concerned with the interrelationships between humans and their environments – both physical (natural) and built (social and cultural) – expressed through spatial practices. For Raju, the major problem of method in geography comes from an intellectual legacy of the discipline as caught between whether it is part of 'science' proper or social science. Underlying assumptions and expectations that come with an understanding of the discipline as part of science include rational and objective viewpoints in terms of an 'enlightened' way of seeing the world as orderly, neatly organised, causally connected, and generalisable. Here, production of truth meant using methods that lead to an overarching theorisation, without recognising any bias and/or the role of the observer's persuasion in the perception of reality. Therefore, traditionally, geography has been seen as a nomothetic discipline concerned with objectively producing knowledge, using methods of scientific rigour, scrutiny, and validation. Raju argues that Indian geography imported this model without introspection and reflection, to the extent that what was not quantifiable and not amenable to statistical testing couldn't be geography at all. While the Anglophone world gradually moved away from the quantitative revolution and positivistic tradition to a more evolved and complex mix of methodologies, the debate, as Raju sees it, remains at a nascent stage in India. She argues that in order to retain its popularity and relevance, geography ought to contend with the subfield of the geography of gender. Raju challenges the hegemony of scientific assumptions in geography by exclusively positioning gender within the orbit of geographical knowledge in terms of research and teaching in India. She defines gender and differentiates it from sex, while also differentiating the responses of men and women to their respective social and natural environments. She also alludes to intersectionality as she admits that caste, class, religion, and ethnicity intercept gender locations. She is concerned primarily with spatiality and how gender is geographically anchored. Raju concludes on an optimistic note, suggesting that despite geography's rootedness in positivist traditions, its dissatisfaction with empiricism is largely palpable. She tells us that recent research has seen the emergence of mix methods whereby both qualitative and quantitative approaches have been adopted, along with visual studies and in-depth case studies.

Yasmeen Arif's speculative essay scrutinises the temporal frame of the contemporary to outline the sets of responses to the question of scientific method in the social sciences. An important entry point for Arif is the recent 'shifts' in sociology/social anthropology towards the study of conventionally 'objective' realms. She problematises the simplistic binary between scientificity and subjectivity, since the question of method in sociology/social anthropology cannot be posed in these limited and pregiven ranges of responses. She points out right at the outset that the dividing lines between positivism, empiricism, objectivity, and arguments of reliability and validity, on the one hand, and intricacies of narrative structure, subjectivity, authenticity, plurality, representation, languages, and modes of writing on the other are increasingly blurred.

With several rich conceptual formulations, Arif explains that in the contemporary moment, research concerns in sociology/social anthropology emerge in the assemblage of the local and the 'planetary' as well as in the convergence of past–present–future. The question of who gets to make meaning and from where becomes more important than whether that meaning is constructed 'scientifically'. Arif goes on to describe what she calls a 'geopolitics of epistemology', wherein the Global South is placed in opposition to the Global North or in another way, postcolonial epistemological positions are asserted. She links such a geopolitics to the decolonising movement in sociology/social anthropology that strives to claim both epistemological and methodological reparation. Arif discusses the methodological divide between sociology and social anthropology in India, exemplified by the distinction between the MSc in Anthropology and the MA in Sociology programs at the University of Delhi. She alludes to a particular epistemological attitude wherein the physical, objective, and scientific aspects of a physical anthropology are separated from a more socio-cultural, subjective, ethnographic, and qualitatively accessed anthropology. She traces the historic developments in sociology/social anthropology in India and argues that the analytical toolkit that accompanied the development of the discipline(s) in India reflected global theoretical environments, including structural functionalism, political economy perspectives, postcolonial critiques, and post-structuralisms. In this context, Arif proposes that to find an appropriate method for research objects of the contemporary moment, one needs to move beyond the postcolonial, for possibilities of opening up predefined limits of epistemological stances. Arif argues that pedagogy now encourages a mixed-method approach, combining qualitative and quantitative toolkits. So the methodological spectrum now ranges from the subjective realms (qualitative) of autoethnography to the mining of large datasets with sophisticated statistical tools (quantitative). In this way, Arif brings to light that method is required to fundamentally acknowledge and respond to the changing social environment in which questions are formulated. With the substantive example of violence, humiliation, and subjugation of Dalit experience, Arif shows us how epistemic inequalities are connected to methodological processes of representation and possibilities of knowing. In the last section, Arif offers the reader a sketch of her own work that brings science and the social in close conversation in the contemporary. Specifically, she attempts to make the reader conscious of the significance of the biosocial during an unprecedented event like the COVID-19 pandemic.

....

Notes

1 While the question of method has opened up in and through interdisciplinary domains like cultural studies, science and technology studies, women's studies, development studies, to name a few, our interest was to look carefully into how canonical disciplines, deeply entrenched in the academia, have explored the question of method. Exploring the method question in interdisciplinary domains would be another volume.

2 It is important to articulate the fact that both of us use/occupy (and also identify with) the space of Anthropology. We recognize the differences in our understanding and approach to the discipline of anthropology – which may be attributed to the early training we received at departments that were our academic homes. What is important to note here is that the disciplines of sociology and anthropology have often combined theories and methods of research. With reference to research methodologies, arguably the big breakthroughs have occurred in anthropology more than in sociology. However, the critical theorisations on science as a social institution date back to Robert Merton. A critical perspective to science as a knowledge-making system dates back to Thomas Kuhn. Science criticism, in a sense, then predates the anthropological work of scholars like Bruno Latour and others in STS.

3 See Houtman, Aupers, and Laermans (2021). For a critique of *The March for Science*, see Sarukkai (2017), Thomas (2017).

4 The terms method and methodology are almost used synonymously. We mean here the methodological assumptions of research, both ontological and epistemological and not just the tools of research.

5 In disciplines such as social anthropology where ethnography began as a method of research is arguably seen today as philosophy and theory, not as method. This is an example of how a 'dominant' method in a discipline can get presented as theory itself because it changes the very assumptions of research. The distinction between method and theory gets blurred here. For an engaging reflection on the relationship between anthropology, theory, and philosophy, see Veena Das, Michael Jackson, Arthur, Kleinman, and Bhrigupati Singh (2014).

6 See Jason A. Josephson-Storm (2017, 2021).

7 For instance, Sasanka Perera notes that the aversion and suspicion towards the 'visual' and photography in Sociology and Anthropology is due to the existing scientism in these disciplines (Perera, 2020).

8 See Tim Ingold (2017), Irfan Ahmad (2021.)

9 For instance, see Renato Rosaldo (2013). Also see Anand Pandian and Stuart McLean (2017).

10 For a recent discussion on scientification in humanities and social sciences, see Axel Michaels and Christoph Wulf (2020).

11 There have been attempts to discuss and problematise the relation between methods and imperialism and the need to decolonize methodologies. See Linda Tuhiwai Smith (1999), also see Mwenda Ntarangwi (2010).

12 Immanuel Wallerstein, in the famous Report of the *Gulbenkian Commission on the Restructuring of the Social Sciences*, traces the origin of the social sciences and the creation of their disciplinary silos in the Enlightenment ideology of the 18th century from where it was also cast in the same mould as the method of the natural sciences. The Report outlines the need that has been felt since the end of the Second World War, for a radical restructuring of the social sciences but, in a way, that takes into account the diversities and pluralities of the social sciences as they are practised across the globe.

References

Ahmad, Irfan (ed.). 2021. *Anthropology and Ethnography Are Not Equivalent: Reorienting Anthropology for the Future*. Oxford: Berghahn Books.

Baber, Zaheer. 1998. *The Science of Empire*. New Delhi: Oxford University Press.

Barnes, Barry. 1974. *Scientific Knowledge and Sociological Theory*. London: Routledge.

Bloor, David. 1976. *Knowledge and Social Imagery*. London and New York: Routledge & Kegan Paul.

Cao, Tian Yu (ed.). 1999. *Conceptual Foundations of Quantum Theory*. Cambridge: Cambridge University Press.

Chadha, Gita. 1998. 'Sokal's Hoax and Tensions in the Scientific Left.' *Economic and Political Weekly*, XXXIII (35): 2194–2196.

Chadha, Gita. 1999. 'Sokal's Hoax: A Backlash to Science Criticism.' *Economic and Political Weekly*, XXXIV (13): 779–784.

Chadha, Gita. 2015. 'Tracking a Consciousness: Questions, Dilemmas and Conundrums of Science Criticism in India.' In Krishna, Sumi and Chadha, Gita (eds) *Feminists and Science: Critiques and Changing Perspectives in India*, Vol. 1, pp. 257–272, xxxvi–xlvi. Kolkata: Stree Publishers.

Chadha, Gita. 2017a. 'Fingerprints and Erasures: Mapping the Creative Process in Science.' In Krishna, Sumi and Chadha, Gita (eds) *Feminists and Science: Critiques and Changing Perspectives in India*, Vol. II, pp. 207–231. Kolkata and New Delhi: Stree Publishers Sage Publications.

Chadha, Gita. 2017b. 'Science, Sociology of Science and Society: Is there a Relationship?' In Dhar, Anup, Niranjana, Tejaswini, and Sridhar, K. (eds) *Breaking the Silo: Integrated Science Education in India*, pp. 114–131. New Delhi: Orient Black Swan.

Cowles, Henry M. 2020. *The Scientific Method: An Evolution of Thinking From Darwin to Dewey*. Cambridge: Harvard University Press.

Das, Veena, Jackson, Michael, Kleinman, Arthur, and Singh, Bhrigupati (eds). 2014. *The Ground Between: Anthropologists Engage Philosophy*. New Delhi: Orient Blackswan.

Daston, Lorraine, and Galison, Peter. 2007. *Objectivity*. New York: Zone Books.

Feyerabend, Paul. 1975. *Against Method*. London: Verso Books.

Feyerabend, Paul. 1981. *Realism, Rationalism and Scientific Method*. Cambridge: Cambridge University Press.

Gower, Barry. 1996. *Scientific Method: A Historical and Philosophical Introduction*. London: Routledge.

Haraway, Donna. 1988. 'Situated Knowledges: The Science Question in Feminism and the Privilege of Partial Perspective.' *Feminist Studies*, 14 (3): 575–599.

Harding, Sandra. 1986. *The Science Question in Feminism*. Ithaca: Cornell University Press.

Harding, Sandra. 1989. 'Is There a Feminist Method?' In Tuana, Nancy *Feminism and Science*. Bloomington and Indianapolis: Indiana University Press.

Hegde, Sasheej. 2014. *Recontextualizing Disciplines: Three Lectures on Method*. Shimla: IIAS.

Horgan, John. 2019. 'We Should All Be Science Critics.' *Scientific American*, August 5. https://blogs.scientificamerican.com/cross-check/we-should-all-be-science-critics (accessed on 25th October 2021).

Houtman, Dick, Aupers, Stef, and Laermans, Rudi. 2021. 'Introduction: A Cultural Sociology of the Authority of Science.' In Houtman, Dick, Aupers, Stef, and Laermans, Rudi (eds) *Science Under Siege: Contesting the Secular Religion of Scientism*, pp. 1–34. London: Palgrave Macmillan.

Ingold, Tim. 2017. 'Anthropology Contra Ethnography.' *HAU: Journal of Ethnographic Theory*, 7 (1): 21–26.

Jasanoff, Sheila. 2019. *Can Science Make Sense of Life?* Cambridge: Polity.

Josephson-Storm, Jason A. 2017. *The Myth of Disenchantment: Magic, Modernity, and the Birth of Human Sciences*. Chicago: University of Chicago Press.

Josephson-Storm, Jason A. 2021. 'Max Weber and the Rationalization of Magic.' In Yelle, Robert A., and Trein, Lorenz (eds) *Narratives of Disenchantment and Secularization: Critiquing Max Weber's Idea of Modernity*, pp. 31–49. London: Bloomsbury.

Kang, Gagandeep. 2021. 'Vaccine Platforms Old and New for SARS-CoV2.' *The India Forum*, March 5. https://www.theindiaforum.in/article/vaccine-platforms-old-and-new-sars-cov2?utm_source=website&utm_medium=organic&utm_campaign=category&utm_content=Covid-19 (accessed on 28th October 2021).

Keller, Evelyn Fox. 1992. 'How Gender Matters, or, Why It's So Hard for Us to Count Past Two.' In Kirkup, G., and Keller, L. (eds) *Inventing Women: Science, Technology and Gender*. Cambridge, UK and Cambridge, USA: Polity Press.

Keller, Evelyn Fox, and Longino, Helen E. (eds). 1996. *Feminism & Science*. Oxford and New York: Oxford University Press.

Maani, Nason, and Galea, Sandro. 2021. 'What Science Can and Cannot Do in a Time of Pandemic.' *Scientific American*, February 2. https://www.scientificamerican.com /article/what-science-can-and-cannot-do-in-a-time-of-pandemic/ (accessed on 28th October 2021).

Michaels, Axel, and Wulf, Christoph (eds). 2020. *Science and Scientification in South Asia and Europe*. London: Routledge.

Nanda, Meera. 1998. 'Reclaiming Modern Science.' *Economic and Political Weekly of India*, XXXIII (16): 18–24.

Ntarangwi, Mwenda. 2010. *Reversed Gaze: An African Ethnography of American Anthropology*. Urbana: University of Illinois Press.

Pandian, Anand, and McLean, Stuart (eds). 2017. *Crumpled Paper Boat: Experiments in Ethnographic Writing*. Durham: Duke University Press.

Perera, Sasanka. 2020. *The Fear of the Visual?: Photography, Anthropology, and Anxieties of Seeing*. New Delhi: Orient Blackswan.

Rosaldo, Renato. 2013. *The Day of Shelly's Death: The Poetry and Ethnography of Grief*. Durham: Duke University Press.

Salk, Jonas. 1979. 'Introduction.' In Latour, Bruno, and Woolgar, Steve (eds) *Laboratory Life: The Social Construction of Scientific Facts*. London: Sage Publications.

Sarukkai, Sundar. 2017. 'The March from Yesterday.' *The Hindu*, August 9.

Shah, Chayanika, and Chadha, Gita. 2015. 'Teaching Feminist Science Studies in India: An Experiment.' In Krishna, Sumi and Chadha, Gita (eds) *Feminists and*

Science: Critiques and Changing Perspectives in India, Vol. 1, pp. 257–272. Kolkata: Stree Publishers.

Shapin, Steven, and Schaffer, Simon. 1985. *Leviathan and the Air-Pump: Hobbes, Boyle, and the Experimental Life*. Princeton: Princeton University Press.

Smith, Linda Tuhiwai. 1999. *Decolonizing Methodologies: Research and Indigenous Peoples*. London: Zed Books.

Snow, C. P. 1959. *The Two Cultures and the Scientific Revolution*. London and New York: Cambridge University Press.

Sokal, Alan D. 1996a. 'Transgressing the Boundaries: Toward a Transformative Hermeneutics of Quantum Gravity.' *Social Text*, 6/47, Spring/Summer Issue 217–252.

Sokal, Alan D. 1996b. 'A Physicist Experiments with Cultural Studies.' *Lingua Franca*, May–June Issue 62–64.

Subramaniam, Banu. 2014. *Ghost Stories for Darwin: The Science of Variation and the Politics of Diversity*. Urbana: University of Illinois Press.

Subramaniam, Banu. 2016. 'Stories We Tell: Feminism, Science, Methodology.' *Economic and Political Weekly*, 51 (18): 57–63.

Thomas, Renny. 2017. 'Can Science and Social Science Really March together?' *The Wire*, August 22 (accessed on 28th October 2021).

Thomas, Renny. 2018. 'Narratives in Feminist Sociology of Science: Contextualizing the Experience(s) of Women Scientists in India.' In Chadha, Gita, and Joseph, M. T. (eds) *Re-Imagining Sociology in India: Feminist Perspectives*, pp. 316–337. London and New York: Routledge.

Thomas, Renny. 2020. 'Brahmins as Scientists and Science as Brahmins' Calling: Caste in an Indian Scientific Research Institute.' *Public Understanding of Science*, 29 (3): 306–318.

Thomas, Renny. 2022. *Science and Religion in India: Beyond Disenchantment*. London: Routledge.

Thurs, Daniel P. 2015. 'Myth 26: That the Scientific Method Accurately Reflects What Scientists Do.' In Numbers, Ronald L., and Kampourakis, Kostas (eds) *Newton's Apple and Other Myths About Science*, pp. 210–218. Cambridge: Harvard University Press.

Todorov, Tzvetan. 1993. *On Human Diversity: Nationalism, Racism and Exoticism in French Thought*. Cambridge: Harvard University Press.

Uberoi, J. P. S. 1978. *Science and Culture*. Delhi: Oxford University Press.

Uberoi, J. P. S. 1984. *The Other Mind of Europe: Goethe as a Scientist*. Delhi: Oxford University Press.

Visvanathan, Shiv. 1985. *Organizing for Science: The Making of an Industrial Laboratory*. Delhi: Oxford University Press.

Visvanathan, Shiv. 1988. 'On the Annals of the Laboratory State.' In Nandy, A. (ed.) *Science, Hegemony and Violence: a Requiem for Modernity*. Delhi: Oxford University Press.

Wallerstein, Immanuel. 1996. *Open the Social Sciences: Report of the Gulbenkian Commission on the Restructuring of the Social Sciences*. Stanford: Stanford University Press.

Weinberg, Steven. 1993. *Dreams of a Final Theory*. London: Hutchinson Radius.

Part I

SHIFTS WITHIN THE SILO: HUMANITIES

INTRODUCTION TO PART I

Most universities and academic institutions divide disciplines into three broad silos: sciences, social sciences, and the humanities. They get listed like this, almost, in the order of their perceived importance to the modern pursuit of knowledge. In fact, if we use science as an identity marker, the humanities, constituted by philosophy and literature (linguistics being somewhat of an outlier even in this classification), are put out of the silos of the sciences – natural and social – which are seen as unified by a method. The 'humanities' are constructed as 'creative' fields while the sciences are not, much to the chagrin of the scientists. Philosophers are perceived as using speculation without empirical verification and the literature *wallahs* are seen as exercising imagination without a sense of the real. Linguistics is of course seen as lying in between being a science and an 'arts' discipline. Since we had decided, in a considered manner, to imitate the disciplinary order in the academic institutions, we placed linguistics in this section as an 'arts' subject. While debating the foundations of methods in any discipline, we inevitably find ourselves excavating its philosophical foundations. Often practitioners of the natural sciences say that the question of method is more to do with philosophy than with science. Paradoxically, we also know that the birth of modern science was born out of generally severing ties with philosophy and more particularly with metaphysics. But how does the discipline of philosophy itself look at the question of method is a fascinating question to explore. It is also interesting to explore how the method question is articulated in the study of both literature and language. Does a 'scientific reading' of a literary text enhance or endanger the core text? Does academic equal to scientific, or are there other ways of marking academic readings of literacy texts? The essays in this section explore some of these questions.

DOI: 10.4324/9781003298908-3

1

METHODS IN SUBSTANTIVIST LINGUISTICS

Probal Dasgupta

Organisation

Disagreements in linguistics and in its classical version, grammar, have always been difficult to resolve. This feature of the discipline goes back to the intellectual landscape of ancient societies. Ancient Romans were used to seeing grammarians debate subtle points without ever reaching a consensus. Horace, in his 'Ars Poetica', famously wrote: 'Grammatici certant et adhuc sub judice lis est', which is Latin for 'The grammarians are contending, and the matter is still under dispute' (line 78 of the 'Ars Poetica', quoted in Williams 2013: 67). Thus, speaking for the discipline to an audience largely composed of non-specialists becomes a fraught enterprise not only because of the interminable contestations within the field but also because linguistics strikes today's general public as a particularly opaque discipline.

This unusual opacity has to do with the fact that linguistics has not been accepted as a science whose authority in the domain of language is uncontested. All societies to this day have their language apparatuses – their systems of language teaching, publishing, editing, etc. – driven by traditional grammar, not by linguistics. In the minefield of ideas and perspectives in linguistics, there is one school of thought, known as substantivism, which treats this unfinished journey of linguistics towards social acceptance not as an irrelevant external difficulty but as part of the core concerns that the discipline has to address. By a fortuitous coincidence, substantivism is also intimately connected to modern India, which makes it a suitable vantage point for an intervention in this volume.

This chapter is organised as follows. In the first section, 'A Science of Language?' we look at considerations that justify a science-of-language enterprise. The second section, 'Compositionalism, Overreach, and Consequences', takes up some factors that have brought this enterprise into a state of malaise verging on a crisis. The third section, 'The Architecture Meets the Thatchwork', flags some resources that help us to diagnose and address the crisis. The last section, 'Situating the Resonances', places the specific crisis of linguistic science in a larger context.

DOI: 10.4324/9781003298908-4

A Science of Language?

The 'common sense' underlying our scientific discourse classifies happenings into actions and mere events.[1] Actions reflect free will. There is no science to them. Sciences study the patterns of necessity and chance, not the actions an agent has chosen to perform. The patterns of 'mere' events that nobody is responsible for fall under the purview of science. Where does this leave language?

My writing these pages for you is an exercise of free will. If it counts as 'language', then there is no science of language. How can acts of writing or speaking be 'mere events'?

Pursuing this question is trickier than one might imagine. Suppose I say to you, 'Listen, the cat really is on the mat.' My saying this, or any other utterance – spoken or written – is a freely chosen action, therefore outside the purview of science. But a science of language treats each language as a super-pattern of possible event-patterns, within which the speaker or writer makes actual choices.

Thus, both (1) *The cat is on the mat* and (2) *The bat is on the hat* are possible sentences in English. But the range of the possible excludes (3) *Bat the is hat the on* (the prefix '*' labels a string of words as formally inadmissible, excluded, ill-formed).

To speak/write English, in a given context, involves choosing a sentence to utter, within the range of possible sentences. The concrete **choices** one makes fall outside science; they count as **discourse**. Studies of discourse cannot be systematised into a science. However, every **super-pattern of event-patterns** within which speakers/writers make their discursive choices is a **language**. Linguistics is the science dedicated to languages. The linguistics of English is a province within that science.

This binary – language (systematic event-pattern-schema) vs discourse (system-eluding personal choices) – is more robust than non-specialists imagine. There has never been a science of discourse per se,[2] nor is such a science about to take off. But the science of language has flourished since the 18th century as a systematic global enterprise. Today's linguistics uses tools and terminology drawn from 2000-year-old grammatical traditions whose systematic writ used to run within regional clusters of languages.

Among these traditions, some carry the labels Indian, Chinese, and Arabic. One tradition – the Greco-Roman legacy filtered through later European accretions – is called 'traditional grammar', treated as the trans-regional default basis for contemporary linguistics. Contrary to post-colonial expectations, linguists from China, India, etc., do not challenge the default status of this Europe-derived tradition. Linguistics plays the North/South game differently (see the sections 'The Architecture Meets the Thatchwork' and 'Situating the Resonances'); for now, let us focus on shared factors rather than those that lead to contestations. Traditional grammar works

44

alongside lexicography, dictionary-making; it is their joint enterprise, call it Traditional Lexicography-and-Grammar (TLG), which **codifies** practices of **writing**, setting standards for them. We have said that a language is a super-pattern of event-patterns defining the range of the 'possible'. TLG codifies written language standards; its notion of language parses 'possible' as validly **writable**, as acceptable in this normative codification.

Normative dictionaries have always specified the **writable *words*** in a given language. But TLG cannot characterise the range of **writable *sentences*** – an infinite range, even within a single language. No classical tradition ever tried to map rigorously the syntactic (sentence-making) system. The 20th-century syntactician who took up this work, Otto Jespersen, is an amphibian figure, significantly contributing to both the old TLG (Jespersen, 1933) and the new linguistics (Jespersen 1937). Is Jespersen guilty of confusing scientific analysis (linguistics) with normative-cultural prescription (TLG)?

On the contrary, his two-pronged approach corresponds to the fact that his object of study, the sentence, is an amphibian creature, equally at home in the water of language and on the landmass of discourse. This chapter is built around the following articulation of the amphibian character of the sentence:

(1) The **sentence** is the point at which grammar/science (linguistics) meets grammar/culture (TLG). As an entity structured by the syntactic system, **a sentence exemplifies the logic that combines words into phrases**, and belongs to **Language**. It invites **exact, rational** scientific analysis and more. As an entity assembled by an individual speaker or writer, **a sentence is a minimal composition**, and belongs to **Discourse**. It invites **prudent, reasonable** normative-cultural editorial advice and more.

'Analysis and more/advice and more' because both the science of language and the cultural pedagogy of writing are compelled to engage with history – to acknowledge the fact that change is a constant.[3] Linguists stress the role of spoken language as a factor directing change in the written language. Phonetics, the branch of linguistics that deals with speech, was accommodated in the TLG apparatus when foreign language pedagogy began to teach learners how to speak the target language. Dictionaries have been using the International Phonetic Alphabet to register the normative pronunciation of words. Thus, the general public is familiar with this alphabet and peripherally aware of linguists who are responsible for it. But modern societies still keep TLG, not linguistics, in charge of the normative machinery of language in education and publishing.

That the public is only peripherally aware of linguistics, as a limited scientific supplement to TLG, reflects the fact that, ever since the modern discipline took off in the 18th century, linguists specialised in the peripheries. British and French authors of monographs on languages spoken in colonies[4] like India and Algeria, or American experts on the languages of indigenous peoples, were marginal figures. But linguists in the 20th century – seeing themselves as scientists who seriously understand language – aspired to gatecrash into the mainstream and depose TLG, just as modern astronomy once unseated the Ptolemaic-Aristotelian orthodoxy. Why, then, did they fail?

While analysing the languages spoken by colonial subjects or indigenous peoples, white linguists engaged in objective 'description, not prescription'. Turning this gaze on their own societies, those linguists were horrified by the undiminished power of prescriptive TLG, which they proceeded to critique from a descriptivist standpoint. Hall's *Leave Your Language Alone* (1950) was a radical overstatement, but even moderate descriptivists failed to convince the public that linguists seriously understand language. It is palpably absurd to treat what a three-year-old toddler speaks as a self-contained, independent system. The public knows that children acquire successive approximations to adult speech. No 'science' can 'describe' this fact away. Embedded in this fact is an implicit or explicit pedagogy. Prescriptive TLG may have committed authoritarian excesses, but throwing the pedagogic baby out with the authoritarian bathwater hardly helps.

As long as linguists and TLG workers confined themselves to the grammar of words, this stand-off continued. When the best-known linguist of our times, Noam Chomsky, took on the grammar of sentences – syntax – he introduced a conceptually significant twist, which ended the stand-off. He proposed (Chomsky 1965) that linguists describing a particular language (PL) should imagine describing a homogeneous community of idealised speakers who all speak the same variety of PL perfectly and unconstrained by limits of memory and attention. The methodological point of this 'perfection idealisation' is to focus sharply on the grammatical core of linguistics – separating it from sociolinguistics (concerned with dialect diversity within PL) and psycholinguistics (which studies the limiting factors that constrain even adult performance and that shape the trajectory of children[5] gradually acquiring PL). By pushing those independent factors off the screen of grammar, Chomsky's move dramatises the fact that any PL, qua language, harbours infinitely many sentences of any length and complexity, which pure syntax must address. The fact that no flesh-and-blood speaker can ever use more than a tiny fraction of the massive range of syntactic options reflects limitations of memory and attention; those fall under psycholinguistics, not grammar.

Why did Chomsky's move end the stand-off between TLG and linguistics? TLG, which revolves around language pedagogy, imagines perfectly proficient speaker-writers as the ideal end-products of this teaching. Generative (Chomskyan) linguistics – even though it inherits and deepens the descriptivist insistence on spontaneity over pedagogy[6] – projects an idealisation akin to the TLG image of pedagogy's ideal end-product. Given their different priorities, practitioners of linguistics and of TLG cannot see eye to eye. But the two tribes found themselves in operational proximity at the level of idealisations and built a tacit coalition. The perfection idealisation was one factor enabling this; decolonisation was another. With the empires dismantled, those peripheral experts of colonial languages came back home and placed white people's mother tongues under fresh scrutiny. A modus vivendi with TLG became a pressing need.

How was this need met? What form did the tacit linguistics–TLG coalition take?

Years after the death of the linguistics-inclined traditional grammarian Otto Jespersen (1860–1943), who had systematically studied syntax for the first time in the history of TLG, a 1961 course taught at MIT by the Linguistics Professor Edward S. Klima flagged his *Analytic Syntax* (Jespersen 1937/1984) as a key resource.[7] Klima and his students proceeded[8] to translate many of Jespersen's formulations into the versatile notation for syntactic description created by Chomsky (1957) and continuously updated by the growing generative enterprise.

MIT's Department of Linguistics and Philosophy did more than an insight-mining exercise involving one book. It urged students to place Jespersen in historical context. Barbara H. Partee (p.c. 2021) informs me that, shortly before her 1964 Ph.D., 'There was actually a Jespersen vs Sonnenschein question on our generals [=*Ph.D. qualifying*] exam, with our task being to translate the debate into Chomskyan terminology and discuss it – if I remember rightly, it was about object-raising or something like that.' By that time, Sonnenschein's (1927) critique of Jespersen was an antiquated reference. The linguists leading the generative enterprise in the early 1960s, by taking Jespersen and his erstwhile adversaries seriously as interlocutors, reconfigured the relation between TLG and linguistics. To be sure, it was a TLG overhauled by Jespersen and a linguistics overhauled by Chomsky that forged such a bond.

But the format of this relationship was a tacit coalition, not a merger.[9] Recall that language (events and event-patterns, inviting scientific analysis) contrasts with discourse (a realm of action, the stuff of history, not science). Grammar/science aka linguistics remains **language**-focused, descriptive, given to exact systematisations (which stem from cross-linguistic comparisons and surveys), and oriented skywards, towards the limitless potential for sentence formation. Grammar/culture aka TLG is wedded to the **discourse** of one particular language. Although its theoretical foundations allow an amphibian Jespersen to do his linguistics and his extraordinary TLG in the same breath, the practice of everyday TLG prescribes a pedagogy, enforces an editoriality, and closely hugs the ground of present-day usage, as well as saliently remembered periods of the past.

Why do we claim, then, that the amphibian character of the sentence makes Jespersen possible? What is the purport of our observation, showcased as '(1)' earlier, that the sentence has a grammatical/scientific structure and is also subject to grammatical/cultural 'editoriality'?

An example or two will dispel the puzzlement. Lessing (1987: 68) writes:

> Or, if I say I think ginger groups, pressure groups, are invaluable because they prevent a society from going sleepy and unself-critical, then that is all right too – no, it is the word "élite" that is suspect. Very well, let's discard it, we live in a time when people may murder for the sake of a word, or a phrase...

Our main point is that, if we take a 'formal sentence' to be any stretch of text starting with a capitalisation and ending with '.', then the first formal sentence of the passage just cited is interrupted on purpose; the dash marks a pause, followed by an afterthought. Exactly like the '...' at the end of the quote (present in the original, and intended to create a special effect), such deliberate self-interruption defies grammatical analysis (of either variety – generative or TLG), but belongs to the toolkit of our 'editoriality'. Lessing writes 'no' and 'very well' to stage an exchange between two figures on her podium; the second voice appears to concede a point to the first voice; but she makes it clear that they are compartments in the author's mind.

Readers who imagine that 'editoriality' might be a side-effect of punctuation are invited to consider another Lessing passage (1987: 68):

> The minority develop this idea, at first probably in secrecy, or semi-secrecy, and then more and more visibly, with more and more support until ... guess what? This seditious, impossible, wrong-headed idea becomes what is known as "received opinion" and is loved and valued by the majority.

If Lessing were to have turned the '...' and the '?' into paired dashes and used lower-case 't' for 'this', her sentence boundaries would have changed minutely. But her **editoriality** – in the sense of this term we seek to illustrate rather than to define – would remain intact. We see here yet another instance of Lessing interrupting herself and rearranging her thoughts after the pause.

Again, no grammatical account of these effects is available, for reasons of principle. Editoriality rides on an intricate interplay between the actions of an author and the sentential structure-schemas that the language makes available to her. Grammar only deals with the syntax of the language, consisting of such schemas. Hence our claim that a sentence is an amphibian entity belongs both to 'structure'/**language** and to 'editoriality'/**discourse**.

An outsider might wonder: why do we call it 'editoriality'? Why not attribute these effects to composition (grammar's twin) and speak of 'compositionality'? This is actually a serious question, requiring a twofold answer. The content-level answer is that one's ability to edit a sentence (to redraft it in the written mode, or take it back and say it again with a correction in the spoken mode) is a specific fact, distinct from composition. Terminologically speaking, we cannot use the word 'compositionality' because it is a technical term. The compositionality principle is defined in (2) below. We are about to see, in the 'Compositionalism, Overreach, and Consequences' section, that excessive application of that principle has landed linguistics in trouble.

(2) **The compositionality principle:** Words combine to build constructions. Constructions combine to build a sentence. **Compositionality** is the principle that the meaning of every construction is a specific combination of the meanings of its component parts; the specification comes from the mode of assembly (*astute thief* and *alleged thief* put the noun and the adjective together in two different ways). However, idioms like *spill the beans* 'reveal the secret' bypass compositionality and call for special treatment.

Compositionalism, Overreach, and Consequences

What we are about to watch is an object lesson in scientific overreach. A reader who reads the trees too closely will miss the forest, hence this prefatory warning.

Generative linguists found compositionality both pleasingly exact and easy to use in syntax, the study of sentence structure. They proceeded to extend its use to the study of word structure – known as morphology. At first sight, many of us fail to see that this move instantiates 'scientism', an overextension of the methods of science. We need to identify it as a separate move for its consequences to become visible. Here is one formulation:

(3) Practitioners of **pervasive compositionalism** ('p-compositionalism', from this point on) segment a word into fragments called 'morphemes', associated with individual meanings. They regard the word as a construction composed of morphemes, which observe compositionality in the sense defined at (2). They admit exceptions analogous to the idioms mentioned in (2), e.g. *cock-tail*.

Certain plausible-looking examples initially make (3) look attractive. When we go from *combine* to *re-combine* to *re-combin-able* to *re-combin-abil-ity*, the outputs look like miniature syntax. A thinker preaching recombinability recruits devotees called *re-combin-abil-itar-ians*, who swear by *re-combin-abil-itar-ian-ism*. (Here the slight rewrite of *ity+ian* as *itar+ian* leaves undisturbed p-compositionalism's triumphal march from sentence structure into word structure.)

49

But (3) lands us in deep trouble. To see how it does, try placing *con-ceive* alongside *de-ceive, re-ceive, per-ceive; con-sist, de-sist, re-sist, per-sist; con-fer, dif-fer, re-fer, pre-fer; con-sume, re-sume, pre-sume*. The system segmenting words into morphemes authorises[10] every hyphen here, creating the **opaque** (individually meaningless) **morphemes** *con-, de-, re-, per-, pre-* on the left and *-ceive, -sist, -fer, -sume* on the right. But we were told – when we were taught how to identify morphemes, on the basis of the p-compositionality idea – to segment words in such a way that every morpheme would carry a meaning of its own, were we not? The sight of these opaque morphemes worries us.

The p-compositionalist, who needs opaque morphemes to save her theory, will dismiss our worries, observing that syntacticians cheerfully accept an **opaque word within a phrase**. We all accept *betwixt* in *betwixt and between*, or *let* in *without let or hindrance*, or the French *fur* in *au fur et à mesure* ('in due course'), or the Hindi *taach* in *puuch taach karnaa* ('to make inquiries'). Why, then – the p-compositionalist asks – do the **opaque morphemes within a word** worry some of us, leading us to contest (3)?

Surely even non-linguist readers appreciate our worry. Within a phrase, an opaque minor receives handholding or chaperoning from a responsible adult – an independently meaningful word. Thus, the **minor words** *betwixt, let, fur, taach* have their hand held by *between/hindrance/mesure/puuch*. But p-compositionalism allows opaque *con-* on the left and opaque *-sume* on the right to magically produce a meaningful 'consume'. How can morphology let two minor morphemes add up to one adult word? Syntax never allows two minor words to add up to one adult phrase.

In response, our p-compositionalist will point to the numerous minor morphemes that have both their hands held, from the left and the right – like *o* in *soci-o-political, psych-o-dynamics, electr-o-magnetic, Sin-o-American*.[11] They ask us to allow a few anomalous members in the broad category of opaque morphemes. But even if we grant this, (3) remains troublesome. The 'both hands held' logic makes us count *al* in *dramatic-al+ly, terrific-al+ly, monotonic-al+ly* as doubly chaperoned (*ic*-word on the left, *ly* on the right). Such an account, however, fails to handle cases where the *al* falls within an *ical*-word on the left (leaving no opaque *al* morpheme) – *magical+ly, lackadaisical+ly, physical+ly*[12] – or cases where the left-hand word allows either an *ic* or an *ical* ending (resulting in a 'poltergeist'[13] opaque *al*) – *analytical+ly/analytic-al+ly, philosophical+ly/philosophic-al+ly, metaphorical+ly/metaphoric-al+ly*. The account can be patched up, by inventing exceptions within exceptions. But even non-linguist readers now see that p-compositionalism keeps getting its adherents into odd and exception-ridden commitments.

Some readers will probably say that the difficulties surrounding (3) stem from a laudable effort to make the study of language rigorous and scientific; that issues that look intractable today will dissolve after fresh breakthroughs; that none of this amounts to a crisis. We agree; our remarks above only showcase for the lay reader some examples of what the extreme move (3) leads to. We see a crisis not here, but in the consequences of the larger project – extending the logic of syntax (inter-word relations in a sentence) into morphology (inter-segment relations within a word). That 'unified grammar' logic sees segments and their combinations everywhere and seeks exact laws governing both syntactic and morphological assembly. This project triggers a crisis on several fronts. Given our audience, we focus only on the relation between rules and 'irregular' forms, often called 'exceptions'. In morphology (the grammar of words), exceptional forms nearly always veto their regular (rule-governed) alternatives. Past of *go* or *come*? Always *went/came*; never **goed/*comed*. A few exceptions, like *learnt/burnt*, do not exercise the veto; they tolerate regular alternatives (*learned/burned*). But the overwhelming majority of exceptions are like *went*, which vetoes **goed*.

In this respect, morphology sharply contrasts with syntax (the grammar of sentences), where irregular forms consistently tolerate regular alternatives. The exceptional constructions *What say you?* and *Suffice it to say that we are terminating the negotiations* happily coexist with their regular versions *What do you say?* and *It should suffice to say that we are terminating the negotiations*. We could multiply examples, but nobody contests this fundamental operational difference between syntax and morphology. The project of unifying them is therefore not merely a frustrating exercise in practice; it lands the theory in a muddle.

Anybody who implements p-compositionalism all the way, attributing 'meanings' to those word-segments called 'morphemes', finds that 'morpheme-meanings' lack the specifiability found in a typical word-meaning. Try chopping *driver* into *drive+er*. In an imaginary world where *er* consistently carried a specific meaning like 'someone who performs the [driving or other] action', p-compositionalism and morphology–syntax unification might have made sense. But, in our real world, *driver*'s *er-* 'morpheme' shares only meaning-like connections with the *er-* 'morphemes' in *astrologers, geographers* (subject specialists); *photographers, biographers* (skilled professionals); *forty-niners, women's-libbers* (thematically identified person types); *farmers, miners* (who get paid); *boarders, lodgers* (who pay); *diners* (where paying happens); *dishwashers, can-openers; page-turners, thrillers; loudspeakers; fivers, tenners*.

Now, this *er* is a typical 'morpheme'. Quite generally, 'morpheme-meanings' are what we might call 'rudimentary'. Readers may wonder if words never have rudimentary meanings. Some do, like *pertain*. But one finds no language that expresses the notions listed above in the format of **phrases** that use a 'rudimentary' **word**. No language translates our list into **astrology pertainants, *geography pertainants, *forty-nine pertainants …, *five pertainants, *ten pertainants*. Once one realises the unimaginability of such a language, one sees that how drastically morphology differs from syntax.

Any residual hope of unifying the rudimentary meanings of 'morphemes' with the specifiable meanings of words perishes on the rocks of a classic paper in which Tirumalesh (1991) explains why one cannot express 'punching Bill again' as *repunching Bill*. The core of his argument highlights cases like *repunch the sheets of paper*, where one's predecessor using the punching machine missed several sheets and one has to redo the job properly. Tirumalesh's point is that doing it over, as a concrete assignment, merits a single-word embodiment in such a context. The word *repunch* in that scenario does not just combine the meanings of the bits *re+punch*. Formatting the combination as a single word dramatises the coherent specifiability of the task.

We can feel free to contradict Tirumalesh's empirical conclusion. Imagining an outlandish scenario – where punching Bill again becomes a second gangster's job after a first thug bungles it – might enable an author to write *Sylvester detested having to repunch Bill, but he did it for Kyle's sake*. Our extension of his reasoning contradicts his empirical conclusion but reinforces Tirumalesh's (1991: 266) methodological point: 'The pragmatic landscape of *re*-words has only certain signposts to guide us and these – we hope to have demonstrated – resist all attempts at semanticization'. Signposts are not rules carved in stone; somewhere in life or art, 'repunching Bill' might make sense.

Tirumalesh places his 'signpost' idea in the context of Aronoff's (1976) thinking, which rejects the p-compositionalist proposal that syntax should annex morphology. Aronoff, Tirumalesh, and other linguists who listen to words carefully for their resonances draw on the resources of semiotics – the study of signs and their functioning (both in language and in painting, film, the performing arts, and everyday life). Semiotics alone cannot rescue linguistics from its crisis, but it certainly helps make sense of it. Let us see just how.

The Architecture Meets the Thatchwork

In a language like English, the verb agrees with its subject for grammatical number (*The **girl is** playing*, singular, vs *The **girls are** playing*, plural); this **architectural** apparatus holds the sentence together. But in Indonesian, these sentences come out as *Gadis* (girl) *itu* (the) *sedang* (in progress) *bermain* (play) vs *Gadis-gadis sedang bermain*. In the plural version, we see *gadis* (girl) appearing twice, 'reduplicated' – the standard marker of noun plurality in Indonesian. But the verb shows no change. Generative linguists, being formally minded, equate the plurality marking performed by Indonesian reduplication with the work done by the *s* in English *girls*. But this formal equation gets muddled in Hindi, where – in contrast to singular 'long pole', *lambaa daṇḍaa* – plural 'long poles' is either *lambe daṇḍe* (where the vowels indicate singular vs plural) or *lambe lambe daṇḍe*, with reduplication

52

vaguely emphasising plurality. Hindi and Indonesian force us to ask how facts of '**thatchwork**', such as reduplication – especially in languages with no agreement, like Indonesian – are aligned with agreement-like **architectural** patterns, whose ubiquity in European languages like English has led linguists to focus on them too exclusively for too long.

The point is not limited to something as obvious a piece of thatchwork as reduplication. In English, the architectural system of pronouns (*I*, *we*, *you*, *she*, *he*, *it*, *they*) differs from the thatchwork properties of demonstratives (*this*, *these*, *that*, *those*) in subtle respects, which interest generative linguists. In sentence (4), the two occurrences of the pronoun *she* may 'co-refer' (indicate the same person):

(4) She is still exactly where she was five minutes ago.

But in (5), the second *this* must carry stress and be associated with a specific pointing. The two demonstratives cannot co-refer – their referents must be distinct:

(5) This is still exactly where *this* was five minutes ago.

The co-referential reading calls for (6), using the pronoun *it* instead of the second *this*:

(6) This is still exactly where it was five minutes ago.

Now contrast English with Hindi, where both (4') and (5') (translations of (4) and (5)) allow the two copies of the repeated word to co-refer:

(4') Vo pããc minaṭ pahle jahãã thii, abhii tak vo vahĩĩ hai.
(5') Ye pããc minaṭ pahle jahãã thaa, abhii tak ye vahĩĩ hai.

The Hindi *vo* is a pronoun and simultaneously belongs to the demonstrative pair *ye* ('this')/*vo* ('that'). Hindi demonstratives, unlike English *this* and *that*, allow co-reference. Generative linguists have been struggling to find resources to handle the interplay between 'thatchwork' (as in demonstratives, as in reduplication) and 'architecture' (as in pronouns, as in *s*-marked English *girls* and *e*-marked Hindi *ḍaṇḍe*). The present section showcases one response to this challenge.

Working with kindred spirits like Tirumalesh and Aronoff, some of us have been running a school of thought within generative linguistics called **substantivism** (Dasgupta et al. 2000; Singh and Starosta 2003).[14] We work with modes of substantive organisation that cut across standard formal boundaries separating morphology from syntax and other components.

Our recognition[15] of architecture and thatchwork as contrasting modes of substantive organisation is one of our tools for what we see as the main task today: acknowledging textures and resonances, and recognising the specific connections and disconnections that obtain between sensitive phenomena of this type. To this end, we refrain from chopping words up into morphemes: we recognise that word-parts encode rudimentary semi-meanings and participate in the build-up of texture.

This obviously means rejecting p-compositionalism. But we go further, pushing the debate right back into syntax, the home ground of compositionality. There, we ask: just how do words forge compositional bonds with each other? The conventional answers rely on the architectural mechanisms of agreement and case-marking noticed from Sanskrit, Greek and Latin onwards: verbs case-mark their objects, adjectives agree with nouns, etc.[16] But the world of thatchwork phenomena we must **now** face does not lend itself to mechanical extensions of those games. Our '**now**' refers to the post-colonial moment, which forces linguists to take all languages seriously, suspending the old hierarchies at least tokenistically.[17] What do the problems of this new era look like on the ground?

For clarity, we focus on the spectrum of noun categorisation. At one end is 'gender', familiar to us all (masculine and feminine in Hindi). At the other end are 'classifiers', unfamiliar to many readers.

Hindi copies masculine/feminine markers onto adjectives agreeing with nouns, or verbs agreeing with nouns. Watch the *aa*'s and *ii*'s in *Lambaa* (tall) *larkaa* (boy) *aayegaa* (will come), *Lambii* (tall) *larkii* (girl) *aayegii* (will come). The noun category in Hindi is subclassified only into two classes, called genders. Gender marking in Hindi is fused with the marking of the two grammatical numbers, singular and plural. The sentences just given, when pluralised, insert a characteristic *n* inside the verb, and change the vowels at the end of the nouns and adjectives: *Lambe larke aayenge* 'tall boys will come', *Lambii larkiyãã aayengii* 'tall girls will come'.

At the opposite end of the spectrum is Indonesian, with a staggering 60 noun classes, each class marked by a characteristic classifier (Dardjowidjojo 1978: 64, cited by Chung 2000: 162). An Indonesian classifier appears just once – boldfaced in the examples *se-orang bocah* ('one-Classifier boy'), *dua buah topi* ('two-Classifier hat') – and is never copied onto any word agreeing with the noun. Indonesian classifiers belong to pure 'thatchwork', devoid of agreement or any other 'architectural' behaviour.

To make some sense of classifiers, try looking at *tiga ekor gajah*, Indonesian for 'three elephants'. The classifier *ekor* here says that elephants belong to the animal class. But *ekor* as an independent noun means 'tail'. The *orang* and *buah* boldfaced as classifiers in the examples above, when used as nouns, mean 'person' and 'fruit', respectively; hats classify as fruit like in Indonesian noun categorisation. Not all classifiers are so transparent;

highlighting the transparent ones helps us understand those that have become obscure.

Unlike gender markers, classifiers do not normally fuse with the plural system. Indonesian grammar says one either counts the elephants (*tiga ekor gajah* 'three elephants') or plural-marks them (*gajah gajah* 'elephants'); doing both produces the ungrammatical *tiga ekor gajah gajah* ('three Classif elephant elephant'). Chung (2000: 169) states this in Note 8 but reports a few counterexamples in her main text. Such hesitancy is typical of thatchwork, which lacks the fixity of architecture.

What turns these phenomena into a domain of interest for substantivists is the middle of the noun categorisation spectrum, where we find Swahili. It has ten noun 'classes', each with a characteristic marker. Swahili is half-way between Hindi (whose gender markers form part of the word and are drawn from a tiny set of two genders) and Indonesian (whose classifiers stand apart and come from a large set of 60 classifiers). Swahili has reasonably many (ten) class markers, but these form part of the word and participate in agreement – exemplified by the boldface material in **Wa**naume (men) **wa**kubwa (big) **wa**lianguka (fell) 'big men fell', **Vi**tabu (books) **vi**kubwa (big) **vi**lianguka (fell), 'big books fell'.[18]

Just as Hindi agreement fuses gender with number, Swahili agreement fuses noun class with number. Thus, the plural class markers *wa/vi* have matching singulars *mw/ki*: **Mw**anaume (man) **m**kubwa (big) **a**lianguka (fell) 'a big man fell', **Ki**tabu (book) **ki**kubwa (big) **ki**lianguka (fell), 'a big book fell'. Swahili arranges all class markers in singular–plural pairs.

Clearly, whenever noun categorisation enters the agreement system, it fuses with number. Most generative linguists have framed their questions about the categorisation spectrum (Hindi to Swahili to Indonesian) in terms of architectural features like number, resulting in an excessively tight approach to the categorisation spectrum. Chierchia (1998) provides an account of this sort. Responding to him, Chung (2000) advocates a selectively relaxed approach, so as to acknowledge the Indonesian flexibility flagged above. But delivering the relaxation she recommends calls for resources unavailable within the mainstream generative linguistics that frames most work. Chung has been chipping away at the problem, along lines akin to ours; see Chung and Ladusaw (2007) for one direction in which she seeks to overcome the rigidity of the standard p-compositionalist architecture.

Substantivism's semiotic package draws on the exact architecture's solidity and the discretionary thatchwork's versatility as twin virtues and helps deliver the selective flexibility required. These resources enable substantivist work on noun categorisation (Dasgupta 2018, 2019) to map the initially surprising contrast in India between the Dravidian and the main[19] Indo-Aryan languages vis-à-vis gender, for example. That Dravidian has 'natural' gender, while Hindi categorises *granth* as masculine and *pustak* as feminine even though both these words mean 'book', falls into place on our approach.

How do these points advance our argument?

Where p-compositionalism took the transparent compositional logic for granted in the sentence and chopped the word into pieces to push this syntactic logic into morphology, substantivism leaves words intact and asks how syntactic composition works in the first place. Pressing this question leads us to see that the agreement systems found in the architecture-prominent languages intensively studied in most grammatical traditions have lulled us into assuming that either agreement proper or mechanisms mimicking it forge the inter-word bonds that drive syntactic composition. That assumption does not survive the serious study of a wider range of languages, which brings to the fore the texture of thatchwork at the Indonesian end of the spectrum, and in the Swahili middle, where the combination of discretionary thatchwork with exact architecture continues to challenge us.

With compositionality itself demoted from 'assumption' to 'problem', linguists find themselves slowly coming to terms with the realisation that linguistics is an area of **precarious exactitude**, half-way between the simple exactitudes of a geometry and the meandering coastlines of a geography.

Did the reader notice that Chung's interlocutors included not just the descriptive enterprise of generative linguistics (represented by Chierchia (1998), whose work Chung (2000) cites and contests) but the normativity of Indonesian TLG as well?

Situating the Resonances

Scientific linguists emphatically distanced themselves, as descriptivists, from authoritarian-prescriptive TLG at the moment of structuralism, flagged in the 'Compositionalism, Overreach, and Consequences' section. But TLG has not stood still. It has kept pace with advances in the culture of democracy and taken a leaf out of the book of linguistic science as well. TLG's prescriptive-authoritarian enterprise that linguists found so upsetting focused on words. The peak of TLG's work on sentences, however, took the form of Jespersen and participated in the process of shaping modern generative syntax itself. When we look at sentences today and ask how they look under the linguist's analysis and under the author's or the proofreader's editoriality, we find that TLG, coupled with its literary-pedagogic extension, is the only site at which the analysis and editoriality can meet. TLG is obliged to constantly monitor empirical realities as part of the work of updating its normative standards – and draws on linguistics for this monitoring.

When substantivism borrows key resources from the semiotic study of resonances in order to align the apparatus of linguistics with the texture of thatchwork, it has to anchor this exercise in the textual space of the speech community. Many resonances come from widely circulated texts, especially from the TLG and the literary pedagogy associated with it. In order to bring out the significance of this localistic emphasis of substantivism, we need to mention Indonesian again, at the risk of annoying the reader.

From a shared 'Malay language' base, two mutually intelligible national languages emerged and were fully codified after Malaysia and Indonesia gained independence. In 1972, leading TLG scholars of the two countries agreed on a shared orthography, ironing out some points of divergence between their British and Dutch colonial legacies.[20] Now, this TLG apparatus did not just happen. The standardisation of Indonesian was carried out under the leadership of the iconic language standardiser Sutan Takdir Alisjahbana (1908–1994), and that of Malaysian followed his example. Alisjahbana found linguists incapable of satisfying any of his needs as he did his work. At one point, he felt compelled to take the whole enterprise of modern linguistics to task for its fundamental failure to understand that a sustainable relationship would have to be built between the analytical science of linguistics and the normative art of TLG. He spoke for the global South as a whole when he formulated his criticism in the following terms (1965: 14–15, cited in Coulmas 2016: 270–271):

> What they [= *the leaders of new nations*] need is not descriptive, but prescriptive linguistics. It is thus very regrettable that precisely in these processes and problems that are crucial for the languages of the developing societies, processes and problems which can be formulated succinctly in the terms standardisation and modernisation, modern linguistics, through its static, formal and micro approach, is least able to contribute.

That substantivist linguistics places the need to build a sustainable TLG–linguistics relationship on its agenda[21] is not surprising if one considers the fact that, like Alisjahbana, the architects of substantivism have either been from the global South or (like Ford and Starosta) focused on third or fourth world languages.

However, there is a larger issue at stake here. Linguistics is not just another discipline; its progress, in tandem with its exoteric twin TLG, is a prerequisite for the health of a democratic society. The enterprise of free speech in the public space, at the heart of the construction and maintenance of the democratic imagination, depends on steady and civil contestation. The dominant current of discourse, with its entrenched defaults,[22] needs to be challenged at multiple points by marginal streams of discourse, which seek to unsettle those defaults. If this process of constant dissent does not flourish, neither does the culture of a democratic society. It is the editoriality – initially installed by the TLG apparatus and by the basic literature pedagogy – that tacitly teaches children and young adults the linguistic and textual defaults while drawing their attention, of course, to salient material that keeps the defaults in the presupposed background.

Social and cultural struggles today have been foregrounding a few of the defaults by labelling them as discriminatory against people of colour, or Dalits, or indigenous peoples, or women, or queers, or some other identifiable category. But overwhelmingly many defaults remain naturalised and invisible, as they are pervasive structural patterns that do not even appear to be directed for or against any category. By the same token, the dissent against those invisible defaults also takes the form of invisible undercurrents unconsciously shaping the criticism or the counter-discourse that the dissident authors and artists themselves articulate only in the aesthetic, ideological, or other widely intelligible terms in which they consciously understand them.

Some Indian linguists, in and outside India, have been at the forefront of the struggle for an accountable, democratic linguistics that will help redraw the cultural and ideological map not just for language but for its cultural and social crucible as well, in the South and in the North. One key step, at the level of reconfiguring the linguistics–TLG equation, is to note the place that Jespersen's TLG has taken in shaping the very terms on which generative linguistics conducts its 'purely scientific' research. Another step is to reiterate the relevance of a critique of linguistics (coupled with the presentation of a positive alternative with semiotic characteristics) articulated by a Jespersen-like figure from the South, who is often seen only as a littérateur and cultural icon – Rabindranath Tagore. In another article (Dasgupta 2019), to which we refer the reader, we have provided an elaborate exposition (with substantivist spice added) of Tagore's critique of linguistics and shown that it applies to contemporary generative linguistics as well.

Once the fact that the exactitude scientific linguistics pursues is a precarious exactitude begins to register, the rest of the discipline's self-understanding will change as well, and a realignment of North–South equations with respect to language and public discourse will follow. We hardly need to add that 'North–South' does not reduce to geography alone. Within every society, located anywhere, the technological hardware, software, and personnel that underpin the search engines, automatic translators, text-to-speech converters – all the way to the social media which alone the commentarial chatter flags as meriting our critical attention – constitute the form that the North takes. Structurally speaking, the North is the techno-scientific foundation, and the South is the 'us' whose vulnerability negotiates with the North and keeps picking itself up and renewing the negotiation after every defeat. When we speak of resonances being located, and the substantivist approach using resonances as the basis for its deployment of semiotic resources so that the thatchwork and the architecture are aligned, we are speaking of 'locations' consisting of these 'everywhere' versions of the North and the South, gentle reader. Please read this chapter in the context of its resonances with the rest of this volume, and with other work by like-minded authors, as we inch our way forward towards a clearer understanding of the issues of discourse, and of the languages in which it is conducted.

Acknowledgements

Conversations with Sandra Chung, Florian Coulmas, Chinmay Vijay Dharurkar, Hannsjoerg Hasche-Kluender, Diego Krivochen, Mélanie Maradan, Ken Safir, Susan Schmerling, and Niranjana Thokchom helped shape this chapter. The standard disclaimers apply.

Notes

1 We do not doubt that this use of the term 'event' is open to contestation. When people speak in commonsensical terms, they do not normally use the word 'event'. We request the reader to allow us to use the term without specialist commentary. Any analysis of the term 'event' would take us into metatheoretical contestations without adding anything useful either to our argument or to any obvious counter-argument.

2 To be sure, a heterodox view of the social sciences might characterise them as, precisely, specialist sciences of subdiscourses. Our point is that this is not a known self-characterisation prevalent in those disciplines. An enterprise known as 'discourse analysis' does exist, has never been claimed to be a science, and has not yet become a stable discipline with internationally accepted concepts and terms of reference.

3 The 1901–1902 Bangla grammar debate (or '*Bengali* grammar debate', for readers who prefer that name for the language Bangladesh shares with India) is a particularly dramatic and well-documented example of the need to engage with history. Our brief account of that debate (Dasgupta 2017) includes a translation of Hara Prasad Shastri's article (1901) that initiated the discussion and summarizes the debate itself, highlighting Rabindranath Tagore's (1973a 1973b) intervention. Tagore argued for a grammar of modern Bangla as a language independent of Sanskrit.

4 Colonies and indigenous peoples are only examples of the 'peripheral' topics that attracted linguists. They also found rural dialects in Britain exciting. But the case of colonies happens to be crucial for our argument.

5 Full disclosure: the topic of children matters because Chomsky's idealisation imagines children learning their mother tongue instantaneously; thus, they too count as perfectly proficient. This part of Chomsky's move emulates the point mass postulation strategy of physics.

6 Chomsky (1986) provides a particularly clear formulation distinguishing the linguist's notion of language from the general public's TLG-based 'teleological' view of a language as a goal towards which the standard pedagogy guides its speakers.

7 James D. McCawley states this in his editorial introduction (Jespersen 1937/1984: xi)

8 Testimony from two of the students in question, George Lakoff (p.c. 1982) and Barbara H. Partee (p.c. 2021; up to the early 1960s she was known as Barbara Hall).

9 We return in the 'Situating the Resonances' section to the question of how to move beyond this inadequate tacit coalition, which became an operational reality in the 1960s and crucially shaped the formal mechanics of generative syntax but was never acknowledged in the official narratives either in TLG or in any school of generative linguistics.

10 To see what logic authorises the segmentations, consider the *conceive–perceive* series. Stating only once that *-ceive* verbs become *-ception* nouns is more eco-

nomical than itemising *conception, perception,* etc. In the morphemic logic, such economising becomes an obligation, not a choice.

11 Sanskritists will see that vikaranas, like *a* in *tud-a-ti* ('hits'), are doubly-chaperoned opaque morphemes.

12 We use boldface to highlight the opaque elements; that these three words show nothing in bold type means that the opaque *al* morpheme is absent here.

13 A poltergeist is an intermittently visible ghost.

14 The title of this chapter is a riff on *Methods in Structural Linguistics*, a book by Chomsky's teacher Harris (1951).

15 Dasgupta (2013) makes this move, concretely deploying what the formal apparatus calls 'phi-features' and 'tau-features'.

16 Readers clued into formal linguistics will recognise that agreement, in this chapter, serves as a metonym for the family of devices that includes trace theory and other tools for tracking co-reference.

17 By this, we merely mean that nobody at a conference today would dare describe some languages as 'civilised' and others as 'savage', which was a common practice in colonial times.

18 Data taken from https://en.wikipedia.org/wiki/Swahili_grammar#Concord, consulted on 9 June 2021.

19 Eastern Indo-Aryan languages like Odia, Assamese, and Bangla lack gender and fall outside this pattern.

20 To be sure, other differences remain; a Wikipedia article we consulted on 11 June 2021, https://en.wikipedia.org/wiki/Comparison_of_Indonesian_and_Standard _Malay, informs us that that the two languages 'are generally mutually intelligible, yet there are noticeable differences in spelling, grammar, pronunciation, and vocabulary, as well as the predominant source of loanwords'.

21 Recall the point, made in the 'A Science of Language?' section, that the valorisation of Jespersen's version of TLG by the linguists at MIT in the early 1960s led to a tacit coalition between the two enterprises. Unfortunately, once generativists had taken what they needed from Jespersen, new recruits to generative linguistics from the late 1960s onwards received the legacy as an undifferentiated 'generative syntax apparatus'. TLG stopped being regarded as a potential coalition partner, and the tacit coalition dissolved even in the mainstream linguistics of the global North. Not a single school of mainstream Northern linguistics has even begun to take on board the task of nurturing the growth of TLG in Southern speech communities, or to realise that an explicit partnership with TLG – on mutually beneficial terms which need to be negotiated – is a prerequisite for the serious linguistic analysis of Southern languages.

22 A default is an automatic, semi-unconscious setting in the discourse. One says 'bread and butter', not 'butter and bread'. People pace 'up and down', not 'down and up'. Unsettling these trivial defaults does not become a mission for dissidents, of course. But dissent may take the form of preferring new formats of verse, new ways of writing fiction, new modes of debate (such as the mode pioneered by Lakatos 1976), instead of the defaults that have been entrenched for centuries.

References

Alisjahbana, Sutan Takdir. 1965. *The Failure of Modern Linguistics in the Face of Linguistic Problems of the Twentieth Century*. Kuala Lumpur: University of Malaya Press.

Aronoff, Mark. 1976. *Word Formation in Generative Grammar*. Cambridge, MA: MIT Press.

Chierchia, Gennaro. 1998. 'Reference to Kinds Across Languages.' *Natural Language Semantics*, 6: 339–405.

Chomsky, Noam. 1957. *Syntactic Structures*. The Hague: Mouton.

Chomsky, Noam. 1965. *Aspects of the Theory of Syntax*. Cambridge, MA: MIT Press.

Chomsky, Noam. 1986. *Knowledge of Language: Its Nature, Origin, and Use*. New York: Praeger.

Chung, Sandra. 2000. 'On Reference to Kinds in Indonesian.' *Natural Language Semantics*, 8: 157–171.

Chung, Sandra, and William A. Ladusaw. 2007. 'Chamorro Evidence for Compositional Asymmetry.' *Natural Language Semantics*, 14: 325–357.

Coulmas, Florian. 2016. *Guardians of Language: Twenty Voices Through History*. Oxford: Oxford University Press.

Dardjowidjojo, Soenjono. 1978. *Sentence Patterns of Indonesian*. Honolulu: University of Hawaii Press.

Dasgupta, Probal. 2013. 'Scarlet and Green: Phi-Inert Indo-Aryan Nominals in a Co-Representation Analysis.' In Shu-Fen Chen and Benjamin Slade (eds) *Grammatica et Verba/Grammar and Verve/Studies in South Asian, Historical and Indo-European Linguistics in honour of Hans Henrich Hock*, pp. 46–52. New York: Beech Stave.

Dasgupta, Probal. 2017. 'Analysis and Modernity: The Language Debate in the Bangiya Sahitya Parishad.' In M. Sridhar and Sunita Mishra (eds) *Language Policy and Education in India: Documents, Contexts and Debates*, pp. 112–125. London/New York: Routledge.

Dasgupta, Probal. 2018. 'Classify or Agree: The View From a Split Classifier Language.' *Kalakalpa*, 3 (1): 15–32.

Dasgupta, Probal. 2019. 'Too Big a Bang for Language: Tagore's Critique Reloaded.' In Partha Ghose (ed.) *Tagore, Einstein and the Nature of Reality: Literary and Philosophical Reflections*, pp. 33–58. London/New York: Routledge.

Dasgupta, Probal, Alan Ford, and Rajendra Singh. 2000. *After Etymology: Towards a Substantivist Linguistics*. Munich: LincomEuropa.

Hall, Robert A., Jr. 1950. *Leave your Language Alone*. Ithaca, NY: Linguistica.

Harris, Zellig. 1951. *Methods in Structural Linguistics*. Chicago, IL: University of Chicago Press.

Jespersen, Otto. 1933. *Essentials of English Grammar*. London: George Allen & Unwin.

Jespersen, Otto. 1937. *Analytic Syntax*. Copenhagen: Levin & Munksgaard. Reprinted 1984. Chicago: Chicago University Press.

Lakatos, Imre. 1976. *Proofs and Refutations: The Logic of Mathematical Discovery*. Cambridge: Cambridge University Press.

Lessing, Doris. 1987. *Prisons We Choose to Live Inside*. New York: Harper & Row.

Shastri, Hara Prasad. 1901. 'Bāṅgālā byākaraṇ.' *Sahitya Parishat Patrika*, 8 (1): 1–7.

Singh, Rajendra, and Stanley Starosta (eds). 2003. *Explorations in Seamless Morphology*. New Delhi/Thousand Oaks/London: Sage.

Sonnenschein, Edward Adolf. 1927. *The Soul of Grammar: A Bird's-Eye View of the Organic Unity of the Ancient & the Modern Languages Studied in British and American Schools*. Cambridge: Cambridge University Press.

Tagore, Rabindranath. 1973a. 'Bāṅglā krit o taddhit' [Primary and Secondary Derivatives in Bangla].' In his *Rabīndra-Racanābalī*, Vol. 12, pp. 382–396. Calcutta: Visvabharati.

Tagore, Rabindranath. 1973b. 'Bāṅglā byākaraṇ' [Bangla grammar].' In his *Rabīndra-Racanābalī*, Vol. 12, pp. 564–578. Calcutta: Visvabharati.

Tirumalesh, K. V. 1991. 'Why You Can't Repunch Bill: An Inquiry into the Pragmatics of Re-Words in English.' *Journal of Pragmatics*, 16: 249–267.

Williams, H. G. D. 2013. 'Shattering Tradition: A Rejection of Analysis by Genre in Horace's *Ars Poetica*.' *Akroterion*, 58: 61–77.

2

'IF NOT PRECISELY A SCIENCE'

The Provocations of Literary Studies

Sharmila Sreekumar

Literary studies was awhirl with change. The hectic decades of the 1930s–1950s saw fervid efforts to re-professionalise English literary studies both in the Anglo-American academia and in the colonies straining towards political independence. Stern boundaries were erected to discretise the discipline. Literature was separated from literary studies; the two were deemed to be 'distinct activities' – the 'one is creative, an art; the other, *if not precisely a science*, is a species of knowledge or of learning', it was announced (Wellek and Warren 1949: 3, emphasis added). This, then, is an interesting moment to catch the kinetic and polymorphic story of literary studies, not least of all because of the curious wobble at the very core of attempts to define and discipline it. Take, for instance, the above statement: what began as an assertive description of literary studies soon teeters on a slew of hesitations; it gets tucked into a parenthesis (hedged by a faltering 'if', a negating 'not'), followed by a sprawling categorisation ('a species …') and a see-sawing conjunction ('or') that simultaneously signal equivalence and uncertainty. Together, they cobble a tremulous simile which imperfectly aligns literary studies with 'science'. Just when critics René Wellek and Austin Warren seem poised to disclose the precise characteristics of this volatile discipline, they get tangled in manoeuvres that involve a darting reference to 'science'. This, we shall see, is symptomatic.

***

In this chapter, I will tease out some strands of the provocative relationship that literary studies establishes with 'science'. I will show that literary studies uses 'science' to clarify and consolidate itself, to explicate methodological and definitional possibilities, and to hazard innovations and transformations.

But first, a set of cautions and disclaimers: while explicating itself, literary studies imbricates 'science' in a web of mis/representations that often bleach 'science' of all particularities (as in the above quote) and/or reduce it to idiosyncratic sets of attributes and methods. This chapter does not adjudicate

DOI: 10.4324/9781003298908-5

what a legitimate invocation of science and scientific method *should* be. Nor does it aim to establish a healthy, mutually sustaining relationship between literary studies and science. Second, it makes no effort to prescriptively fix the business of literary studies or to 'cover' every variant/school of literary criticism and theory. It is but *a* story of the shifting dis/affiliations between literary studies and 'science'; other traversals are likely to animate appreciably different aspects. Finally, it is all too likely that this story resonates in other disciplines of the humanities, social sciences, and, even, the broadly classified sciences. The attempt here is to delineate impulses that have characterised literary studies, not necessarily those unique to it.

What follows is unabashedly episodic. Foraying across decades and continents, I tarry at select waypoints to examine how (English) literary studies invokes 'science'. Interspersing these are segments where I risk propositions – about what these invocations *do* and what 'species of knowledge or of learning' literary studies sets itself to be.

Improvisation I (in Two Acts)

Act I

The year: 1959. Delivering the Rede lecture at Cambridge University, scientist turned novelist C.P. Snow indicted literary scholars for partitioning intellectual life into 'two cultures' – 'arts' and 'sciences'. The debates he triggered blew into a full-sized controversy when F.R. Leavis, the literary critic, blitzkrieged Snow and his formulations. Besides setting the terms for describing the breach between sciences/arts, this controversy also signalled a charged moment when literary criticism was being positioned as the vital core of academia.

By 1929 Cambridge critic I.A. Richards had already published his epoch-making book *Practical Criticism*, where he detailed his 'experiments' in the close reading of poetry. Together with the literary critical methods propounded by F.R. Leavis, these signalled complex transactions with 'science'. Leavisian criticism also insisted on the centrality of literature to civilisation. Refusing the separation of art from life, the aesthetic from the moral, Leavis tasked the critic with educating the public, tending the mind, and shaping cultural–intellectual life.

Act II

The year: again, 1959. Another blistering dispute. Across the Atlantic, in an elite school in the USA, an apparently humdrum class on poetry was underway. A student read aloud from the textbook: 'If the poem's score for perfection is plotted on the horizontal of a graph and its importance is plotted on the vertical ....' Just as the class had settled into familiar distractions,

Mr Keating, the English teacher, enunciated, 'Excrement!' – startling the class to attention even with the long-winded synonym of the four-letter word. Mr Keating, it turns out, was incensed by the essay's postulations and wanted its writer, a certain J. Evans Pritchard, expunged from the textbook. 'Leave nothing of it. Rip it out! Rip! Be gone, J. Evans Pritchard, Ph.D. Rip. Shred. Tear. Rip it out!', he commanded. For Mr Keating, poetry had to be savoured, its talismanic properties cherished, not eviscerated by analysis and intellectualism. Mr Keating's was a righteous war; his enemies were the '[a]rmies of academics going forward, measuring poetry'. The arid J. Evans Pritchard, with his caricaturable scientism, embodied all that blighted the proper appreciation of literature.

Postscript: Readers would recognise this iconic scene from the movie *Dead Poets Society* (Weir 1989).

New Criticism and Its Dis/contents

Notwithstanding its fictional vintage, Mr Keating's classroom is historically resonant. The offending essay, 'Understanding Poetry', all too obviously references an influential textbook and poetry anthology of the time. The eponymous *Understanding Poetry* (1938) codified the ideas of New Criticism and moulded it into a coherent classroom approach, transforming the teaching of literature in the USA and, indeed, the world. Mr Keating's fusillade against the fictitious J. Evans Pritchard was, in fact, directed at the presumptuous New Critics, the progenitors of what sometimes gets called 'scientific' literary criticism.

Who were the New Critics and what did they propose?

In the 1930s, a crop of young scholars sought to wrest literature away from the gentlemanly persuasions to which it had been consigned in the US academia: the fawning considerations of writers' lives, rapturous 'appreciation' of 'masterpieces', ecstatic declamations of aesthetic passages and paraphrases of ennobling thoughts.[1] The New Critics announced the need for an immediate 'change of policy' by which 'literary criticism' could possess 'its own charter of rights' and 'function independently' (Ransom 1937). John Crowe Ransom, a founder member, further stipulated that criticism 'must become more scientific, or precise and systematic', though he also freely conceded that it 'will never be a very exact science, or even a nearly exact one', and further, that it 'does not matter whether we call them sciences or just systematic studies' (Ransom 1937). Aside from offering another instance where initial certainty knots into ambivalence, this also reveals that rather than science itself, what Ransom championed were specific qualities associated with science – namely precision and systematicity. Indeed, Ransom was not the only New Critic to tactically plumb 'science' for attributes and analogies which would recast 'literary criticism' as a legitimate, robust academic practice.

In the rest of this segment, I shall reveal some of the intimacies that New Criticism forged with 'science' as it ventured to (1) reconfigure the object and objectives of literary criticism, (2) re-professionalise literary studies, and (3) systemise methods of literary enquiry.

* * *

Audaciously, New Criticism sought to reconstitute the object and concomitantly the objectives of disciplinary enquiry. It isolated the literary work from other kinds of writing and severed it from concerns that had previously imbricated it: the literary work was disengaged from the writer, the reader, and socio-biographical histories. The responses a work elicited – whether it caused 'oblivion of the outer world, the flowing of tears, visceral or laryngeal sensations, and such like' – were no longer consequential (Ransom 1937). All such considerations were deemed extraneous and fallacious as the New Critics refashioned the literary work into an autonomous object.[2] They focused attention solely on the compositional specificities of the literary work and the facticity of words-on-the-page, thus aligning themselves with I.A. Richards and F.R. Leavis who also concentrated intellectual energies on the literary work.[3]

The empiricism undergirding this enterprise becomes clearer when read in the light of Mary Poovey's thickly detailed analysis of the 'model system' of criticism. Unpacking the metaphor of the 'organic whole', Poovey argues that modern biology governs 'the practices of critics as apparently diverse as New Critics, deconstructionists, and New Historicists' (Poovey 2001: 409). Alerted to this long-standing metaphor, we rediscover the New Critics' affirmation of the 'simple truth ... that a successful work of art is a whole in which the parts collaborate and modify one another' (Wellek 1978: 617). Further, that a literary work is 'a highly complex organisation of a stratified character with multiple meanings and relationships', even as they also fret that the 'usual terminology ... of an "organism", is somewhat misleading, since it ... leads to biological parallels not always relevant' (Wellek and Warren 1949: 17). The New Critics, it would seem, preferred the built and the constructed – metaphorising poetry as a 'well-wrought urn' (Cleanth Brooks) and a 'verbal icon' (Wimsatt).[4] But whether they worked with the artifactual or the organic, it is clear that the New Critics accorded the literary work with indwelling identity.

Poovey also discloses the 'functional equivalence' of 'genre' and 'genus', arguing that the '*model system* serves the same function in biology that *genre* does in literary criticism' (Poovey 2001: 410, emphasis in original). Although the New Critics did not explicitly use the term 'genre', they grouped together and taxonomised a range of 'highly complex' literary objects into categories based on perceived relations of similarity.[5] Duly 'object'ified and classified, the literary work was primed for positivist observation and methical study.

It is important to clarify that this was not, by any account, the 'invention' of literary criticism.[6] What this moment consolidated was the officialising of literary criticism within the university-system and its professionalisation as a knowledge community. Not surprisingly, when Ransom declared the independence of 'Criticism Inc.' and instituted it as the primary business of literary studies, he mandated the professor of literature as its sole, legitimate exponent. All other claimants were summarily disqualified: the artist, because 'his' 'intuitive' understanding could not satisfactorily explain the 'theory of the thing'; the philosopher, because the sweep of 'his theory' disallowed a 'persistent and intimate' acquaintance with 'particular works of art' and 'their technical effects' (Ransom 1937); the journalist, the reviewer, the amateur reader – for comparable reasons. In contrast, the professor, uniquely credentialed for this specialised activity was to train 'strong young minds ... to sharpen their critical apparatus into precision tools ... as never before' (Ransom 1937).

Meanwhile, there was an accretion of machismo and militarism around New Criticism, partly because it grew in the shadows of the World Wars. In fact, it is believed that New Criticism's escalating influence was in part because it made literature determinate, object-centred, and amenable to pedagogical transmission at a time when returning soldiers sought entry into the university-system. The aura of rigour and scientism which settled upon New Criticism made it attractive to many, while others denounced it for capitulating to the capitalist-military-industrial-technological complex (Graff 1974: 73). However that may be, as the New Critics reorganised literary studies into a robust, professionalised discipline within the academia, they found it convenient, even necessary, to confer it with a species resemblance to 'science'.

The methods systematised by New Criticism were crucial to consolidating the claims and knowledge regimes of literary studies. The most influential of these methods, the practice of 'close reading', centred the literary work for microscopic observation. The trained eye of the New Critic penetrated the surface of the literary work, probed depths inaccessible to the amateur eye, and revealed layers of artistic truth. In Britain, I.A. Richards and F.R. Leavis had already privileged observation, the latter even founding a journal appositely called *Scrutiny*. This specialised seeing was to be further reinforced by New Criticism's stance of dispassionate objectivity. While its ardent scientism cannot be missed, new critical methods drew its authority from other quarters as well. More particularly, it leaned upon the older Arnoldian tradition of 'disinterestedness', by which criticism was to eschew 'ulterior, political, practical considerations about ideas ... which criticism has really nothing to do with' (Arnold 1865). This lofty aloofness, transfused with poet-critic T.S. Eliot's idea of 'depersonalisation', secured the protocols of New Criticism.

Meanwhile, in his magisterial fashion, Leavis had opined that the 'business of the critic was to perceive for himself (sic), to make the finest and sharpest relevant discriminations and to state his (sic) findings as responsibly, clearly and forcibly, as possible' (Leavis 1936: 8). What was getting incubated in this period was a mode of writing that confirmed literary criticism's authority as well as enacted its distance from non-specialist writings and its own fuzzy lineages. Here, the *method* of literary criticism, the *knowledge* it produced, and the *writing* of literary criticism were all crucially imbricated within each other.

This new genre of description-explication (not yet widely labelled 'argumentation') allowed the critic committed to a certain kind of formalism to isolate constituent parts of a literary work without undermining its integrity. These were then closely scrutinised and analysed through well-defined objective categories to excavate reserves of meanings. It was in/through such writing that criticism clarified its scope, established procedures, criterial justifications, and standards of adequacy; it also preserved itself from being scorned as a 'frivolous form of self-indulgence' (Graff 1974: 89). With this criticism was well on the way of being both a mode of analysing literature and a distinct kind of writing *about* literature.

Curiously, the language that New Criticism mobilised for this purpose was performed as transparent and planar, without unseen depths or poetic intensities. Its denotative lucidity was achieved, in part, by customising a theoretical–analytical vocabulary for specialist use. Let us pick a term that New Criticism thus elevated: 'paradox'.[7] As against its use in philosophy, where 'paradox' stood for statement/s that were overtly innocuous but, on deeper examination, militated some foundational law, here, 'paradox' represented statement/s which at first seemed wrong, even, absurd, but lead, potentially, to unexpected insights. New Critics rendered 'paradox' inalienable to literary works and 'inevitable to poetry' (Brooks 1956: 1).[8] As opposites clashed and incongruities brimmed over, 'paradox' was believed to push against the limits of everyday language, startling readers to unaccustomed perceptions and intuitions. It was thought to estrange readers from conventional wisdom, convey incandescent emotions and ideas and provoke new ways of apprehending the world. Far from being a frilly, figurative accessory in poetry, 'paradox' distilled its character and spirit, constituting its very ontology. Despite its overt contrariness, 'paradox' welded form-and-content, holding within it the principle of order and unity.

Crucially, New Criticism also re-purposed 'paradox' as a tool for meticulous analysis. In fact, the close reading of poetry came to be compacted with the scrutiny of taut paradoxes embodied in statements, situations, and structures. Because 'paradox' was neither straightforward nor self-evident, critical expertise was required to prise open anomalous juxtapositions of contrary ideas and reveal rich veins of subterranean significance. 'Paradox' as method, effectively seized the structure, meanings, and dynamics of a poem. It invested literary criticism with well-defined conventions and procedures, aligning it, if incompletely, with 'science'. Simultaneously, it also underscored literary criticism's distinction. For, unlike science, which required the 'purg[ing] of every trace of paradox, the truth which the poet utters can be approached only in terms of paradox' (Brooks 1956: 1).[9] 'Paradox', therefore, was not to be eradicated, but nurtured and used. It, thereby, dramatised the impossibility of distilling a set of true–false statements about literary works. And in this, we encounter the ambivalences of literary 'scientific' criticism – to which I turn in the next section.

Questions and Propositions I

Many believed that New Criticism's 'scientific pose, conscious or unconscious' was its main strength: 'here, for the first time, was "art for art" with teeth in it, with a precise method and scientific aplomb' (Raleigh 1959: 23). Without doubt, New Criticism had recruited dispassionate observation and procedural rigour to dispel vagueness and uncertainty. But did this critical repertoire sufficiently constitute it as 'science'? Equally moot, did New Criticism *intend* itself to be 'properly' scientific? New Critics have spoken in a babel of tongues and commentators have added to the muddle. Some have maintained that the very 'allegation that the New Critics want to make criticism a science' was 'preposterous' (Wellek 1978: 618).[10] Others have suggested that New Criticism abstracted select methods and dispositions from 'science' as 'protective coloration' (Raleigh 1959: 23). Still others have condemned 'objective' literary criticism for its servile emulation of positivist science and its stance of value-free neutrality. In contrast, are those who have commended New Criticism for rescuing objectivity from narrow positivism (Graff 1974: 89); even of entirely decanting scientific method from 'science', so as to free it for new uses. In the ebb and flow of these debates, we find that New Criticism is just as likely to be censured for abducting criticism into the scientific as for not being scientific enough.

Our brief examination of 'paradox' permits us to unpack the problem somewhat differently. Besides exposing the limits of treating the literary work as an integrated object with stable coordinates, 'paradox' as a mode of explication and analysis also highlighted the seeming perversity of New Criticism's techniques. As a critical tool 'paradox' neither methodised referential, testable truths nor standardised interpretation. It neither guaranteed procedural objectivity (where critics can arrive at the same, invariant conclusion by following specific analytical procedures) nor, indeed, convergent objectivity (where a critic can reach the same result as other critics following different analytical routes).[11] At best, 'paradox' allowed a professional community to conduct lively explications and debates on a shared terrain.

Even as conceived by New Criticism, literature was not a closed system that could be seized by its specialist critical terms. While 'paradox' served to extract component entities of a work (imagery, symbol, tone) and establish relations (causality, correspondence, potentiality) between them, it did not lead to the formulation of general laws from which, in turn, stable truths could be deduced. More often than not, literary criticism was simply content to arrive at robust explanations and expositions; it was neither invested nor overly exercised about predictions or principle-making. To consider these as evidence of literary criticism's failures would, however, be overhasty. We would be better served to re-address both literary criticism's invocations of science and its claim to being a 'species of knowledge'.

* * *

Here is how T.S. Eliot 'defines' his influential idea of 'depersonalisation', a term that has been seen to mirror neutral, dispassionate scientism:

> It is in this depersonalization that art may be said to approach the condition of science. I shall, therefore, invite you to consider, *as a suggestive analogy*, the action which takes place when a bit of finely filiated platinum is introduced into a chamber containing oxygen and sulphur dioxide.
>
> (Eliot 1919, emphasis added)

It is important to diagnose that for Eliot, depersonalised criticism was neither platinum in a gas chamber nor the catalysing process. Instead, his invitation was to consider this vivid, sensual scenario as a 'suggestive analogy'.

And so, the proposition I advance here: 'science' is interpolated in literary criticism primarily as a 'trope', as a way 'to turn, direct, alter' (*tropos*), and to figuratively characterise the shifting object/ives of literary criticism. This tropisation of 'science' is somewhat distinct from how other disciplines, especially of the social sciences, strategise their relation to 'science'. The latter, usually, attempt to challenge and alter the descriptive-definitional latitudes of 'science', often, in order to authorise their own practices and protocols. They seek admission and/or recognisability, in some measure, within the revamped sciences. In contrast, literary studies rarely strives to reform 'science' or to quarrel with its foreclosures. It marshals 'science' less as a historically rooted entity and more as compelling gestures that entice, repel, and clarify from the 'outside'. Rather than intervening *in* 'science', literary studies ferrets out impulses for its own distinctive knowledge-making enterprise.

Therefore, to seal 'science' and criticism into separate silos of 'two-cultures', or endlessly debate whether criticism is adequately scientific – is to miss the serious work that science-as-trope does in literary studies. We have already seen, albeit in passing, some of the varied labours that it was summoned to perform. It was enlisted, for instance, to disrupt prevailing epistemes and knowledge regimes, to trouble current methodologies, systematise emergent practices, and furnish 'organising metaphors' (Poovey 2001: 410). It helped contour the concerns and contradictions of literary studies; sharpen its analytical tools and heuristics; choreograph aspirations and adjacencies (where criticism 'approach(ed) the condition of science'); manoeuvre departures, contrasts, and hostilities. It also goaded the formulation of new problems and possibilities. Consequently, to evaluate literary criticism *as* scientific method is to mis-recognise both the practice of literary criticism and the complex ways in which science folds into its enterprise.

It is important to understand that literary criticism elaborated itself primarily in its 'doing', by enacting its theoretical frames through specific acts of interpretation (starting, arguably, from I.A. Richards's *Practical Criticism*). So, when called upon to give an account of itself, it frequently turned to 'science' (though not exclusively to 'science', as we shall see) for 'suggestive analogies' and enabling metaphors. Furthermore, when literary criticism confronted problems that it had not previously encountered or been appointed to address, it plucked at notions of 'science' to renew ways of seeing, searching, discovering, explaining, and analysing.

But just when literary criticism seemed to affirm 'science', it also reinvigorated ways of exceeding, defering, and defying it. This is at least partly because literary criticism stumbled between its fitful craving for the definitional certitudes and precisions of 'science' and its clearheaded understanding of their impossibility and inadequacy, especially as it tackled connotative pluralities and textual ambiguities. It would be instructive to pursue how literary criticism ditheringly piloted science-as-trope to innovate new modes of doing hermeneutics, poetics, and interpretation. But these are directions that this chapter can only flag before it swings to another juncture of literary studies.

'An Indian Response'

Academic trade-routes brought both the Richard–Leavisite schools of criticism and New Criticism to India, where they took root and mutated in ways that allow us to re-examine the heuristics of 'science'. One way to pick these possibilities is to forage the work of C.D. Narasimhaiah (popularly, CDN).[12]

When CDN returned to newly independent India after back-to-back fellowships in the Anglo-American academia, he brought with him novel ideas and ideals. His attempts to overhaul literary education may, at first, seem to mirror developments in England and the USA. CDN aspired to intellectualise the discipline, discretise literary works, and subject them to critical scrutiny. The British teachers who had previously taught him in Mysore, despite their individual generosities, had sought primarily to inculcate the language and awe of English-ness – assuming that 'Indian students shouldn't concern themselves with higher criticism' (Narasimhaiah 1991: 29). In contrast, CDN was determined to place criticism at the centre of the university-system and, indeed, of national culture. He not only presented literature as 'the criticism of life', quite like Mathew Arnold, but also invested the academic-critic with

71

specialist roles and messianic purposes reminiscent of F.R. Leavis. It was the critic's mission to reveal literature's aesthetic–moral grandeur, and there-from, move organically and expansively towards dispelling socio-cultural corruptions and consolidating the civilisational core.[13] He strove to make criticism a teachable–learnable exercise, working to revise the curriculum and render it rigorously testable.[14] These institutional changes were pow-ered by notions of objective criticism. However, it will soon become evident that 'science' did not emerge as either an important interlocutor (as with Leavis) or a fertile trope-making resource (as with the New Critics). It is this crucial difference that I shall attend in the rest of this section.

Throughout his oeuvre, CDN elaborated exacting standards and criteria of literary analysis; he established principles for probing the internal com-position of literary works and for arbitrating in/appropriate criticism.[15] But to see these moves as imitative and/or derivative is to miss a vital aspect of CDN's initiative. If anything, he was anxious to consolidate a substantive 'Indian' response. He lamented that education had 'made us beggars at oth-ers' doors picking up … rags and bones at their dustbins' (Narasimhaiah 1986a: 4). And again, that we had 'forgotten the antiquity of … our criti-cal tradition rooted in the Upanishads and looked pitifully … for European models' (Narasimhaiah 1970: 4). As remedy, CDN advocated an 'Indian aesthetics' and an 'Indian response to Literature'. In effect, he foregrounded select strands of Sanskrit poetics, which were presumed to be readily accessi-ble and uniformly relevant. 'Culture' emerged in his writing as an undivided, shared, national possession that was internally uncontested. Intriguingly, the cultural specificity of 'Indian aesthetics' did not appear to subtract in any manner from conceptions of the universal. I shall highlight below two ways in which CDN plotted the purported universality of 'Indian genius'.

First, let us consider the near emblematic mode by which CDN under-took to a richer synthesis between 'Indian sensibility' and British thought. In his book, *The Swan and the Eagle*, he contended that the two birds, despite their unconcealed contrasts, were 'a fruitful conjunction' that could embody compatibility without collapsing differences (Narasimhaiah 1969: xii–xiii). In this vein, he proposed that the concepts of *rasa* and *dhvani* were congruent with Gerard Manley Hopkins's idea of 'inscape' and 'instress' (Narasimhaiah 1986b: 49); and that '*akhandacarvana*', the 'endless chew-ing' of great works of art, approximated F.R. Leavis' 'close reading' of min-ute particularities (Narasimhaiah 1986b: 51).

Besides synthesising compatibilities, CDN also posited Indian culture as the originary fount of modern literary criticism. Take the crucial case of dispassionate objectivity which the New Critics had affiliated with 'science'. CDN dramatically realigned this by highlighting its 'original' characterisa-tion in Mathew Arnold's 1857 essay, where he, almost passingly, referred to it as 'the Indian virtue of detachment'. By insistently foregrounding this description and repeatedly annotating detachment as a withdrawal from material investments and interests, CDN shifted the lineages of detachment from the scientific to the cultural, from positivism to 'Indian' spiritual make-up and 'Indian' sensibility.

It is not uncommon to find similar declarations of independence from 'western ideals' and assertions of national self-sufficiency anchored by a rarefied Hindu–Sanskrit 'tradition'. What is noteworthy for the purpose of this chapter is the heuristic shift from science to sensibility. It is evident that CDN did not aim to supplant or radically alter Leavisian/New Critical conceptions of literature; nor did he challenge the need for adjudicable literary criteria. Precisely because his attempts were additive – seeking to inflect and enrich, rather than disrupt and challenge – the displacement of scientific registers becomes conspicuous. It showcased them for what they were – heuristics that served strategic purposes. After all, CDN's revised 'Indian' response also nurtured identical intentions – i.e., to construct a robust body of knowledge through critical methodologies that were recognised as consistent, procedurally meticulous, and publicly sharable.

Literary Theorisation: Approaches 'That Consent to Remain Metaphorical'?[16]

By the late 1950s, other crises were brewing in literary studies. Criticism came to be challenged and re-fashioned in the Anglo-American academia by what came to be labelled 'literary theory'. Saussurean Linguistics, continental philosophy, overlapping waves of feminist, Marxist, psychoanalytic, critical race and postcolonial analyses – with their varying histories – convulsed the academia. New segments of people entered universities. Social movements – against racism, war, patriarchy – mobilised new inducements. These churnings impelled fresh rounds of questions in literary studies. The question, 'what is the study of literature', again necessitated the allied question, 'what is literature'. In this segment, instead of anatomising the various 'schools' of literary theory, I shall pursue certain shared tendencies and explicatory trajectories.

Roland Barthes' manifesto-like essay, 'From Work to Text', is a good place to begin as it emphatically centred a new 'object' of study – the 'Text'. Unlike the literary 'work', which was a delimited object, Barthes described the 'Text' as expansive. Whereas the work was 'a fragment of substance, occupying a part of the space of books (in a library for example), *the Text is a methodological field*' (Barthes 1971, emphasis added). Notwithstanding the prepositions deployed in the title, what Barthes formulated was not quite a historical swing – *from* a prior entity, the 'work' *to* a new formation, the 'Text' – as much as a methodological shift (although the 'work' and the 'Text' also possessed differing ontologies).

This move from 'an object that can be computed' (the work), to that which can be *experienced only in an activity of production*' (the Text), had several implications (Barthes 1971). Importantly, it dismantled the solidity and autonomy of the well-wrought literary object, which the New Critics had produced. Barthes offered this suggestive analogy:

> Just as Einsteinian science demands that *the relativity of the frames of reference* be included in the object studied, so the combined action of Marxism, Freudianism and structuralism demands, in literature, the relativization of the relations of writer, reader and observer (critic). Over against the traditional notion of the *work*, for long—and still—conceived of in a, so to speak, Newtonian way, there is now the requirement of a new object, obtained by the sliding or overturning of former categories. That object is the *Text*.
>
> (Barthes 1971, emphasis in original)

According to Barthes, the combinatory network of the 'Text', unlike the structured unity of a Newtonian object, 'tries to place itself very exactly *behind* the limits of the *doxa*', hence making the Text 'always *paradoxical*' (Barthes 1971, emphasis in original).

Different 'schools' of literary theory innovated distinct ways of pushing beyond the taken-for-granted *doxa* and activating the 'paradoxical'. Vigorous initiatives were launched to disrupt the received canons of literature, reformulate aesthetic criteria, and foreground writings of people from different genders, races, classes, geographies, and sexualities. In doing so, they animated Barthes' formulation that 'the Text does not stop at (good) Literature; it cannot be contained in a hierarchy ... even in a simple division of genres. What constitutes the Text is, on the contrary (or precisely), its subversive force in respect of the old classifications' (Barthes 1971). Simultaneously, there were also attempts to re-read familiar texts with new analytical optics. Adrienne Rich, lesbian feminist, poet and critic, famously called this the act of 're-vision', emphasising that this 'act of looking back, of seeing with fresh eyes, of entering an old text from a new critical direction' was in effect, an 'act of survival' (Rich 1972: 18).

The 'paradoxical' could also be seen in attempts to unearth what lay suppressed: to uncover 'the assumptions ... which ... drenched [us]' (Rich 1972: 18).[17] This became an abiding preoccupation for a vast range of theoretical schools. Thus, the Marxist critic, Fredric Jameson, worked to excavate the 'latent meaning behind a manifest one', to slough away the 'mystification or repression', and seek out what was 'unrealized in the surface of the text' – i.e., its 'political unconscious' (Jameson 1981: 60, 48). The hermeneutic phenomenologist, Paul Ricoeur, in his significantly different undertaking, proposed that 'to interpret is to understand a double meaning' to demystify illusions through a 'hermeneutics of suspicion' (Ricoeur 1970: 8). For Gayatri Spivak, a critic associated with Marxism, feminism, deconstruction, and postcolonialism, 'theory' was the 'reading of a hidden ethical or ideological agenda in a literary text' (Spivak 1981: 673).

Piquantly, these diverse theoretical persuasions converged upon similar tropes of probe-and-discovery, excavation-and-revelation – which, though not the monopoly of 'science', grooved easily with it. Critical energies were trained towards mining the deep structures of the text, divulging hidden genealogies, uncovering overlooked artistries, and demystifying meanings and effects. Spivak, among others, steered the need to read 'ideology' and to fill 'the vision of literary form with its connections to ... history, political economy—the world' (Spivak 1981: 673, 671).[18] Precisely what New Criticism had prised apart as extraneous to the literary work was now brought back and adjudged as constitutive of the critical enterprise. From being dismissed as inert socio-historical backdrops to literary works, these became crucially interwoven into the methodological field of the text, inaugurating thereby new species-mutations of literary studies and its relation to life and literature.

It would be instructive to trace an arch from New Criticism's focus on the *paradox* (which inhered within the literary 'work') to the *paradoxical* (which was animated in various ways by literary theory). Against New Criticism's resolve to fortify the discipline of literary studies, here we find the breaking down of 'the solidarity of the old disciplines ... in the interests of a new object and a new language' (Barthes 1971). Barthes regarded this as producing new, exciting possibilities, signalled by the term '*interdisciplinarity*'. Various schools of literary theory, separately and compositely, participated in the 'sliding or overturning of former categories', the autonomous literary work, and discrete disciplinary boundaries. In its place, they set up carnivals of interaction, circulation, and exchange.[19] Take for instance New Historicism, which enthusiastically brought together 'metaphors, ceremonies, dances, emblems, items of clothing, popular stories—previously held to be independent and unrelated', in order to ply the mutual imbrications of the literary and the non-literary (Veeser 1989: xii).

Clearly, this was a new episode of what could (still) be called literary studies. Militating against New Criticism's commitment to the unity of the literary work, Barthes maintained that the Text 'can be broken' (Barthes 1971). What changed consequently was not only the object but also the methods and functions of literary studies. The primary project of literary studies was no longer the exegesis and evaluation of particular works/genres. Rather, it initiated more abstract explorations and 'theorisations'. The literary theorist devised new modes to diagnose, describe, and analyse; to disassemble and demystify; and to reconstitute polyvalent unbounded texts and produce new epistemic structures. In short, criticism mutated into *critique*, the latter became the new object(ive) of literary studies.

***

Meanwhile, distinct urgencies in the Indian academia threw up other ruptures. The convulsions of Mandal and Masjid and the changing economic
regimes in the 1980s–1990s, conjoined with the entry of new social segments into the university-classroom and the challenges posed by 'literary
theory', led to the crisis of literary studies in unprecedented ways. Its object/
ives came to be substantially re-framed: areas that were earlier 'defined
out of the discipline as "political" or "administrative" or ... on the wrong
side of the art/non-art divide' now became included *within* literary studies
(Tharu 1997: 2).

Susie Tharu, surveying the field in the late 1990s, picked the term
'alienation' to analyse this 'paradigm shift', noting that alienation also
described the situation of English/literary studies in India (Tharu 1997: 3,
2). Attentive to socio-historical vagaries, Tharu composed an edgy, dimen-
sional understanding of 'alienation' which included in its sweep the mundane boredom of students in the English classroom, the estrangement of
subjects (like 'woman') from the project of liberal humanism, differential
access to English (language and literature), and also the role of institutions
and ideologies governing English/literature in India. What this meant was
that as an analytical term 'alienation' was chronically inconstant. It was also
inextricably meshed with the problem it sought to study, clearly interrupting protocols of scientificity. I shall not stay with the details of Tharu's mul-
tiscalar discursivemethodology, but move on to consider the dis/continuities
between Tharu's conception of 'alienation' and the apolitical disinterestedness of Leavisite/New Criticism.

According to Tharu, English/literary studies, through its chequered history in the subcontinent, has been deployed to subdue the alienations of the
(post)colonial subject and thereby expunge the '*threat to existing institutions and structures of power*' (Tharu 1997: 25, 9, emphasis in original). In
contradistinction, she seeks to leverage 'alienation'

> as an effect of the power relations structuring the discipline, its
> curricular theory and classroom practices .... *Alienation is there-
> fore—and that is its magic rub—also a means of wedging open,
> interrogating and engaging with these power relations.* It is not some-
> thing either to be overcome or to be set aside. As a mark of exclu-
> sion or subjugation, a border line, it is something to be confronted,
> elaborated and engaged with, politically and epistemologically.
>   (Tharu 1997: 28, emphasis in original)

Delineated in this fashion, 'alienation' inserted an analytical distance with
the object of study, 'wedging (it) open' for scrutiny, serving, thereby,a purpose that was somewhat congruent with what dispassionate objectivity did
in New Criticism. Except that, unlike the latter, the estrangements instituted

by 'alienation' implicated the enquirer within the field of enquiry; it was embodied, socio-historically located, and ideologically mediated, allowing us to think and resonate it with notions of 'situated knowledges'.[20]

While we can see the oblique shadows of science-as-trope and as heuristics, and more prominently, the drive of the 'paradoxical', what we find here is another reshaping of literary studies. The impulse here is not 'to reduce our "alienation" but to transform it into a sign of our relationships with the dominant ideologies of literature', ensuring that 'we turn to our world, asking its questions, and searching out its dimensions, in order to articulate the visions and engage in the battles of our times' (Tharu 1997: 31–32).

Questions and Propositions II

It would be no exaggeration to say that literary studies has been in a near-permanent state of critical change since at least the 1930s. Its object/s of study, functions, objectives, methods, and theories have all transmuted in complex conjunction with changes in university-spaces, socio-political currents, and epistemological dispositions. This has fed enduring self-descriptions of being a discipline in 'crisis'. While each 'moment' of crisis has a 'take' on earlier regimes of doing literary studies, they are rarely subjected to rigorous refutation. More often than not, previous object/ives are discarded or left behind because they are deemed effete and unproductive. Their formulaic applications are not seen to yield new insights, worse, these are seen to obfuscate, even actively thwart new knowledge possibilities. Accordingly, each 'crisis' re-frames the object/ives and re-draws the coordinates of literary studies without necessarily negating the old or effecting decisive 'paradigm shifts'.

In fact, literary studies rarely possesses a consensual paradigm which unites its disciplinary community. Rarely, again, is an old paradigm wholly superseded by a new. What obtains, instead, are swirls where the old, the new, and the renewed eddy together. Take for instance Mr Keating's theatrics: Do we have here the last, beleaguered stand of the old order (of literary 'appreciation') against the conquering forces of New Criticism? Or, do we find revitalised claims for a spontaneous, unmediated relation with poetry?[21] Does Mr Keating evidence the silent victories of New Criticism (because, despite his hostility, he reads many of the poems that New Criticism centred into prominence)? Or, do we encounter a New Criticism that is flailing, outworn, and reduced to caricaturing itself (in/as J. Evans Pritchard)? Did New Criticism die in the 1950s, or did it merely withdraw into the shadows to re-surface from time to time? Or, did it quietly and uninterruptedly continue to underwrite literary studies?[22]

In so far as it is difficult to conclusively answer these questions, we have to reckon with the chronic volatility that undergirds this discipline. The notion of 'progress' has little play in literary studies; it does not chart triumphal journeys from error to (increasing) truth. This points us, yet again, to the epistemic configurations of literary studies and their difference with dominant conceptions of 'science'.

Customarily, the explanations/knowledges produced in this discipline defy empirical testing and measurement. They do not respond well to the question: 'Is it true?'; neither are they readily falsifiable. How then does evaluation, accountability, and responsibility get arbitrated? The adequacy of an analysis or of a theoretical proposition is commonly gauged by its density, its persuasiveness, the rigour of its argumentation, and the insights that it sustains. What literary studies commonly examines is the internal coherence of formulations within domains of contestation that it has established. Hence, we can ask if 'paradox' is a category that is useful for analysing a particular work; rarely do we ask if 'paradox' in itself is true or false. In other words, literary studies is invested in exercising the explanatory-analytical range of its categories. Typically, an enquiry could press to further investigate a well-studied field, or take a familiar question to a new discursive field, or ask if a known field/text/debate can be re-approached from a fresh direction, or if hitherto discrete strains of scholarship can be productively cross-pollinated, and so on. Knowledge-making usually produces patterns, connections, collisions, linkages, frameworks, potentialities, and/or arguments, which are suitably exemplified and substantiated. While doing so, it might also advance new concepts, methods, and protocols

Meanwhile, no large theories are disproved. Instead an explanatory grid or a set of propositions are floated and demonstrated to be convincing/sustainable. For instance, Gayatri Spivak's postulations in 'Three Women's Texts and a Critique of Imperialism' do not render false previous studies on 19th-century British literature. Rather, it establishes an initial condition: 'It should not be possible to read [19th century British literature] without remembering that imperialism ... was a crucial part of the cultural representation of England to the English' (Spivak 1985: 243). Her readings of select novels establish the validity of this contention and show how such an understanding offers new, persuasive analyses of these novels. Furthermore, how these propel us beyond and beneath what was already known and understood (the 'paradoxical', as it were).

Even when there are overt disputes, as for instance, when K. Satyanarayana explicitly quarrels with another critic, D.R. Nagaraj's theoretical approach to contemporary Dalit literature, what we find is an enquiry into explanatory frameworks and a critique of what these do not allow us to describe–explain (Satyanarayana 2019: 9). Thereupon, there is a substantial re-framing of the problem and within its revised terms, the furnishing of a better explanation.

Besides being occasions to renovate analytical vocabularies, methods, and practices, such re-framings also mediate criterial conditions of coherence and responsibility. The analyses and theorisations which ensue do not necessarily crystallise conclusive 'facts'. Instead, the 'findings' proposed are usually in the nature of interpretations and arguments that become preludes to further searches, contestations, and reformulations. What gets marshalled are also new objects of scrutiny. Because in re-formulating the analyses-interpretations-explanations to an existing problem, scholars frequently engineer the re-fashioning of the problem itself.

Improvisations II (In Lieu of a Conclusion)

A quick montage of some recent shifts (randomly picked, eccentrically arranged): each 'shot' assembles incipient shapes in/of literary studies; each sparks its own provocations with 'science'.

- The reclamation of Literary Darwinism: on the eve of Darwin's bicentennial, the journal *Style* produced a double issue titled 'An Evolutionary Paradigm for Literary Studies', exploring a controversial 'movement' whose resurgence can be tracked from Joseph Carroll's 1995 study. The same year Jonathan Gottschall argued for a 'new humanities' founded on Arnoldian disinterestedness, the evolutionary paradigm, and the emulation of scientific method. Sailing on the question: 'What exactly are the sciences doing so right that we are doing so wrong?', he called for an 'upheaval; for new theory, method, and ethos; for paradigm shift' (Gottschall 2008: xi, xii).

- The revaluation of quantitative research and 'literary data': especially since the consecration of digital humanities, a whole host of possibilities have been tested and named – data visualisation, 'big data' analysis, literary mapping, algorithmic literary criticism, and computational formalism. Among the most conspicuous is the work of Franco Moretti and the Stanford Literary Lab that he co-founded. Bringing together a range of materialistic, empirical, statistical, digital, neo-Darwinist approaches and tools, Moretti has popularised fields such as literary geography and methods such as 'distant reading'. The latter steps away from the close reading of individual texts to examine, instead, micro and macro details that enable an understanding of 'the system in its entirety' (Moretti 2000: 57).

- The recuperation of literary history, a domain long consigned to the fringes: a collection of essays – first incubated in a conference – worked to re-construct India's literary history in the 19th century (Blackburn and Dalmia 2004). As the essays *did* literary history, they also reflexively examined the making/writing of the field. In the process, they drew from various disciplinary practices to innovate conversations between culture, economy, expressive traditions, literary forms, and technological landscapes.

- The institution of postcritique, 'surface reading', 'close but not deep reading':

Trenchant criticisms have erupted against 'critique'. The latter (as we saw through the 'paradoxical') read texts 'suspiciously'; it laboured to demystify, defamiliarise and dredge ideological under-pinnings. In sharp contrast, several scholars have maintained that nothing lies withdrawn from immediate scrutiny and that it was the very surface of texts that had to be read productively using observational social sciences, affect theories and also 'some of the methods of science'.

(Best and Marcus 2009: 17)

- The renovation of lowly 'methods' (rather than theoretically freighted 'methodologies'): venerable journals have produced special issues on seemingly humble practices like 'description' and the 'case'; they have traced how these methods have been deployed in disciplines as various as law, medicine, psychoanalysis, humanities, and sciences (*Representations* 2016; *Critical Inquiry* 2007). Such considerations have subsequently been looped back into literary studies.

Because the fast-paced sequence of the montage condenses time, space, and information, it allows us to skip individual details and highlight juxtaposi-tional patterns. The initiatives fast-cut above differ from Barthesian notions of interdisciplinarity in more ways than one. Unlike the 'paradoxical' which was a site of easeful borderlessness, in several of the above transactions, scholars from diverse disciplines were invited to purposively (and provi-sionally) bend their enquiries towards particular aspects/questions/debates. Domains of 'science' appear here not (only) as trope but (also) as partici-pants/interlocutors. Interestingly, several of these exchanges were convened as edited volumes, conferences, or journal (special) issues. It is not acciden-tal that these exchanges have occurred in venues of literary studies that are at the limits of the university-habitat. Implicitly, but also often explicitly, many have staged the inadequacies and obduracies of the current univer-sity-system that purportedly disallow it from hosting these enquiries. As new economies, ideologies, and transnational and transdisciplinary opera-tions thicken and reshape academia, we are perhaps already in the throes of new modes of professionalisation, object/ive-making, knowledge-seeking. These will undoubtedly unfurl new crises and provocations of/for literary studies.

Notes

1 I thank Probal Dasgupta, Gita Chadha, and Renny Thomas for their generous and improving comments on this paper. Also, precious friends and students who have ignited discussions that have nourished this paper in more ways than they recognize. Accounts detailing similar practices can be found in other cultural geographies as well; see for instance Narasimhaiah (1991: 59).

2 W.K. Wimsatt and Monroe Beardsley famously formulated (i) 'Intentional Fallacy' (the error of judging a literary work by assuming the intent/purpose of its writer) and (ii) 'Affective Fallacy' (the error of judging a work based on the emotional effect on its readers). These were not just tendencies to be avoided, but *fallacies* that undermined knowledge production.

3 Though by no means collapsible with the New Critics, these Cambridge scholars have often been uneasily allied with the latter, some even conceding that '[d]iscussion of The New Criticism must start with Mr. Richards' (Ransom 1941: 3).

4 This is also affirmed by Poovey who footnotes that the New Critics 'supplemented [the metaphor of the organic whole] ... with the metaphor of structure' (Poovey 2001: 433).

5 For instance, the influential textbooks of Cleanth Brooks and Robert Penn Warren were titled *Understanding Poetry* and *Understanding Fiction*.

6 Its 19th century lineages have been elaborated, among others, by Poovey (2001) and Moore (2013).

7 'Irony', 'tension', and 'ambiguity' were all foundational tools of New Criticism, each with distinct trajectories. Irony, for instance, has a variegated history from Ancient Greece to postmodernism, during which course it has been a rhetorical device, a narrative mode, a figure of speech, and an analytical tool. Though not quite as wanton, 'paradox' too was prolific, as we shall see ahead.

8 See also Brooks (1947). For a criticism of Brooks' 'monism', see Crane (1948).

9 Perhaps it bears reiteration that this paper does not adjudicate whether Brooks (or others) mis-characterise 'science' nor indeed whether literary criticism is un/ like 'science'. It seeks merely to show how 'science' gets invoked variously – as foil, ally, kin, adjacency – when literary studies sets about explicating itself.

10 Wellek launches a long tirade against science, roundly chastising it for being 'the villain of history' and declaring that the New Critics were the 'enemies of science' (Wellek 1978: 618).

11 For a useful unpacking of objectivity, see Douglas (2004).

12 I thank Dr Vijay Boratti for helping gather a cache of CDN's works and for being a stimulating interlocutor.

13 I thank Probal Dasgupta for urging me to strengthen this line. More clearly needs to be done to trace the changing transactions between literature, criticism, and life, or as Edward Said terms it later 'The World, the Text, and the Critic' (1983).

14 His autobiography is a treasure-trove of finely observed details of English disciplinary practices in the 1940s–1970s. For an account of how he instituted changes in the question-paper format so that they reflected new methods of literary criticism, see Narasimhaiah (1991: 128–133).

15 Significantly, the scholarly journal that he founded in 1952 was called *The Literary Criterion*.

16 Barthes (1971).

17 This set of readings have also been called 'suspicious reading', critical reading etc.

18 To Spivak, ideology is 'the loosely articulated sets of historically determined and determining notions, presuppositions, and practices [...] which goes by the name of common sense or self-evident truth [...]—a conception not very far from Barthes' "doxa"' (Spivak 1981: 673).

19 It would be useful to track interdisciplinary efforts involving literature/literary studies in other disciplinary formations (to explore, for instance, the 'linguistic turn', the 'anthropological novel') and examine how these re-cast relation with 'science'.

20 Donna Haraway proposed 'situated knowledges' as a way to think outside the duality of objectivity–relativism. This term, which pleats together epistemological–ethical–political possibilities, has acquired a portability outside 'the science question in feminism' within which Haraway first postulated it.

21 Some years after the fictional Mr Keating, Susan Sontag, in *Against Interpretation* (1966), re-articulated the discontent with rule-bound, intellectual exegesis which ignored the sensuous impact of art. More recently, 'postcritical reading' (briefly mentioned in the last montage-segment) has suggested that the proper response to art is, in fact, to be moved by it. Curiously, impulses that have been dismissed as outmoded have shown a tendency to re-surface as lively new initiatives within literary studies. The last segment gestures to a few such dis/continuous tracks.

22 Gayatri Spivak, when reconsidering literary studies in the 1980s, described her recuperations of 'practical criticism' because 'its strategies are extremely useful in interpreting and changing the social text' (Spivak 1981: 678).

References

Arnold, Matthew. 1865. 'The Function of Criticism at the Present Time.' http://public-library.uk/ebooks/24/100.pdf (accessed June 12, 2020).

Barthes, Roland. 1971. 'From Work to Text.' http://worrydream.com/refs/Barthes%20-%20From%20Work%20to%20Text.pdf (accessed September 24, 2020).

Best, Stephen, and Sharon Marcus. 2009. 'Surface Reading: An Introduction.' *Representations*, 108 (1): 1–21.

Blackburn, Stuart, and Vasudha Dalmia (eds). 2004. *India's Literary History: Essays on the Nineteenth Century*. Delhi: Permanent Black.

Brooks, Cleanth. 1947. *The Well Wrought Urn: Studies in the Structure of Poetry*. New York: Harcourt, Brace & World.

Brooks, Cleanth. 1956. 'The Language of Paradox.' http://leonidas.pbworks.com/w/file/fetch/114564013/Brooks%20(1971)%20The%20Language%20of%20Paradox.pdf (accessed July 14, 2020).

Brooks, Cleanth, and Robert Penn Warren. 1938. *Understanding Poetry: An Anthology for College Students*. New York: H. Holt and Co.

Carroll, Joseph. 2008. 'An Evolutionary Paradigm for Literary Study.' *Style*, 42 (2–3). https://www.jstor.org/stable/10.5325/style.42.issue-2-3 (accessed July 8, 2020).

Crane, R. S. 1948. 'Cleanth Brooks; Or, the Bankruptcy of Critical Monism.' *Modern Philology*, 45 (4): 226–245. https://www.journals.uchicago.edu/doi/abs/10.1086/388774 (accessed May 15, 2020).

Douglas, Heather. 2004. 'The Irreducible Complexity of Objectivity.' *Synthese*, 138 (3): 453–473.

Eliot, T. S. 1919. 'Tradition and Individual Talent.' https://www.poetryfoundation.org/articles/69400/tradition-and-the-individual-talent (accessed January 8, 2020).

Ferguson, Frances, and Bill Brown. 2007. 'Special Issue: On the Case 2007.' *Critical Inquiry*, 33 (4). https://www.journals.uchicago.edu/toc/ci/2007/33/4 (accessed November 12, 2019).

Gottschall, Jonathan. 2008. *Literature, Science, and a New Humanities*. New York: Palgrave Macmillan.

Graff, Gerald. 1974. 'What Was New Criticism? Literary Interpretation and Scientific Objectivity.' *Salmagundi*, 27: 72–93.

Haraway, Donna. 1988. 'Situated Knowledges: The Science Question in Feminism and the Privilege of Partial Perspective.' *Feminist Studies*, 14 (3): 575–599.

Jameson, Fredric. 1981. *The Political Unconscious: Narrative as a Socially Symbolic Act*. Ithaca: Cornell University Press.

Leavis, F. R. 1936. *Revaluation: Tradition and Development in English Poetry*. London: Chatto & Windus.

Marcus, S. 2016. 'Special Issue: Description Across Disciplines.' *Representations*, 135 (1): 1–21.

Moore, Gregory. 2013. 'Literary Criticism and Models of Science.' In M. A. Rafey Habib (ed.) *The Cambridge History of Literary Criticism, Vol. 6: The Nineteenth Century, c. 1830–1914*, pp. 565–587. Cambridge: Cambridge University Press.

Moretti, Franco. 2000. 'Conjectures on World Literature.' *New Left Review*, 1: 54–68.

Narasimhaiah, C. D. 1969. *The Swan and the Eagle: Essays on Indian English literature*. Simla: Indian Institute of Advanced Studies.

Narasimhaiah, C. D. 1970. 'How Major Is Our Literature of the Past Fifty Years?' In C. D. Narasimhaiah (ed.) *Indian Literature of the Past Fifty Years (1917–1967)*, pp. 1–10. Mysore: Prasaranga, University of Mysore.

Narasimhaiah, C. D. 1986a. 'The Function of Criticism in India.' In C. D. Narasimhaiah (ed.) *The Function of Criticism in India: Essays in Indian Response to Literature*, pp. 1–42. Mysore: Central Institute of Indian Languages.

Narasimhaiah, C. D. 1986b. 'Towards a Formulation of a Common Poetic for Indian Literatures Today.' In C. D. Narasimhaiah (ed.) *The Function of Criticism in India: Essays in Indian Response to Literature*, pp. 43–58. Mysore: Central Institute of Indian Languages.

Narasimhaiah, C. D. 1991. *N for Nobody: Autobiography of an English Teacher*. New Delhi: B.R. Publishers.

Poovey, Mary. 2001. 'The Model System of Contemporary Literary Criticism.' *Critical Inquiry*, 27 (3): 408–438.

Raleigh, John Henry. 1959. 'The New Criticism as an Historical Phenomenon.' *Comparative Literature*, 11 (1): 21–28.

Ransom, John Crowe. 1937. 'Criticism, Inc.' *The Virginia Quarterly Review*, 13 (4): October 1, 1937.

Ransom, John Crowe. 1941. *The New Criticism*. Norfolk, Connecticut: New Directions.

Rich, Adrienne. 1972. 'When We Dead Awaken: Writing as Re-Vision.' *College English*, 34 (1): 18–30.

Ricoeur, Paul. 1970. *Freud and Philosophy: An Essay on Interpretation*. New Haven, CT: Yale University Press.

Satyanrayana, K. 2019. 'The Political and Aesthetic Significance of Contemporary Dalit Literature.' *The Journal of Commonwealth Literature*, 54 (1): 9–24.

Spivak, Gayatri Chakravorty. 1981. 'Reading the World: Literary Studies in the 80s.' *College English*, 43 (7): 671–679.

Spivak, Gayatri Chakravorty. 1985. 'Three Women's Texts and a Critique of Imperialism.' *Critical Inquiry*, 12 (1): 243–261.

Tharu, Susie J. 1997. 'Government, Binding and Unbinding: Alienation and the Subject of Literature.' In Susie Tharu (ed.) *Subject to Change: Teaching Literature in the Nineties*, pp. 1–32. New Delhi: Orient Longman.

Veeser, H. Aram (ed.). 1989. *The New Historicism*. London: Routledge.

Weir, Peter (dir). 1989. *Dead Poets Society*.

Wellek, René. 1978. 'The New Criticism: Pro and Contra.' *Critical Inquiry*, 4 (4): 611–624.

Wellek, Rene, and Austin Warren. 1949. *Theory of Literature*. Harmondsworth: Penguin.

3

PHILOSOPHY AND METHOD

Sundar Sarukkai

The notion of method is essential to any form of systematic thinking and systematic action. Thus, it is an important term not just for philosophy but for all other disciplines and activities which have well-defined aims. Method can be understood as a series of steps that must be performed and/or a set of rules to be followed. These steps and rules are chosen for some specific reasons and they function as a strategy for reaching some desired goals. So any talk of method needs to begin with the aim or goal and the tools at hand to reach that goal. This is true not just for our everyday actions but also for discourses. The tools that we use in discourses are constrained by their dependence on the objects of discourse. Thus, we could say that in its most general sense the idea of method is a systematic way of achieving some desired objectives with the help of certain tools that depend on the nature of the objects of interest, including 'objects' of discourse, in achieving that goal. The issue of method becomes important because it is often the answer to the question 'How does one do philosophy?' rather than 'What does philosophy mean?' Obviously, it would not be possible to do something without first knowing what it is that one wants to be done. Thus, method is always in a constant engagement with the aims and goals of a particular activity.

However, the objects of study and the goals of study are not really independent of each other. The objects of study are as much a product of the discourse. For example, the aim of natural science is to describe and explain the natural world and so the methods in science are chosen so as to lead to these goals. However, although nature is the object of study for the natural sciences, each of the sciences has to define and create the object called nature so that it is amenable to that particular method of studying it (Daston 2014). The conception of nature in physics is different from its conceptualisation in chemistry and biology, for instance. Similarly, 'society' and the 'social' are the 'objects' of interest in the social sciences and each of the social sciences constructs its own meaning of these terms (Guru and Sarukkai 2019). Thus, one could say that the objects of study are discursively constructed; if this is

DOI: 10.4324/9781003298908-6

a view that alarms some, we can say that objects of study in any discipline are 'discursively prepared'.

The idea of method has a long history starting from the philosophical traditions of both the Indian and the Greek traditions. This intellectual history has influenced many of the contemporary approaches to method in different disciplines such as science and the humanities. This is to be expected since all modern disciplines are, to various degrees, derived from philosophy. When new disciplines were formed, the concept of method became one important characteristic to distinguish one discipline from another. For example, in the early days of social science, an important challenge to its legitimacy came from literature (Lepenies 1988). In response, method became an important way by which one could distinguish these two activities even though there was some overlap with their ultimate goals. So although, in some sense, both literature and sociology aimed to describe society, they could be distinguished in terms of the methods of studying/understanding humans and society.

We distinguish disciplines by their methods but often without sufficient recognition that the methods are not independent of other elements of that discipline, including their ontologies and foundational assumptions. Philosophy in Europe, even up until Newton's time, was called 'Natural philosophy', a philosophical study of nature. Newton in his *Principia Mathematica* refers to four 'Rules of Reasoning in Philosophy', which are a methodological way of approaching phenomena. The creation of modern science post-Newton is not just a methodological question but also a question of ontology – the kind of objects that a discipline decides to study. Once the objects of the study get delineated, then the methods of studying these objects get clarified.

So to do philosophy is to know the goal of what we are doing and the method we use to approach this goal. But one of philosophy's goals is method itself – so the notion of method itself becomes an object of philosophical enquiry. Philosophy's enquiry on method is self-reflexive because it uses philosophical methods to study the notion of 'method'. We can contrast this with the use of methods in other disciplines where a general understanding of method would serve as a blueprint for the way one 'performs' that discipline. This self-reflexivity is captured by a more general view of philosophy as 'thinking about thinking'. The enquiry into method in philosophy is precisely of this character: it reflects on the very nature of what method is even as it aims to use these methods to 'do' philosophy.[1]

Methods in Philosophy

There are very few texts that deal explicitly with method in philosophy. Baggini and Fosl (2003) offer what they call a toolkit for doing philosophy. They give a list of the toolkit of concepts and methods used in philosophy

– but as is so common in the field today, they equate philosophy with western philosophy. I want to start with some examples from this list to give an idea of the range of the methods of doing philosophy, but my approach in this paper will differ from theirs in two ways: (1) I will focus specifically on the idea of method and (2) I will offer a more inclusive understanding of philosophy by drawing on examples from Indian philosophy.

Baggini and Fosl break up the list of tools into a few major categories: Basic and Advanced Tools for Argument, Tools for Assessment, Tools for Conceptual Distinctions, Tools of Historical Schools and Philosophers, Tools for Radical Critique, and Tools at the Limit. Firstly, we can note that in their view philosophy is primarily to do with making arguments, assessing the validity of these arguments, creating and clarifying concepts (and this can depend upon and/or be reduced to historical analysis), and articulating a critique of traditional philosophical claims. The tools to analyse arguments are the standard ones in logic, including the different types of logic, notions of validity and soundness, the role of contradiction, and the various kinds of fallacies involved in reasoning. Logical analysis using these concepts thus can be seen as an important philosophical tool.

A method may have more than one such tool. For example, not each of the fallacies is a method, but fallacies as a particular framework can be seen as a method to evaluate arguments. The aim of a philosophical argument in many cases may not be to reach 'truth' as such or even knowledge about something but only to establish the consistency of a position or an argument.

One of the important tasks for philosophy is to clarify how concepts, which might otherwise seem similar, are distinguishable. In western philosophy, there are a set of standard distinctions that have been influential. They are typical markers of a philosophical approach and can be seen as part of a methodological approach. Baggini and Fosl describe a variety of tools, which interestingly are described in terms of a binary structure, for conceptual distinction: a priori/a posteriori (and related to this, analytic/synthetic), de re/de dicto, entailment/implication, essence/accident, necessary/contingent, objective/subjective, types/tokens, and so on. These are more like frameworks which supply a philosophical vocabulary that enable a philosophical description.

Many of these elements are commonly understood to be a part of methods in philosophy. For example, Daly (2010) discusses six major categories: common sense, analysis, thought experiment, simplicity, and science. Some of these ideas are more a listing of the terms that appear in philosophy rather than a sustained argument on why these steps have to be seen as a method.

A more elaborate discussion is found in an edited volume on philosophical methods (Daly 2015). In the summary of the methods in the Introduction, Daly mentions the following elements of method in philosophy: logic and formal methods, common sense, the linguistic turn (analysis of the structure of the language of the philosophical questions), conceptual analysis, verificationism, ordinary language philosophy, intuition, reflective equilibrium, epistemic conservatism, naturalism, thought experiments, and experimental philosophy. The book does not engage with the kinds of methods that arise in continental philosophy including disciplines such as Phenomenology, Existentialism, and Marxism. Although some of the chapters in the book deal with certain conceptual terms in some details, not all of them are useful in informing students as to why these methods are chosen nor how the students should begin to 'do' philosophy. I would argue that an important requirement in any discussion on method is an explanation as to why some approaches should be seen as method.

It is disheartening that even when there has been a large amount of literature on philosophical traditions in Asia and Africa, any discussion on philosophy gets restricted to a small group of 'western' philosophers. This is ironic considering that many of the themes, concepts, practices, and methods that are mentioned above are present in other philosophical traditions. And there are many other elements related to methods that are special to these global practices of philosophy. For example, many of the common ideas related to logic, reason, and argumentation are very much part of Indian philosophy. All Indian schools discuss the nature of inference in great depth and the method by which they do it includes the role of contradiction and the nature of fallacies.

There are many other methods used in logic in Indian philosophical schools including sophisticated methods such as the method of sublation (Vedanta), apoha and catuskoti (Buddhists), syadvada (Jain), and the use of delimiters (Navya Nyāya) (Matilal 1985, 1999). The extremely rich and complex use of negation is used as a methodological device in many Indian philosophical schools. It would be apt to call syadvada (seven-valued syllogisms) and nayavada (standpoint view) of the Jains as methods of philosophising since they apply these methods to every situation (Bharadwaja 1982). One can apply these methods of standpoint and the seven-valued syllogisms to questions about the soul as well as the reality of the table in front of us. These genuinely qualify as methods because they are the steps in a philosophical analysis through which a particular result is obtained.

Although these are examples of what could be seen as part of methods in philosophy, what I want to do in this paper is to explore the conceptual idea of method in philosophy. Through this exploration, it may be possible to understand why some of these concepts mentioned above are seen as part of a method. Moreover, I do not subscribe to the hegemonic interpretation of philosophy primarily through 'western' philosophical categories and hence will analyse the idea of method through examples from both western and Indian traditions, which is also an example of a methodological approach that in the final analysis should be able to include many other global philosophical traditions.

Themes That Underlie Methods

Methods in every discipline are based on certain underlying themes which dictate not just the worldview of those disciplines but also the assumptions underlying their function and goals. I believe that we can get a deeper understanding of why certain methods are used in different disciplines if we identify these basic structures underlying them.

We can start with the standard examples of philosophy and analyse the methods used in 'doing' philosophy. While there are debates on whether the standard distinction used in western philosophy is applicable to Asian philosophies, we can nevertheless start with the standard classification since these themes are part of most philosophical traditions. The classification of philosophy into disciplines such as Metaphysics, Ontology, Epistemology, Logic, Ethics, Aesthetics, and various Philosophies of language, science, mind, society, etc., can be a starting point. Typically, a philosophical approach would be to enquire into concepts such as reality, knowledge, truth, mind, consciousness, perception, inference, and language.

What philosophy does in these activities is to produce 'visions', 'perspectives', 'worldviews', 'frameworks', etc., which do the task of 'uncovering', 'visualising', 'ordering', 'understanding', and so on. Thus, one of the most important goals of philosophy is to clarify the foundational structures of the world, our experience, our thoughts, our language, and almost anything else that we can 'think' about. The description of Indian philosophies as **Darsana** implying perception rather than explicitly knowing suggests that philosophy could well be described as 'perception of truth' and not 'love of knowledge/wisdom', a distinction which would immediately influence the required methods. I give a brief description of the different themes that are common to an activity that could loosely be classified as philosophy. They explain why the certain methods are chosen as philosophical methods.

Perhaps the most important theme that influences philosophical methods is the distinction between **appearance and reality**, leading to the methodological suspicion about what is perceived. The appearance/reality distinction is not specific to philosophy (and is most important in both science and art), but the way philosophy addresses this problem is indeed special. While the atomic theory in science is an answer to the appearance/reality question (and it is important to note that ancient philosophical traditions in India as well as Greek philosophers proposed the idea of atom within the ambit of philosophical reflection), it is nonetheless different when compared to the way philosophy addresses the same issue. The underlying reality of objects is atomic for science, whereas the underlying reality of objects is the metaphysical categories for philosophy. Metaphysics as a discipline offers answers and methods to address this problem of the nature of the real. The postulation of universals, for example, would be a way of finding the truly 'existent' entities, which are common to objects. In a sense, the metaphysical categories are like the atoms in that they are the final constituents of what makes objects real, but they are not materially reducible like the atom. They are the final constituents as far as the question of their status as real is concerned.

The Descartian project of **methodological doubt** – doubt everything until we reach the point where we cannot doubt ourselves doubting – is part of the methodology of suspicion of appearance (Sarukkai 2015). We can note an earlier emphasis on doubt in Nyāya where doubt is an essential category in the steps leading to certainty. In both these cases, we could say that doubt catalyses enquiry – one in a positive sense and the other as a critical method. Suspicion and doubt by themselves are not enough since they may only lead to scepticism. Questions and interrogation become the next methodological tool for a positive project where reality can come to be known. So also the methodological approach of giving an account of **sameness and difference**, a theme deeply related to nature of perception and metaphysics. This methodological approach of recognising sameness-difference as a two-sided binary leads to important philosophical ideas of essence, continuity, change, temporality, identity, and so on.

One of the common methodological principles in Indian and Greek philosophies is that of **debate** (Matilal 1999; Raghuramaraju 2007). All Indian philosophical traditions recognise debate as one of the most important aspects of doing philosophy, as did the Greeks. The detailed classification of the types of debates, the various fallacies, clinchers, and so on offers a specific way of conducting debate and thus constitutes an important methodological principle for philosophy. This is starkly exemplified in Indian philosophy and is also present – even if in more dilute forms – in contemporary philosophy also. Indian philosophical texts also use a particular method which is related to the way philosophical discussions or debates should be conducted. The texts begin with the opponent's position, which is then systematically dismantled. But the insistence on presenting the opponent's view leads to the introduction of ethical elements into philosophy as debate. There are certain core principles that should be followed in 'correctly' presenting the opponent's view although this may not always have been followed. Cort (2000) points out that the Jain philosophers were most scrupulous about this aspect of correctly presenting their opponent's views and sees this as part of a larger idea of 'intellectual ahimsa' which is associated with the method of Jain philosophy. I believe that this can be extended into recognising this practice as a particular methodological practice of Jain philosophy.

Although not explicitly discussed as part of method, it is important to recognise the **commentarial tradition** as a method in philosophy. One of the first ways of learning to do philosophy is to begin with commentaries of texts. While this is sometimes classified under the method of hermeneutics, there are different types of commentarial methods which are used in philosophy. There are methods on how to commentate on texts, ranging from religious texts to philosophical ones. The hermeneutic method draws upon many deep concepts of language, interpretation, action, and so on. A commentary on a text leads to many new philosophical concepts and new arguments. One may argue that this method is not specific to philosophy since interpretation in the larger sense is essential to any discipline. However, although interpretation is a general method used across disciplines (as well as in ordinary everyday language), there are specific elements that are special to philosophy. Scholasticism was always seen as part of philosophy, but modern science begins by rejecting scholasticism. One could safely claim that the removal of the excessive engagement with texts and scholarship paved the way for a new discourse of science post Bacon. The carry-over of such practices left over from this scholastic method has also been highly criticised by many social scientists. Within philosophy itself, the analytic school has departed to a great extent from this practice although in continental philosophy this practice continues. Interestingly, Indian philosophical texts are also dependent on scholastic method but there are important differences between the types and methods of commentarial tradition in these schools and that of the hermeneutic tradition, particularly those related to Biblical and religious studies. Moreover, the different methods of commentarial tradition are exemplified in the different types of texts in Indian philosophical practice.

The emphasis on language and **etymology,** popular with some strands of continental thought, is part of this methodological impulse of mining the written text in order to generate philosophical insights. One major consequence of emphasising this method is that philosophy tends to be seen largely as a history of philosophy. So, unlike other discourses like the sciences, philosophy still harks back to ancient and medieval texts as their starting point. There are reasons for doing this: one of them is that the subject matter has remained quite stable and another reason is the nature of philosophical knowledge, aspects of which we discussed earlier. But this methodological approach has also led to serious problems in the conduct of philosophy today, especially in terms of its capacity to meaningfully interpret and influence contemporary thought within the domain of philosophy per se.

Hermeneutic methodology was primarily about language and the rules and 'methods' for interpretation. There was another methodological movement in philosophy which had to do with language but quite different from traditional hermeneutics. Keeping the question of language as the core aspect of philosophy, the method of analytically understanding the linguistic structure of philosophical questions was influential in approaches such as ordinary language theory and speech act theory. This also had an influence on analytic philosophy as an important method that helped to define what kind of questions were meaningful, what kind of approaches were allowed, and so on. There was also a methodological shift in the way language is understood in relation to philosophy and in much of these views, one can see a resemblance between the way natural science viewed the role of language within its discourse and the way analytical philosophy began to view the problem of language in the 'doing' of philosophy.

Interestingly, this was also influenced by a project in the 19th century which proposed to model philosophy on science and was called '**scientific philosophy**' (Richardson 1997). Modelling on science dominantly means modelling philosophy on the methods of science such as simplicity, parsimony, verifiability, and avoidance of grand metaphysical themes. In this view, philosophy's task was not to resolve grand questions but to solve little problems leading up to the big themes. In particular, the 'meaninglessness' of metaphysics was a focus of criticism. Although this might seem to be exclusive to analytical philosophy, the members of this movement included the early Husserl and Heidegger. But when Husserl and Heidegger moved away from this way of 'doing' philosophy, we begin to see the emphasis towards understanding philosophy as a historical enterprise.

In the western tradition, following this move towards scientific philosophy, the philosophical project of phenomenology began to develop. It is of interest here because of the explicit invocation of what is called the '**phenomenological method**'. Being most famously associated with Husserl, this method consists of steps to understand the foundational structures of our experience. Some of the major elements of the Husserlian method include bracketing and reduction (the phenomenological attitude), use of a free variation of the imagination leading to 'intuiting essences', and description of the essence. The elements of this method include the **eidetic** method. The point to remember is that these are seen as methodological elements leading a human conscious subject to analyse experiences at a deep level. Although phenomenology is quite dispersed in terms of the types of approaches and philosophers who 'fall' into this classification, it nevertheless has influenced methods in anthropology (particularly in qualitative methodology), psychology, and psychiatry. It is also important to note that phenomenology is related to Indian philosophies quite naturally since Indian philosophical traditions were fundamentally a description of consciousness and cognitions. Quite independently of the Husserlian method, advaita has to be understood primarily in terms of phenomenology, including the invocation of 'maya' as a phenomenological attitude as well as **sublation** as a method to discover the reality behind the world of appearance. Given that there are many other techniques to help reach the philosophical goal in schools like the advaita, it is a pity that they are not seen to be relevant to the analysis of method in philosophy today.

Nyāya 16-Step Process as One Model for Philosophical Method

Many of the points described above are neatly captured in a succinct description of what it is to do philosophy in a primary text, the Nyāya Sūtras (circa 550 BC), of the Nyāya tradition of Indian philosophy. The book begins with a 16-step process that well captures the definition, as well as the doing, of philosophy. The text is written in sutra form and the first sutra of the book is this:

> Supreme felicity is attained by the knowledge about the true nature of sixteen categories, viz., means of right knowledge (pramāna), object of right knowledge (prameya), doubt (samsaya), purpose (prayojana), familiar instance (drstānta), established tenet (siddhānta), members (avayava), confutation (tarka), ascertainment (nirnaya), discussion (vāda), wrangling (jalpa), cavil (vitanda), fallacy (hetvabhasa), quibble (chāla), futility (jāti), and occasion for rebuke (nigrahasthana).
>
> (Nyāya Sūtras, p. 1)

Any philosophical task, as we can see from this list, begins with attaining right knowledge through the right means. Vidyabhushana explains the meaning of true knowledge in the following manner: 'Knowledge about the true nature of sixteen categories means true knowledge of the "enunciation", "definition" and "critical examination" of the categories' (Nyāya Sūtras, p. 1). So one can immediately note that the purpose of this philosophical action is to attain true knowledge and the ways of doing this include the methods of enunciation, definition, and critical examination. Many of the 'tools' and 'methods' described in the books on philosophical methods mentioned earlier in this chapter can actually be classified into these larger categories.

Following the first sūtra, there is a description of various kinds of means of knowledge, objects of knowledge, objects of senses, classification of the activities of the body and mind, and then the notion of doubt. In the 23rd sūtra, doubt is classified into five different kinds. Then there is a definition of purpose as something that catalyses action, and purpose can be towards attaining something or avoiding something. What is translated as 'Members' in this list of 16 categories refers to the steps of a syllogism that leads to a conclusion. The nyāya describes a five-step syllogism. The structure of this syllogism includes that of deduction and induction. Its structure also suggests how this syllogism functions as a model of explanation and communicating inferences to others (Sarukkai 2005). Tarka which is translated as confutation here has also been translated as reductio ad absurdum, hypothetical reasoning, etc. It specifies a particular kind of reason in this process. The next three categories of discussion, wrangling, and cavil refer to three types of debate that are possible. There are five fallacies of a reason that are then described in detail. Quibble refers to discursive practices dealing with meaning in relation to critique or countering the claims made in this process and the text defines three types of a quibble. The last two categories are examples of communicative practices in the establishment of a thesis.

As is clear, the 16 steps start from perceiving something to deriving the right knowledge of the right object of perception, related to the clearing of doubt and catalysed by the purpose. The next step in this process is the function of reason starting with logical analysis (inference) which is regulated by rules related to a communicative praxis which demands that the examples used in the argument have to be commonly understood, it should follow certain methods of debate, and if it is a form of debate that is called 'discussion' in this list, it should include the syllogism to demonstrate what one is claiming. As a continuation of the work of reason here, five types of fallacies of reason are pointed out as ones to be avoided. The last few categories take into account the semantic and pragmatic considerations of language constituting the philosophical text.

One can see how this 16-step process gives a broader framework for a methodology describing what it is to do philosophy. It is important to recognise that this process views philosophy as an activity which is closely related to cognitive, communicative, and pragmatic human actions. It is a set of steps or rules to follow which would lead to a philosophical resolution – in short, a set of methods to 'do' philosophy. One such resolution is the removal of doubt – this is included under the category 'ascertainment'. Within these categories, many of the practices of philosophy can be discovered. A commentary of this text achieves precisely these descriptions and the tradition of nyāya is nothing more than the extensive and multiple commentaries and responses to this original text. Unlike a mere listing of philosophical concepts as part of the philosophical method, this process illustrates the activity of doing philosophy. The concepts that are used in philosophy must arise as consequences of understanding what it is to do philosophy and why we do philosophy.

Method and the Human Subject

There is an important aspect of method which I think should be discussed separately. Why is there a need for a method at all? I would argue that for any systematic accumulation and articulation of knowledge, it is necessary to have a 'method' to deal with the presence of the human subject or the knower. Many of the ideas of method that we have talked about before can actually be understood as different ways of taking into account the role of the human subject in the description of the world, which ranges from the world as being completely independent of human cognition to the view of the world as being completely a product of human consciousness. These

94

extremes of absolutely independent reality of the world and the complete immanence of the world lead to the need for methodology as a set of tools/processes needed to justify these positions and/or the many intermediate ones in between.

Scientific method is based on the belief that scientific knowledge is knowledge of properties and other qualities of objects that exist independent of human cognition. However, one cannot just assume that these objects are outside us since there is a constant suspicion that they might be products of our imagination. Method arises as a way to deal with this problem. Aware that all knowledge is essentially produced through the medium of the human through the processes of perception, inference, and the like, how can one remove the effects and the influence of this medium? One response to this doubt leads to the methodological suspicion about perception as mere appearance.

In contrast, phenomenology is an attempt to discover 'objectivity' not by assuming it as a starting point and by assuming the possibility of erasing the presence of the human subject, but through the immersion in the 'subjective' experiences of the subject. Whatever can be discovered must be discovered in the data of experiences, whereas for science whatever has to be discovered is to be found in the data of an external world. These two apparently polar methods – science and phenomenology – paradoxically are performing a similar philosophical task. Science, based on the recognition that all scientific ideas are filtered through human perception and human thought, has to find methods to systematically erase this presence of the human in the description of the world. Phenomenology, based on the recognition that everything has to reside in the human experience, has to find methods to systematically bring the world into these experiences. Scientific method is thus an attempt to erase what is ubiquitously present (subject) while the phenomenological method is an attempt to include what is apparently absent (world). In a deeper sense then, it should not be a surprise to note that for Husserl, phenomenology was the science of the sciences.

It is well known that Kant's philosophical journey was an attempt to find a resolution to the opposition between rationalism and empiricism. The transcendental method, famously associated with him, is a mode of enquiry that attempts to address this binary. Here too the central question is the role and the influence of the a priori structures of human reason in any claim to knowledge of the world. It is also a response to a deeper question as to whether it is possible to know something at all without the influence of the human cognitive subject without giving into full-blown idealism. One can understand the basic structure of Kantian thought as finding a methodological solution to the way one can address this question.

A similar preoccupation is also at the core of Indian philosophical schools. As was mentioned earlier, for all of them cognitive states are the most basic unit of description. Given that, how does one move from the description of one's cognition into statements about an external world? Even a simple statement, 'I see fire' is most commonly interpreted as a description of an individual who sees fire within her field of vision. But there is nothing in the statement 'I see fire' which *necessarily* implies the existence of something called fire that exists '*outside*' and '*independent*' of me. The conclusion that 'There is fire which is outside my body and which I can see' is not an obvious inference and has to be justified only through specific methodologies. I would go to the extent of saying that almost all the methodologies that we commonly use are based on this process of trying to move from a subjective state to an objective claim based on the goal of the enquiry. Another way of stating this is to say that to analyse what we know about the world, it is important to analyse the knower. Many of the modern formulations of knowledge are often based on an unstated distinction between knower and the known.

For some, philosophy is not a mere description of the world nor an intense engagement exclusively with texts. Philosophy is also an active process that generates new ideas, new concepts, and new ways of understanding ourselves and the world. In this view, philosophy's primary task is its capacity to clarify concepts as well as create new concepts. The goal of philosophy is this creative engagement with conceptualization, and for such a goal, the methods needed are truly interdisciplinary in that they draw upon analytical rigour as much as creative practices such as literature. This method of doing philosophy can be found more easily in many contemporary practices, where concept creation becomes a process related to a free-flowing creative process coupled with deep philosophical and historical engagement with ideas. At its foundation, this process is one that explicitly brings and acknowledges the presence of the human subject in the task of doing philosophy.

One can also understand 'history of ideas' as a methodological move within this project of creating concepts. As against 'history of philosophical texts' or 'history of philosophers' or even the broader 'history of philosophy', history of ideas is a methodological way of mining the source of concepts as well as opening up the space of interpreting, understanding, and applying them. It functions as a method because it informs the project of doing philosophy – one way to 'do' philosophy is to work with concepts in this historical manner. Ideally, the exploration of the histories of a concept should not be limited to particular authors or particular schools. Consider the example of a concept like 'causality'. In contemporary philosophy across the world, philosophical issues around causality revolve around standard questions, moving from Aristotle to Hume, with Kant as an important addition. Then some contemporary questions are discussed based on the nature of the question and the philosophical 'affiliation' of the philosopher. A history of ideas method, on the other hand, would be able to take up the analysis of causality independent of whether one's affiliation is to western philosophy, analytical, continental, or Asian philosophies. After all, this concept has been so integral to all philosophical reflections and occurs in other human contexts too. Unfortunately today, the allegiance to philosophical schools has become far more important than philosophical concepts, methods, and goals. A truly useful history of ideas method should enable one to explore and analyse concepts that have been discussed in different philosophical traditions in a more egalitarian manner.

This method is a specific way of acknowledging the presence of the subject in all human activities but at the same time is a methodological way to deal with that presence. Comparative philosophy exemplifies some such attempts. In the last century, one can see how a similar approach led to a broader and more inclusive presence of the human in the articulations of philosophy through the incorporation of the historical, cultural, and social aspects of the knower. This approach added the elements of gender, class, caste, and various other social markers into the process of doing philosophy and in some cases developed a resistance to the erasure or the removal of the presence of the subject in this activity of 'doing'. It has produced a plethora of approaches such as feminist and standpoint epistemologies, many of them arising from and located in departments outside philosophy. These practices constitute a challenge to certain established groups within philosophy and do raise certain fundamental challenges to certain received views of doing philosophy. Rather than take pro- or anti-positions against such philosophical practices, it would be more useful to understand the methodological impulse behind such practices, namely, how they should be seen as a particular way of dealing, engaging, and negotiating with the 'elephant in the room', namely, the subject at the centre of all human articulations.

Philosophy as a Method for Living and Action

The above discussion might seem to be too much focused on aspects of knowing and in general questions of epistemology and ontology. But we know that the goal of philosophy right from ancient times has always been more than a mere analysis of knowing and reality. One could even say that these questions of epistemology and ontology were themselves paths towards another goal that has to do with the way we live our lives and the way we act. A common goal in all philosophical practices in different civilisations was the use of philosophy as teaching us methods for living – living well, living properly, and living ethically. One could argue that almost everything else that came to be associated with philosophy such as specific questions of epistemology and ontology were in response to the more basic understanding of philosophy as a method for teaching us how to live.

This is most explicit in Ethics as a branch of philosophy. But the tenor of this approach influences almost all disciplines that are somehow related to human conduct including sociology, political science, and anthropology. In science, the situation is different. Modern science's dissociation from philosophy, particularly metaphysics, is well known but I think it is important to recognise that one of the most profound breaks between philosophy and science occurs in the removal of the presence of the human subject from the processes of producing and using scientific knowledge. There is nothing in the activity of science which is concerned with questions of right living, living ethically, and so on. Traditional philosophy saw everyday life as part of the method of doing philosophy. This practice of living a life consistent with the philosophical tradition one belongs to can be seen today only in Christian seminaries, Buddhist monasteries, Islamic religious schools, Hindu mutts, and other religious institutions. Thus, the daily practices of good living become a methodological practice in these philosophies, something which was very much a part of knowledge-producing enterprises before the radical break with modern science.

The kind of philosophical concepts that become important for philosophy are also driven by this goal related to living. Concepts such as self and soul have been extremely important in all philosophical traditions. Questions about the purpose of life have influenced the formation of philosophical discourse to a great extent. Similarly, the notion of marga (path) is an explicit invocation of method in Indian philosophies: for example, in the Gita there are three margas that are described, jnana marga (the route through knowledge), karma marga (through proper action), and bhakti marga (through devotion). The use of 'path' as the term for method makes the end goal explicit since all these are paths towards a particular purpose. In the example of the Gita, these are the three ways towards salvation. In the same way, the path in philosophy is related to the goal of philosophy among which are certainty, knowledge, truth, and so on. Similarly, for the Buddhists, the way to enlightenment is the eightfold path which consists of correct action, correct thoughts, etc. At the root of this approach is the methodological principle of taking the middle path and eschewing extremes of one end or the other. This is truly a universal methodological principle in that taking the middle path is the first step moving towards a philosophical discourse. We can see this most famously in Kant when he takes the middle path between rationalism and empiricism.

The goals of living, leading to specific methods of everyday practice and behaviour, are not only meant for personal life. They are also relevant for acting in a society, acting socially as well as a subject matter for speculative introspection. (So even though questions of right action can often be found in the context of our everyday action, there is also a methodological impulse in philosophy to deal with them purely as a speculative enterprise.)[2] In western philosophy, Existentialism can be seen as emphasising the shift from phenomenology as 'pure' philosophy to phenomenology as individual and political practice. Philosophies are different not just in their subject matter but also in the way in which they are 'done'. One such important distinction in the case of Existentialism is the relation of philosophy to everyday living and everyday practice, similar to many of the earlier philosophies of Asia. Similarly, philosophies related to social action become prominent in Marxism and Critical Theory. The methods in these philosophical approaches are dependent on goals related to human action in the social domain. In the Indian context, the philosophical practices inherent in the 'lived discourse' of Phule and Ambedkar offer a philosophical methodology dealing with questions of oppression, humiliation, exclusion, and freedom in terms of caste experiences. Keeping them as being 'outside' philosophy is to misunderstand the plural methodologies of philosophy.

Mimicry, Rituals, and Translation as Methods

In the practice of philosophy, the concept of method is often used heuristically, pragmatically, and many times rhetorically. One of the most powerful consequences of this heuristic use is the ritualisation of method – that is, one 'does' something by merely duplicating the methods defining that act. One might argue that this is the only way of collectively doing something which is ambiguously defined in any case. But the line between habit and ritual is very thin and discourses often exhibit a tendency to function as ritual practices. One important marker of this ritualisation is mimicking, copying something without really knowing why one is doing so. The use of mathematics in the sciences is a classic case of mimicking without the self-awareness of the implications of applying mathematics. More generally, when disciplines in the social sciences mimic the use of mathematics, like in economics, then this ritualisation is of a particular form, what I have referred to as 'epistemological sanskritisation' (Sarukkai 1995). This is a process by which disciplines mimic the practices of other disciplines which are seen to be hierarchically higher to them in order to pull themselves up in the hierarchy. In such cases, the very use of mathematics and mathematical modelling becomes similar to ritual acts.

In a philosophy in which body practices are as important as mental activity, we can see various other modes of method including the practice of yoga and meditation as a methodological necessity for doing philosophy. In theological traditions, religious practices become a path towards understanding philosophical truth and thus are essential as method. In fact, it could be argued that rituals have been grossly misunderstood in modern scholarship and they can actually be seen as methods in some cases; equally, what are often called as methods are more like ritual actions (Sarukkai 2012).

Mimicking and copying are also powerful methods in the doing of philosophy, particularly exemplified in the writing of philosophy. For many students, their first initiation to philosophy is in the forms of writing associated with different philosophers and different 'traditions'. Since the notion of method in philosophy is rarely discussed as part of their pedagogy, mimicry becomes the easiest way to 'do' philosophy. Perhaps this only reflects a truism that is common to many human actions: It is easier to act like something rather than be one. This is as much true of philosophy as it is for other disciplines. I do not think that this practice in itself is wrong because it is an essential mechanism for collectively producing any discourse. This should alert us to the social nature of these disciplines and how a social community is formed through such practices that expand the discipline. Without a community to read what is written, to evaluate and ascribe values, there is no possibility of any discourse. There are deep methodological issues arising from the social nature of these disciplines and mimicry and copying is one such.

Philosophy is deeply engaged with writing, not just the material production of writing but also in thinking about the nature of writing. However, in discussions of method and the process of doing philosophy, the nitty-gritty aspects of writing are not seen as being essential to philosophy. Language is understood as a way to represent philosophical thinking although language itself is one of the most important objects of study for philosophy. But we also know that how one writes is as important in philosophy as what one writes. Unlike the empirical disciplines like the sciences which have a healthy suspicion of language and try to use language in a minimal sense so as to represent the truths of their utterances without distortion, philosophical writing cannot be so reduced. One of the most important and contentious differences between analytical philosophy and continental philosophy (as well as other forms of literary expression of philosophy) is the way language is used in these philosophical practices. Analytical philosophy is not just about the type of problems or the concepts it prefers to use in the act of doing philosophy; it is also based on a belief that language has to have clarity, avoid unnecessary obfuscation, and state things as they are. One can see that the use of language in analytical philosophy is quite similar to that in scientific discourse. In contrast, continental philosophy comes from a tradition which not only uses language to its 'full' capacity but allows language a rein of its own. There is an autonomy to language which must be understood as a methodological move with respect to this style of doing philosophy. Such complex writing is part of the philosophical struggle with language itself. Unfortunately, when this method is copied without much thought, it leads to other kinds of problems which have justly come under criticism.

Why would we consider writing as an important aspect of copying? As I mentioned earlier, doing philosophy is to work with the concepts that are seen to belong to the domain of philosophy. It also includes certain types of argumentative strategies, scholarship, taking into account views which are against one's own, and so on. When somebody who is writing a philosophical text starts with these practices, it is not a ritual act but more of a methodological step. All these practices become rituals when there is a lack of

100

understanding the following: why certain concepts and argumentative strategies are used, why only certain philosophical texts are referred to, what are the contexts in which they can be used, what are the limit points of their use, etc. Mimicking writing strategies and borrowing concepts without due reflection to their origins or the scope of their use are very common practices which border on ritual action. Both in the use of these discursive strategies and in the criticism of one philosophical school by another, we can see the tension between method and ritual. In the case of continental thought, there has been a lot of public criticism on the way some philosophers write, mostly based on a criticism from the analytic camp about the function of language in expressing philosophical thoughts. But the use of language in continental thought is itself the point of their philosophical exercise and they resist reductive understanding of language to express something which is assumed to be 'outside' language. So criticism without understanding why certain philosophical texts are produced in a particular way only reinforces the ritualisation of one's own philosophical traditions. On the other hand, when writers reproduce the kind of writing in many continental philosophers, some of them do it ritualistically. They believe that writing something in a particular way would mark their text as being philosophical. So this ritual copying becomes a methodological tool for them in their understanding of what it is to do philosophy, but the moment they do this, they are no longer 'doing' philosophy.

This aspect becomes a problem particularly in the application of philosophy to study other disciplines. A unique aspect of philosophy as a discipline is the growth of fields which are described as philosophy of X, where X could be science, mathematics, language, art, music, and society. Almost anything can be a subject matter for philosophising.

In the cases where one does 'philosophy of X', there is an interesting relation with the idea of method. It is the distinction between 'doing' philosophy and 'using' philosophy. It is the difference between understanding methods in philosophy as against looking at philosophy itself as a method. So when we do philosophy of mathematics, for example, it is not as much the methods of doing philosophy that are important as the applicability of philosophy to examine mathematics. I am using mathematics here as an example because of the natural connection between 'philosophy of' and mathematical modelling. When a mathematician 'does' mathematics, she is doing something quite different compared to a physicist who is 'using' mathematics. In the case of the former, the mathematician is engaged with the methods of doing mathematics, whereas in the latter the physicist is using methods about applying a given mathematical structure to a physical phenomenon. When science is used to study society, such as in the social sciences, then the methods used in these disciplines are not to be confused with the methods used in doing science. Sometimes there may be overlaps, but in principle they are qualitatively different. When doing science one might use *methods of 'doing science'*, but when doing social science one will be using *methods of 'applying science'*. So when some demand that social processes have to be scientific, they make a mistake in confusing methods *in* science with science *as* a method. Similarly, the methods that describe the way one does metaphysics or epistemology are very different from the methods used when philosophy is applied to understand the nature of scientific epistemology or metaphysics of films. In general, in all these texts which produce philosophy of something, the methodological impulses are very different from 'pure' philosophy. Typically, results or claims from prior philosophical texts are used to reflect on other topics. In this exercise, there is a greater chance of ritual mimicry but at the same time a greater chance of challenging certain established methods in philosophy.

Related to this problem of language, an extremely important aspect of method in philosophy is that of translation. Unlike science, where there is an attempt to codify science within a universal language, philosophy as a form of thinking is present in every society and is very closely aligned with the language of these cultures. Thus, Eurocentric attempts to claim that Greek or German (and perhaps other European) languages are the only ones capable of being philosophical is another way of claiming certain languages as being 'apt' or 'ideal' for philosophy, thus isolating it from the challenge of the diversity of languages and cultures. And this is a challenge because any talk of language already invokes the notion of translation (Chippali and Sarukkai 2018). The only way to deal with this most important problem that is endemic to all philosophical traditions around the world is to look at translation itself as a method (Sarukkai 2013). There are many important aspects to this approach of using the act of translation in a methodological manner, and I believe that philosophy, more than any other discipline, has to pay utmost attention to this urgent problem.

Acknowledgements

I am grateful to Vivek Radhakrishnan, Varun Bhatta, Meera Baindur, and Srajana Kaikini for their critical questions and responses. Thanks are also due to K. Sridhar for his suggestions. Special thanks to Gita Chadha and Renny Thomas for constructive discussions on this topic.

Notes

1 One quick answer to what is philosophy and what it is to do philosophy is to say that – borrowing from similar debates in art – philosophy is what philosophers do and a 'philosopher' is a social title given by a group of people who have the right to accord this title. Although this explanation is appealing, as seems starkly true in the hegemony of western philosophy today, I will not discuss it further in this paper.

2 I thank VivekRadhakrishnan for emphasising this point.

References

Baggini, J and P. S. Fosl. 2003. *The Philosopher's Toolkit: A Compendium of Philosophical Concepts and Methods*. Hoboken, NJ: Wiley-Blackwell Publishing Ltd.

Bharadwaja, V. K. 1982. 'The Jaina Concept of Logic', *Indian Philosophical Quarterly* IX(4): 362–376.

Chippali, M and S. Sarukkai. 2018. 'Conceptual Priority of Translation over Language' in Rita Kothari (Ed.), *A Multilingual Nation: Translation and Language Dynamic in India*, pp. 309–324. New Delhi: Oxford University Press.

Cort, John. 2000. '"Intellectual ahimsa" Revisited: Jain Tolerance and Intolerance of Others', *Philosophy East and West* 50(3): 324–347.

Daly, Chris. 2010. *An Introduction to Philosophical Methods*. Buffalo: Broadview Press.

Daly, Chris (Ed.). 2015. *The Palgrave Handbook of Philosophical Methods*. Basingstoke: Palgrave-Macmillan.

Daston, Lorraine. 2014. 'The Naturalistic Fallacy is Modern', *Isis* 105: 579–587.

Guru, Gopal and Sundar Sarukkai. 2019. *Experience, Caste and the Everyday Social*. New Delhi: Oxford University Press.

Lepenies, Wolf. 1988. *Between Literature and Science: The Rise of Sociology*. Cambridge: Cambridge University Press.

Matilal, B. K. 1985. *Logic, Language and Reality: Indian Philosophy and Contemporary Issues*. New Delhi: Motilal Banarsidass.

Matilal, B. K. 1999. *The Character of Logic in India*. J. Ganeri and H. Tiwari (Eds.). New Delhi: Oxford University Press.

Raghuramaraju, A. 2007. *Debates in Indian Philosophy: Classical, Colonial and Contemporary*. New Delhi: Oxford University Press.

Richardson, Alan. 1997. 'Toward a History of Scientific Philosophy', *Perspectives on Science* 5(3): 418–451.

Sarukkai, Sundar. 1995. 'Mathematisation of Human Sciences: Epistemological Sanskritisation?' *Economic and Political Weekly* 30(52): 3357–3360.

Sarukkai, Sundar. 2005. *Indian Philosophy and Philosophy of Science*. New Delhi: Motilal Banarsidass.

Sarukkai, Sundar. 2012. 'Rituals, Knowledge and Method: The Curious Case of Epistemological Sanskritisation' in Axel Michaels and C. Wulf (Eds.), *Emotions in Rituals and Performances*, pp. 104–114. London: Routledge

Sarukkai, Sundar. 2013. 'Translation as Method: Implications for History of Science' in B. Lightman, G. McOuat and L. Stewart (Eds.), *The Circulation of Knowledge between Britain, India and China*, pp. 309–329. Leiden: Brill Press

Sarukkai, Sundar. 2015. 'To Question and Not to Question: That is the Answer' in R. Thapar (Ed.), *The Public Intellectual in India*, pp. 41–61. New Delhi: Aleph.

The Nyāya Sūtras of Gotama. 1913. Trans. S. C. Vidyabhusana. Allahabad: The Panini Office.

Part II

SHIFTS WITHIN THE SILO: NATURAL SCIENCES

INTRODUCTION TO PART II

While many scientists are comfortable giving their opinions on things beyond their expertise, it is also hard to find, amongst natural scientists, people who are willing and capable of writing on subjects beyond those of their expertise, especially since a lot of discursive writing is styled as per the norms of the social sciences and the humanities. This was itself a question of method of academic writing: why we do not train our natural scientists to write discursively. Given the fact that the question of method is taken for granted in the natural science silo, an academic engagement with method is not organically produced within these disciplines. Given the other fact that the wider universe of academia is divided into academic silos, practitioners of natural sciences do not find themselves naturally engaging with those in other silos, where the question of method is a matter of self-reflective debates. Further, due to the scientism that pervades our academic spaces, attempts at dialogues are often silenced and overpowered by the natural scientists, thus leading to uncompromising closures. Sometimes, the arrogance of the 'other side' is equally harmful to dialogue. Given this scenario, we were fortunate to find practitioners in the natural sciences who were willing to take the risk and write about the question of method in their respective disciplines. While the general understanding of method in natural science is that method largely been fixed and unified over time, largely reinforcing and establishing the standard view of science, we know that in practice the *sotto voce* has always marked the shifts and changes from the accepted discourse. In this section, we present essays on the queen of sciences, mathematics; physics, the most coveted and hardest of the natural sciences; and biology, a discipline that is marked by greater methodological questions than any of the others.

The status of the biological sciences on the positivist pyramid that organises the disciplines according to the level of abstraction they are capable of is rather low. Though in recent times, the recognition of complexity has elevated the status of biology on this pyramid, it is still seen as 'softer', less rigorous, and less theoretical than, say, physics. We also present an

DOI: 10.4324/9781003298908-8

interview that deals with chemistry as a discipline that lies in between study-ing the organic and inorganic matter. The interview also engages with wider questions of science and science education in India that have an impact on the question of method. Each essay attempts to open the respec-tive discipline to methodological pluralism while retaining the rigour of the scientific method.

4

THE METHODS OF MATHEMATICS

Amber Habib

Introduction

Mathematicians tend to believe that our subject provides a model for how human knowledge should be structured and validated. The mathematician in the street may even dispute the qualifying 'human'. The structure that we advocate is a hierarchical one based on the adoption of certain foundational relations, which we call axioms, along with specified rules of logic. We view these axioms as the bedrock on which we build our results. Validation of a result consists of a demonstration, or proof, that it can be reached by applying the specified rules of logic to the axioms. Once a statement is proved, it is called a theorem and can serve as a stepping stone to further theorems and proofs. The daily life of a mathematician consists of conjecturing results and converting them to theorems by finding their proofs. As all mathematicians work with a common choice of axioms and rules of logic, they agree on what is correct and what is not, to a degree not found in any other discipline. Not that we are infallible, errors do occur, but once an error is found there would not be further dispute.

What we have stated so far is the commonly held view of mathematics and mathematical activity. It is quite remarkable that even non-mathematicians subscribe to this view that mathematics stands alone in offering certainty, and an increasing number of disciplines value mathematisation as representing increased rigour. In this chapter, we shall try to flesh out what really happens in mathematical activity and investigate whether the 'common view' offers a true picture or even a decent outline of reality. We can identify three issues right away. First, the description in our introductory paragraph is static and seems to leave little space for history. If the content of mathematics is purely a consequence of the axioms, then the only interesting historical questions would pertain to the choice of axioms. Second, it gives no hint as to the subject matter of mathematics. It is all about process. One may wonder whether this description can even distinguish between mathematics and theology. Third, it does not address why

DOI: 10.4324/9781003298908-9

mathematics is considered important. One justification for mathematics is its usefulness for other disciplines such as physics and economics and its consequent impact on society. Another is that it is worthwhile in itself. Like art, it stimulates the human mind. Neither of these fits comfortably with our description of mathematical activity. It is not obvious why logical deduction from some basic principles should lead to practically useful results, while it also carries the connotation of mechanical routine rather than of creativity and art.

We shall take up these issues in the rest of this chapter. Our main concern is that descriptions of mathematical activity should be tested against evidence and should reflect the actual practice of mathematicians. Philosophers of science have contested the idea that science grows by accretion of verified facts and theories. Popper (1959) asserted that a theory is abandoned when it is no longer able to explain natural phenomena, and that falsifiability rather than verifiability is the test of being scientific. Kuhn (1962) claimed that at any time there is a dominant paradigm which the scientific community subscribes to. When this paradigm starts falling short, a period of crisis results and lasts until a 'revolution' in which a new paradigm replaces the previous one. These descriptions of how the natural sciences evolve are suggestive for mathematics as well, although they do not apply directly. Mathematics is not tested against natural phenomena, and its crises come from within the discipline, when some notions turn out to be more subtle than expected and lead to a period of doubt and uncertainty.

Mathematics in India

Our very first observation is that there have been long-lasting and significant traditions of mathematics that have not subscribed to the view that the essence of mathematics is logical deduction from axioms. In India, for example, there was no search for universal axioms, while there was speculation about different systems of reasoning. Indian mathematicians sought an expanding web of knowledge rather than a tower rising ever higher on firm foundations. Nilakantha, one of the leaders of the 'Kerala School' in the 15th century, is quoted as writing that 'logical reasoning is of little substance, and often indecisive'.[1] The Indian tradition produced a rich body of mathematics over two millennia, covering number theory, algebra, geometry, trigonometry, and the initial steps of infinite series and calculus. Many of the original texts have terse presentations of the final results, and this makes it difficult to identify the underlying method. Indeed, sometimes, it is difficult to be absolutely certain about the precise statements of the results. Techniques are usually explained by working out specific examples. (One of the interesting features is the use of well-selected examples that cover all the cases a technique may encounter, so that mastering them is practically the same as knowing the general approach.) These features have led some to conclude that Indian mathematics was based on empirical observations and lucky guesses and did not value logical reasoning.

However, another aspect of Indian mathematics is the presence of commentaries on the original texts. For example, our understanding of Aryabhata's work in the 5th century relies to a large extent on the commentary written a century later by Bhaskara. Aryabhata's work was especially inspiring for later generations and commentaries were written on it repeatedly, including by Nilakantha. These commentaries make it clear that Indian mathematicians sought general arguments and were not satisfied with working out examples. Rather than providing proofs in the narrow sense of deductive logic, they aimed to create assent and remove confusion through multiple means, which included the use of previously accepted results, intuition, and observation. Mathematical reasoning was not distinguished from other reasoning.

One of the key proof techniques in deductive logic is 'proof by contradiction'. To establish a result, we show that its failure would lead to statements that contradict each other. In the Indian tradition, this approach was sometimes used to show the absence of an object. For example, it was used to justify the statement that a negative number cannot be a square. However, it was never used to establish a presence. Presence had to be directly demonstrated. Thus, the claim that positive numbers do have square roots could only be justified by showing how to find the square roots.[2]

This desire to show existence by direct arguments expressed itself as an emphasis on algorithms and a philosophy that has been called 'computational positivism' (Narasimha 2003). Computational positivism abjures axioms within mathematics and physical models in its applications. It uses circles in astronomy because they fit the data and not because of a belief that they are perfect forms which the universe must conform to. Its flexibility allowed medieval Indian astronomers to make calculations that were much more precise than the ones being made in the Europe of those times.[3] The emphasis on computation also freed the mathematician from having to assign meaning to every concept, and perhaps made it easier to accept the zero and negative numbers. It is worth noting that the Arab scholar Al-Khwarizmi, who is credited with bringing zero and the Indian decimal place notation to the attention of Europeans, did not pick up negative numbers from his sources. This prejudice was maintained in Europe, where negative numbers were seen as lacking reality.[4]

The Indian example shows that the axiomatic approach is not the only way of doing mathematics. Yet, as the Indian tradition petered out by modern times, it could still be argued that the axiomatic approach describes the final or ideal state of modern mathematics. To assess this claim, we must take a closer look at the axiomatic method.[5]

The Axiomatic Method

The creation of the axiomatic method is credited to the Ancient Greeks. It has its genesis in a crisis centring on the use of numbers to measure geometric quantities such as length and area. If we begin with the counting numbers 1, 2, 3, and so on, it appears self-evident that we can reach arbitrarily large numbers just by adding 1 a sufficient number of times. Once fractions are introduced, we can create arbitrarily small numbers by means of the unit fractions 1/2, 1/3, 1/4, etc. This gives us confidence that we can use fractions to describe all lengths and areas. At the same time, the Greeks knew the so-called Pythagorean theorem, which can be stated purely geometrically: The squares standing on the sides of a rectangle equal the square on its diagonal. Euclid gave a proof that dissects and rearranges these squares into each other and thus does not involve any numbers. If we associate numbers to the lengths and areas of the involved shapes, we get the numerical formula that we are familiar with from school: $a^2 + b^2 = c^2$. Applied to a square whose sides are of unit length, this gives $1^2 + 1^2 = c^2$ or $2 = c^2$. That is, squaring the length of the diagonal gives the number 2. Unfortunately, some unknown troublemaker then proved that no fraction can become 2 on squaring! This crisis forced the philosophers of the time to choose between a result and a fundamental principle. It is noteworthy that no one thought of abandoning the result. Instead, Eudoxus delinked 'number' from geometry and created a parallel theory of 'quantity' for treating concepts like lengths and areas. Euclid preserved this distinction in the 13 volumes of his *Elements*, which are dated to about 300 BC (Heath 1908). Volumes I to V consider those aspects of geometry which do not involve measurement. Only in Volume VI does he introduce Eudoxus' theory of quantity and later applies it to further problems of geometry such as the classification of the platonic solids.

The success of Euclid in reducing the geometry of his time to a few fundamental axioms created the model for deductive systems, in which 'self-evident' axioms combine with indisputable laws of reasoning to take us to certain truths. So strong did its hold become that Archimedes, one of the greatest creative mathematicians in all history, hid the methods by which he first obtained his results and only publicised them after recasting in the Euclidean mould.[6]

As Greek science and mathematics dissipated, their key elements were picked up and expanded upon by the Arabs. They carried out extensive translations of the Greek originals, and it was through these translations that Europeans learnt about much of their Greek inheritance. In a repeat of history, Euclid again came to dominate the intellectual discourse, so that even the religious sought comfort from 'proofs' of the existence of God, rather than from pure faith in the received word.

A Change of Attitude

In spite of the success of *Elements* as a model, the work itself started to create unease on two fronts. The first issue was the separation of number from geometry. It is much easier to manipulate numbers rather than geometric entities and as mathematicians took on more challenging problems, they were irked by the bounds put by Euclid. Courant and Robbins (1941) have described the rebellious attitude of the founders of calculus in the 17th and 18th centuries in the following words:

> 'Logically precise reasoning, starting from clear definitions and non-contradictory, "evident" axioms, seemed immaterial to the new pioneers of mathematical science. In a veritable orgy of intuitive guesswork, of cogent reasoning interwoven with nonsensical mysticism, with a blind confidence in the superhuman power of formal procedure, they conquered a mathematical world of immense riches.'

It is doubtful, however, that these pioneers saw themselves in this manner. We judge them now by the standards we have achieved, yet in their time they were no less certain of their correctness than we are of ours.

The second issue that emerged out of Euclid had to do with the requirement that axioms be self-evidently true. Most of Euclid's axioms, such as 'All right angles are equal' or 'Things which are equal to the same thing are also equal to one another' seem to satisfy this. But the notorious Fifth Postulate is far more complex, and thus hardly self-evident: 'If a straight line falling on two straight lines makes the interior angles on the same side less than two right angles, the straight lines, if produced indefinitely, will meet on that side on which the angles are less than two right angles.' It seemed to mathematicians that this axiom needed to be either derived from the others or replaced by one that was genuinely self-evident. For centuries, there was no progress on the first front. Many substitutes were obtained, but no consensus formed for any of them to be taken as self-evident. Eventually, Janos Bolyai and Nicolai Lobachevsky showed, in the early 19th century that one could replace the Fifth Postulate by axioms that contradicted it, yet still led to a consistent deductive system. This revealed that there was not a single geometry mandated by the universe but a variety of geometries. A little later, Bernhard Riemann gave a general framework and pointed out that the geometry of the universe needed to be determined by experiment rather than pure thought. Thus, the status of the axioms of geometry changed from self-evident truths to convenient principles to be adopted as per our needs. Riemann also pointed out that the generalised 'geometry' could be applied in situations other than pertaining to physical space, for example it could be applied to the mixing of colours.[7]

113

This demotion of axioms had a liberating effect on mathematics. Mathematicians began to seek basic principles within purely mathematical concepts as opposed to ones based on our concepts of space or time. They also began to view their subject as being primarily concerned with relations rather than with objects. Consider Euclid's geometry. In his formulation, there are certain fundamental objects (such as points and lines) and relations between them (For example, 'Two distinct lines cannot meet at more than one point.'). The fundamental objects are described by *definitions* while their relations are captured by *axioms*. Now take Euclid's definition of a point:

A point is that which has no part.

A 'part' is a more complicated notion than a point. As a definition, this is not very satisfying. Indeed, one of the remarkable things about *Elements* is that the definitions of the fundamental objects are not used in the proofs – only the axioms are. The implication of this fact went unnoticed for almost 2000 years. It tells us that an object is described by its relations with other things. Or, as Courant and Robbins put it,

> For all purposes of scientific observation an object exhausts itself in the totality of possible relations to the perceiving subject or instrument.
>
> To renounce the goal of comprehending the 'thing in itself', of knowing the 'ultimate truth'… may be a psychological hardship for naïve enthusiasts, but in fact it was one of the most fruitful turns in modern thinking.

Once axioms were no longer seen as self-evident truths about the world, the certainty of mathematics became questionable. The risk is that the axioms may not be consistent with each other and may lead to contradictions. Philosophers first sought certainty in the rules of logic, which govern the formulation of valid arguments. In the late 19th century, Gottlob Frege initiated a programme of reformulating logic so that it could include mathematics, a programme that is now called *logicism*. Frege's attempt failed due to a contradiction discovered by Bertrand Russell. Russell then attempted his own version of logicism but found that keeping it free of contradictions made it difficult to subsume mathematics! To achieve his goals, he had to allow assumptions about the infinite that would not be considered purely logical. There have been other attempts at logicism, but none has succeeded in subsuming mathematics without allowing some rule of logic pertaining to numbers or infinity.

Logicism was followed by a related, yet different, programme called *formalism*. Formalism arose primarily from the work of David Hilbert. Hilbert's first achievement in this direction was to modernise Euclid. In his work *The Foundations of Geometry* (Hilbert 1902), first published in 1899, he first identified certain concepts as the primitive ones which could not be defined in terms of simpler ones. These included both objects (such as points and lines) and relations (such as between and congruent). In the second stage, he provided the axioms, which described the connections between the primitive objects and relations. Then the theorems were obtained as consequences of the axioms. Hilbert also completed Euclid. Euclid had occasionally made assumptions which were not justified by his axioms, and Hilbert added the necessary axioms that removed these gaps. Finally, Hilbert was able to show that his system was consistent, provided that the system of natural numbers was consistent. Thus, the consistency of geometry was reduced to that of arithmetic.

The formalist project had the goal of extending this approach to cover all of mathematics. Moreover, not only were the primitive objects and relations to be deprived of meaning, so were the logical operations. Thus, mathematics would become a game played with symbols, mathematical statements would be certain allowed strings of symbols, and the rules of the game would dictate how one allowed string of symbols may be converted to another. The game would not be arbitrary, however, as it needed to contain a recognisable version of arithmetic.

The Reaction

By the end of the 19th century, mathematicians interested in foundations had settled on the concept of a collection or 'set' as the most primitive notion and had shown that much of mathematics could be built using only this notion. Yet there was enough confusion so that a mathematician might violently oppose a certain proposal only to find that he had been using it himself in another guise. Georg Cantor was the virtuoso of the new Set Theory and he opened undreamt of vistas, so that David Hilbert was moved to remark, 'No one will drive us from the paradise which Cantor created for us.'

Entrance to Cantor's paradise required a willingness to incorporate the infinite into mathematics. Earlier generations had been unwilling to make this step. They accepted infinity only as a marker of an unending process, not as an actual object that could be subjected to mathematical operations. Galileo had noted that the natural numbers could be paired with their squares, and thus one could say that the collection 1, 4, 9, 16, ... had as many elements as the collection 1, 2, 3, 4, ... For Galileo, this was a ridiculous conclusion that showed that one must not talk of notions such as the size of infinite collections. For Cantor, on the other hand, this was merely an example to illustrate that the infinite would have quite different behaviour from the finite.

The modern opposition was led by mathematicians such as L.E.J. Brouwer, who held that mathematics is 'purely the result of the constructive

mental activity of humans rather than the discovery of fundamental principles claimed to exist in an objective reality.' (Iemhoff 2020) In the approach
of the formalists, the existence of a mathematical object is established if its
non-existence would lead to a contradiction with the axioms. For 'intuitionists' such as Brouwer, existence required actual construction, a harder
task. (One is reminded of the attitude towards proof by contradiction in the
Indian tradition.)

Thus, the intuitionists had a higher standard of rigour, in that there were
arguments the formalists would accept but the intuitionists would not. As
an example, they rejected the 'Axiom of Choice', which allows making infinitely many simultaneous choices.[8] If mathematicians truly prioritised certainty over other aspects, one would expect intuitionism to be popular. Yet
it did not turn out that way. Mathematicians have retained certain regard
for intuitionism, yet on the whole have preferred the formalist route. As
happened earlier with the Pythagorean theorem, or with the creators of calculus, mathematicians have been reluctant to give up their theorems over
concerns about basic principles. Purity has repeatedly lost to pragmatism.[9]

The Limits of the Axiomatic Method

These debates led to a clearer understanding of the foundational issues
underlying mathematics, and to startling results about the limitations of
our methods. Kurt Gödel proved in 1931 that any consistent system of logic
that was able to describe the natural numbers would necessarily generate
meaningful statements that could be neither proved nor disproved within
that system. This put paid once and for all to the formalists' hopes of finding
the right set of axioms from which all of mathematics could be obtained.

What was the impact of Gödel's results on the mathematics that followed?
Mathematicians worry about foundations when their normal practice has
led them to the discovery that some fundamental notions are not as straightforward as had been thought. Thus, the Greeks were led to seek fundamental
axioms when geometry and arithmetic suddenly came into conflict. Similarly,
the interest in foundations in the late 18th and early 19th centuries had its
roots in the explosion of mathematical activity following Newton, which
eventually made clear that the earlier notions of curves and numbers were
inadequate beyond a point. Despite Gödel, the work of the logicists and formalists did convince most mathematicians that their discipline currently stood
on solid ground. Indeed, we still do not have any questions of independent
mathematical interest that have been shown to be undecidable. Moreover,
the positive interpretation of Gödel's work is that mathematics would never
be completed, as there would always be undecided (and undecidable) statements.[10] After Gödel, mathematicians' interest moved away from attempting
to show that everything could be proved to the more pragmatic task of actually proving theorems. The Bourbaki collective, a group of leading French

mathematicians, undertook to write a series of texts that would bring some uniformity to the gamut of pure mathematics. To the extent that this meant providing a firm foundation to all of mathematics, they were only partly successful. They did well in several important areas, especially in algebra, but the overall enterprise of even pure mathematics was too varied and messy to be given a common framework. They did not so much fail as run out of steam. However, they did succeed in establishing a new perspective on how mathematics is to be practised, by propagating the notion of 'mathematical structure' as the key concept that mathematical work could focus on.[11]

Gödel's work showed that the axiomatic method could not attain its own goals. One may also question the goals themselves. If we survey the modern mathematical literature, we find that the greater part is the presentation of proofs of theorems. Certainly, one must have a proof to justify a theorem, but this is only one aspect of mathematics. Lakatos (1976) has pointed out that the symbols of formal mathematics are only a partial representation of the intuition and knowledge that is in the mathematician's mind. He has traced the evolution of a formula due to Euler that relates the number of vertices, edges, and faces of a closed polyhedron. The formal description of the shapes to which it applies has been refined several times to remove confusion. However, the formula itself has never been questioned. Mathematicians have always understood that it was the formal representation that needed improvement, and not the underlying mathematical understanding.

William Thurston has also discussed the limitations of the formalist approach, from the viewpoint of one who has extended the frontiers of mathematics. In Thurston (1994), he describes a mathematician's pursuit as being the acquisition and communication of *human understanding* of mathematical knowledge. Proofs are an essential part of this process, but according to Thurston, the notion of proof depends on social context and is not reducible to formal symbolic proof. For example, a proof may be accepted by the community of mathematicians merely on the basis of experts' opinions. A fully formal version of the proof will most likely never come into existence. Further, the formal symbolic proof represents a loss of human understanding of mathematical knowledge, as it is divested of the 'mental models' which led to the proof. The correctness of a proof is not enough by itself to make it valuable to the community. What mathematicians value and seek is the human understanding that is associated with the proof. That is why they react with unease to computer-generated proofs that mechanically check a large number of cases.[12]

Mathematics in the Age of Computers

One significant development in mathematics over the last century, and perhaps especially in the last fifty years, has been the increasing length of published proofs. This has happened in two ways, with differing consequences for the discipline. First, some proofs generate a very large number of cases, which have to be checked individually. Second, we have proofs that introduce multiple new concepts and preliminary results, and these interact in complex ways to yield the main theorem. These new proofs have been very successful and have resolved many famous outstanding problems. However, their history also raises uncomfortable questions about the fallibility of mathematical rigour and the human spirit (Horgan 1993; Krantz 2011).

Proofs of the first kind, involving a very large number of cases, are now facilitated or even completely carried out via computer. The most famous of these pertains to the Four Colour Theorem, which asserts that any map can be coloured with just four colours such that countries with the same colour never border each other (subject to some conditions like every country consisting of a single connected piece). In 1976, Kenneth Appel and Wolfgang Haken announced a proof that required the verification of 1834 configurations. The reduction to these configurations took 400 pages and then computers were used for the verification for each configuration. Errors were pointed out and fixed and, over time, the number of cases was brought down. By now, multiple groups have independently carried out computer-based proofs of the result, so there is general confidence in its truth. Nevertheless, we still do not have what could be considered a 'human' proof – one that would resolve the issues by conceptual insight rather than the checking of a multitude of cases.

Another instance of a computer-assisted proof is Thomas Hales' work on the Kepler Conjecture. Kepler conjectured in 1611 that a particular arrangement of spheres of equal radii would have the greatest density. Hales announced in 1998 that he had carried out a proof that used computers to solve about 100,000 linear programming problems! He submitted his work to the *Annals of Mathematics*, the premier research journal for mathematics. A dozen referees worked together on his submission yet after four years could not absolutely certify its correctness. Hales then embarked on another project – reformulating his proof so that it could be processed by standard proof-checking software. This was completed and the new formal proof published in 2017.

Mathematicians are divided on the value of such work. Initial objections included concerns about programming errors and the difficulty of locating them in lengthy programmes. These concerns have eased with the emergence of standardised proof-checking software. The deeper issue is that while it is reassuring to know a result is correct, we would also like to have a sense of why it is correct. Users of mathematics, such as physicists or computer scientists, may be satisfied with verification, but to a mathematician a proof is an act of communication and a clear proof opens possibilities beyond the specific result that it proves.

In the second category, we have proofs devised by humans but which are of daunting length and complexity. In this category, we have Andrew Wiles' proof of Fermat's Last Theorem, whose final version occupied the entire May 1995 issue of the *Annals of Mathematics*. Fermat's Last Theorem became famous for its difficulty rather than its intrinsic value or utility. By the time Wiles tackled it, about a 150 years of work had gone into creating abstract structures in which it could fit and become a 'standard' problem, and success in achieving that had occurred very recently. Only a small number of mathematicians could understand the techniques that Wiles created, and everyone else had to rely on their judgment. This seems far from the ideal of a proof being something that any rational being can check. Yet mathematicians reacted to the proof with enthusiasm, even when they could not personally verify its correctness. They saw that Wiles had done more than prove a theorem. He had initiated 'an open-ended dialogue that is too elusive and alive to be limited by foundational constraints that are alien to the subject matter' (Harris 2019).

A not-so-publicised issue is the prevalence of errors in mathematical publications. Vladimir Voevodsky became famous in the late 1990s for his work in motivic cohomology, one of the reasons being that it helped circumvent a fatal error in an earlier path-breaking paper. He has described the shock of finding, years later, an error in his own work. This error went undetected for seven years, despite the paper being intensely studied by 'multiple groups of mathematicians' and being used in their work (Voevodsky 2014). Combined with another incident, this convinced Voevodsky that work on the frontiers of his interest could not be deemed reliable without an effective process for checking proofs. The most popular system for reducing mathematics to a few axioms is the ZFC system. Enough of the basic structures and concepts of mathematics have been expressed in this system that most mathematicians believe that it is completely sufficient for their needs. Yet reducing even simple arguments to ZFC formalism requires so much work that it is never done in practice. Voevodsky found that ZFC was inadequate for his plans and sought a way of expressing the foundations of mathematics that would be amenable to computerised checks. This led to a proposal called Univalent Foundations and Voevodsky succeeded in encoding this so that it could be implemented by an existing proof-checking software called Coq.

Three possibilities may be seen for the interaction of computers with mathematics. First, as tools for computation and experimentation, especially for large computations or the visualisation of complex geometries. In this role, computers are already indispensable to mathematicians, even in highly abstract and 'pure' work. The second possibility is as assistants that can check and verify our work, and potentially replace fallible human referees. The work of Voevodsky and the creators of Coq indicate that the primary issue here may be whether mathematicians value this assistance enough to retrain themselves in a new vision of the foundations of mathematics and learn how to write computer-friendly proofs. Possibly, this could happen at some stage simply by a new generation taking over. The third role is as equal partners in the creation of mathematics, able to make conjectures, and find their own proofs. Progress on this front so far is too limited for one to predict the future with confidence. Perhaps mathematics will turn out to be like chess and powerful computation will more than compensate for insight. Or, if mathematics has an essential element of beauty as in poetry or music, then a mathematician computer may at best be competent rather than creative.

Mathematics as a Language

Admirers of mathematics sometimes compare it with poetry. (The other natural sciences presumably play the role of prose.) Hardy (1940) wrote that while poetry involves the patterns formed by words, mathematics is concerned with the patterns formed by ideas. Of course, the word patterns of poetry also represent patterns of ideas, but Hardy's thesis is that the word patterns are more fundamental than the ideas they represent. A trite idea portrayed by a striking arrangement of words may still lead to a memorable poem. If we pursue this comparison further, we have to ask: how does mathematics capture patterns of ideas? To a large extent it uses natural language, but what distinguishes modern mathematics is its use of symbols to stand for abstract concepts and their relationships.

The representation of mathematical concepts by symbols goes back to our earliest days, with the use of scratch marks for counting. As mathematical concepts developed, the rules of natural language were often an impediment to their expression. In India, the rigid conventions of Sanskrit verse made it difficult to incorporate numbers. A convention of 'object-numbers' was developed where certain words stood for corresponding numbers. For example, 'Moon' stood for one because there is one Moon, while 'vedas' stood for four since there are four vedas. General numbers could be represented by placing such words in sequence together with a place-value convention. Since a number could now be represented in different ways, it became easier to satisfy the rules of metre. This is one example of how it became necessary to create special conventions for mathematics. With time, another issue became dominant. Mathematical concepts often have their roots in ordinary problems but then take on a form which is not easily captured by natural language, and the associations carried by that language may mislead us when we use it for mathematics. The use of dedicated symbols helps protect us from such errors. Well-chosen notation also eases the mathematician's work and can even be suggestive of further possibilities. It is widely believed that the advantages of the notation created by Leibniz helped continental Europe make great advances in calculus, while England was held back by its adherence to the less helpful notation of Newton.[13]

Symbolic notation cannot be allowed to proliferate indiscriminately as that would make it unusable. Symbols must be re-used in different contexts and therefore a symbol or expression will usually have multiple meanings. As simple an expression as 2 + 3 stands on the one hand for a procedure of counting (starting with two things and then counting three more), and on the other for the result of the procedure (five). The result of 5 can be obtained in other ways as well (such as 3 + 2, 4 + 1, 6 − 1) and the experienced mathematician sees 2 + 3 as simultaneously representing these as well. Tall and Gray (1994) have described how the ability to easily move between multiple meanings is a key differentiator between students who find mathematics easy and those who find it impossible. These observations also strengthen Hardy's description of mathematics, for ambiguity and multiplicity of meaning is a key aspect of poetry, and we see that it is essential to mathematics as well. The general perception of mathematics as sterile manipulation of symbols may have its roots in our school and university education, which mostly emphasise the procedural aspects and certainly do not portray mathematics as kin to poetry and music.[14]

The Effectiveness of Mathematics

One reason we care about what mathematics is and how it is to be carried out is its evident usefulness. Over the last four centuries, mathematics has come to acquire a central position in the pursuit of knowledge. The mystery is that during this time the core of the discipline has withdrawn inwards and away from direct contact with its applications. How is it that a subject that claims to be independent of perceived reality has become essential to descriptions of that reality?

In a famous essay (Wigner 1960), the physicist Eugene Wigner took up this issue of the 'unreasonable effectiveness' of mathematics in the sciences, particularly physics. He noted that a partial explanation is that if one chooses to use mathematics to describe an aspect of nature, then naturally mathematics will appear useful. This may not be due to any special fitness of mathematics for the task. On the other hand, the mathematics that turns out to be useful often does not have an immediate interpretation that indicates its suitability. Moreover, it may turn out to be useful well outside the domain for which it was initially intended. Wigner gives instances of such 'miracles' but leaves them unexplained. He does note that the known laws of nature 'relate only to a very small part of our knowledge of the world'. It may be that mathematics appears so effective in formulating and applying these laws simply because of their restricted domains. In a meditation on Wigner's article, Hamming (1980) emphasises that we are searching for human understanding and that our evolution determines the way we understand the universe.

> Why then, given our brains [are] wired the way they are, does the remark, 'Perhaps there are thoughts we cannot think', surprise you? Evolution, so far, may possibly have blocked us from being able to think in some directions; there could be unthinkable thoughts.'

Wigner and Hamming are part of a tradition that goes back at least to Galileo and holds that mathematics provides the correct language for formulating science. Sarukkai (2003, 2005b) has asked whether mathematics is really the only language suited to science. After all, large amounts of science are expressed in natural languages like English and even mathematics needs a natural language to convey its concepts. Thus, mathematics may be essential for expressing modern science, but it needs the assistance of natural languages. Moreover, we should not think of mathematics as a single language but as a collection of 'sub-languages' corresponding to divisions like geometry, algebra, and analysis. Each sub-language offers its own narrative possibilities for the mathematisation of a problem in science.

Every description of the effectiveness of mathematics in science creates its own understanding of what mathematicians do that makes it useful. Wigner views their task as creating concepts on which human ingenuity and logic can act. Hamming adds to this that mathematicians constantly extend and generalise their concepts in ways that are governed more by aesthetics than by necessity. Sarukkai wishes them to 'proliferate narratives' to increase the probability of a fit to requirements. In these instances, mathematicians are not advised to do anything specifically aimed at increasing effectiveness, and it appears that as long as the discipline expands so will its applications. But there are also those, even among mathematicians, who worry that the 'pure' side has become too ignorant of the 'applied' side, and that this could lead to a mathematics which has lost its moorings. This tension began in earnest in the 19th century, and while its actual intensity is debatable, it is certainly one more illustration of how mathematics is not a unified activity that can be described in a simple way.

Mathematics and Education

We have been discussing mathematical activity from the viewpoint of the professional mathematician. We should also consider what mathematics means to the wider public, and especially in the context of school education. An important issue is the manner in which beliefs about mathematics impact classroom practice and student attitudes.

Let us start with the positive effects of an emphasis on deductive logic. One is that when equipped with this framework, mathematics is perhaps the only subject where a student can be completely sure of his ground, and can even defend it against his teacher's attacks. Thus, it can lead to a sense of empowerment. Another is that once the rules are sufficiently well understood, they can form the basis of creative explorations. By efficiently structuring knowledge, they can free the student from a clutter of information and allow a focus on the essential principles. For such reasons, an education in mathematics is marketed as essential to developing critical thinking.

On the other hand, an emphasis on deductive logic is not the same thing as full-blown axiomatism. The historical role of the axiomatic approach has been to consolidate the gains of a period of adventurous expansion when they are threatened by paradoxes and confusion. If axiomatism *always* dominates, there would be stagnation. A child's experience of mathematics could be considered in this light. Rules and order should follow exploration and not precede it. Thus, there is a need for balance as well as timing.

Stigler and Hiebert (1999) have surveyed eighth-grade classrooms and described how teachers' beliefs vary by country. In the United States, school teachers generally believe mathematics to comprise fixed definitions and procedures, and so their teaching focuses on repeated drills with the goal of perfecting the execution of these procedures. In Japan, teachers 'act as if mathematics is a set of relationships between concepts, facts, and procedures' and view their role as helping the students to see new relationships between mathematical ideas. Consequently, they offer more challenging problems and allow students to struggle with them, refraining from providing solutions directly. Stigler and Hiebert found that none of the US sessions featured proofs, but 53% of the Japanese ones did, reflecting the greater importance of reasoning in Japanese classrooms.

Ma (2010) compared the content knowledge of primary school teachers from the USA and China and found a greater depth of knowledge in the Chinese teachers. For example, many US teachers calculated $1\frac{3}{4} \div \frac{1}{2}$ incorrectly, while Chinese teachers had no difficulty with it. Exploring the history of mathematics education in the USA, she remarks that in the 19th century reasoning had been made part of teaching arithmetic, but this was abandoned in the early 20th century. She suggests this was due to its presentation in an abstract style which did not take into account children's cognitive development: 'Mathematicians, of course, might approve and appreciate the rigor and parsimony of this presentation. But, from the perspective of lay people, it was dull, tiresome, and meaningless.' Unfortunately, the remedy was to reduce arithmetic to a set of procedures to be memorised. In China, on the other hand, the teaching community persisted with the integration of reasoning in arithmetic and found ways to teach it to school children. The Soviet mathematician Khinchin (2000) has also analysed the disconnect between mastery of formal manipulations and understanding of the corresponding mathematical content, with examples such as

A student will speedily and correctly answer the question 'what is a logarithm', while after the ample time given to him, he is unable to find, without use of tables, $10^{\log 7}$.

These examples illustrate the difficulty of matching the needs of school mathematics with those of the discipline, all the more so if the discipline's needs are cast in a narrow manner. They highlight the necessity for the query 'What is Mathematics?' to be answered in as broad and inclusive a manner as possible.

Notes

1 'Logical reasoning' is interpreted here as deduction from axioms, not merely the use of logical steps while reasoning (Narasimha 2003).
2 Srinivas (2005) gives examples of justifications provided in Indian mathematical texts and analyses how they differ from the proofs of deductive logic, including the avoidance of proofs by contradiction.
3 See Sriram (2005) for examples of algorithms in Indian mathematics and Ramasubramanian (2005) for their uses in Indian astronomy.
4 See Mumford (2010).
5 Plofker (2009) gives a compact presentation of Indian mathematics during ancient and medieval times. For a global history of mathematics, see Boyer and Merzbach (1989).
6 See Heath (1912).
7 The book (Pesic 2007) collects the classic essays by Riemann and others, up to Einstein, covering the evolution of the foundations of geometry and the relationship with sensory experience.
8 Alfred Tarski has described how his proof of an equivalence involving the axiom of choice was rejected by one editor because he objected to the axiom, and by another because it was pointless to derive a true axiom! (Moore 2013: 215).
9 We have painted the three schools of thought (logicism, formalism, and intuitionism) with very broad strokes. In reality, they have interacted with each other and evolved to address the others' concerns; see Horsten (2019) and Davis and Hersh (1981) for a more detailed picture.
10 See Feferman (2006).
11 Mathias (2014) is a very critical take on Bourbaki's knowledge and attitude regarding set theory and its place in mathematics. Corry (1992) distinguishes between the formal and nonformal senses of 'structure' in mathematics and Bourbaki's evident failure in bridging the two. By contrast, Ramanan (2019) offers a personal glimpse of how the Bourbaki texts made it possible for a young student to get a coherent view of a large part of modern mathematics. Several members of Bourbaki took an interest in nurturing Indian graduate students and helped establish the Tata Institute as a leading centre for contemporary mathematics.
12 The development of mathematical logic, as part of the efforts of the logicists and formalists, did lead to success in a different direction – the creation of computer science (Davis 1988).
13 See Sarukkai (2005a) for a detailed consideration of the role of symbols in mathematics.

14 Lockhart (2009) has a dramatic depiction of a world in which music is taught the way mathematics is in ours, with an all-consuming obsession with procedure and repetition.

References

Boyer, Carl B. and Merzbach, Uta C. 1989. *A History of Mathematics*. Second edition. New York: John Wiley & Sons.

Corry, Leo. 1992. 'Bourbaki and the Concept of Mathematical Structure', *Synthese*, 92(3): 315–348.

Courant, R. and Robbins, H. 1941. *What is Mathematics? An Elementary Approach to Ideas and Methods*. London: Oxford University Press.

Davis, M. 1988. 'Influences of Mathematical Logic on Computer Science' in R. Herken (ed.), *A Half-Century Survey on the Universal Turing Machine*, pp. 315–326. New York: Oxford University Press.

Davis, Philip J. and Hersh, Reuben. 1981. *The Mathematical Experience*. Boston, MA: Birkhäuser.

Feferman, Solomon. 2006. 'The Impact of the Incompleteness Theorems on Mathematics', *Notices of the American Mathematical Society*, 53(4): 434–439.

Gray, Eddie M. and Tall, David O. 1994. 'Duality, Ambiguity, and Flexibility: A "Proceptual" View of Simple Arithmetic', *Journal for Research in Mathematics Education*, 25(2): 116–140.

Hamming, R. W. 1980. 'The Unreasonable Effectiveness of Mathematics', *The American Mathematical Monthly*, 87(2): 81–90.

Hardy, G. H. 1940. *A Mathematician's Apology*. Cambridge: Cambridge University Press.

Harris, Michael. 2019. 'Why the Proof of Fermat's Last Theorem Doesn't Need to Be Enhanced', *Quanta Magazine*, June 3, 2019.

Heath, Thomas L. 1908. *The Thirteen Books of Euclid's Elements*. Cambridge: Cambridge University Press.

Heath, Thomas L. 1912. *The Method of Archimedes: Recently Discovered by Heiberg*. Cambridge: Cambridge University Press.

Hilbert, David. 1902. *The Foundations of Geometry*. Translated by E. J. Townsend. Chicago, IL: The Open Court Publishing Company.

Horgan, J. 1993. 'The Death of Proof', *Scientific American*, 269(1993): 93–103.

Horsten, Leon. 2019. 'Philosophy of Mathematics' in Edward N. Zalta (ed.), *The Stanford Encyclopedia of Philosophy*, Spring 2019 Edition. https://plato.stanford.edu/archives/spr2019/entries/philosophy-mathematics (Accessed on 9 August 2021).

Iemhoff, Rosalie. 2020. 'Intuitionism in the Philosophy of Mathematics' in Edward N. Zalta (ed.), *The Stanford Encyclopedia of Philosophy*, Fall 2020 Edition. https://plato.stanford.edu/archives/fall2020/entries/intuitionism (Accessed on 9 August 2021).

Khinchin, A. Ya. 2000. 'On Mathematical Formalism in High Schools Curricula', *The Teaching of Mathematics*, 3(1): 1–14.

Krantz, Steven G. 2011. *The Proof is in the Pudding: The Changing Nature of Mathematical Proof*. New York: Springer.

Kuhn, Thomas. 1962. *The Structure of Scientific Revolutions*. Chicago: University of Chicago Press.

Lakatos, Imre. 1976. *Proofs and Refutations: The Logic of Mathematical Discovery*. Cambridge: Cambridge University Press.

Lockhart, Paul. 2009. *A Mathematician's Lament: How School Cheats Us Out of Our Most Fascinating and Imaginative Art Form*. New York: Bellevue Literary Press.

Ma, Liping. 2010. *Knowing and Teaching Elementary Mathematics*. Anniversary Edition. New York: Routledge.

Mathias, A. R. D. 2014. 'The ignorance of Bourbaki', *The Mathematical Intelligencer*, 14(3): 4–13.

Moore, Gregory H. 2013. *Zermelo's Axiom of Choice*. New York: Dover Publications.

Mumford, David. 2010. 'What's so Baffling about Negative Numbers? – A Cross-Cultural Comparison' in C. S. Seshadri (ed.), *Studies in the History of Indian Mathematics* (pp. 113–143). New Delhi: Hindustan Book Agency.

Narasimha, R. 2003. 'Axiomatism and Computational Positivism', *Economic and Political Weekly*, 38(35): 3650–3656.

Pesic, Peter (ed.). 2007. *Beyond Geometry: Classic Papers from Riemann to Einstein*. New York: Dover Publications.

Plofker, Kim. 2009. *Mathematics in India*. Princeton, NJ: Princeton University Press.

Popper, Karl. 1959. *The Logic of Scientific Discovery*. London: Hutchinson and Co.

Ramanan, S. 2019. 'Bourbaki and his Legacy', *Bhāvanā*, 3(3). http://bhavana.org.in/bourbaki-and-his-legacy/

Ramasubramanian, K. 2005. 'Algorithms in Indian Astronomy' in Gerard G. Emch, R. Sridharan and M. D. Srinivas (eds.), *Contributions to the History of Indian Mathematics* (pp. 183–208). New Delhi: Hindustan Book Agency.

Sarukkai, Sundar. 2003. 'Applying Mathematics', *Economic and Political Weekly*, 38(35): 3662–3670.

Sarukkai, Sundar. 2005a. 'The Use of Symbols in Mathematics and Logic' in G. Sica (ed.), *Essays on the Foundations of Mathematics and Logic* (pp. 99–120). Monza, Italy: Polimetrica International Scientific Publisher.

Sarukkai, Sundar. 2005b. 'Revisiting the 'Unreasonable Effectiveness' of Mathematics', *Current Science*, 88(3): 415–423.

Srinivas, M. D. 2005. 'Proofs in Indian Mathematics' in Gerard G. Emch, R. Sridharan and M. D. Srinivas (eds.), *Contributions to the History of Indian Mathematics* (pp. 209–248). New Delhi: Hindustan Book Agency.

Sriram, M. S. 2005. 'Algorithms in Indian Mathematics' in Gerard G. Emch, R. Sridharan and M. D. Srinivas (eds), *Contributions to the History of Indian Mathematics* (pp. 153–182). New Delhi: Hindustan Book Agency.

Stigler, James W. and Hiebert, James. 1999. *The Teaching Gap*. New York: Free Press.

Thurston, William P. 1994. 'On Proof and Progress in Mathematics', *Bulletin of the American Mathematical Society*, 30(2): 161–177.

Voevodsky, Vladimir. 2014. 'The Origins and Motivations of Univalent Foundations', *The Institute Letter – Institute of Advanced Studies*, Summer 2014: 8–9.

Wigner, E. P. 1960. 'The Unreasonable Effectiveness of Mathematics in the Natural Sciences', *Communications on Pure and Applied Mathematics*, 19(1): 1–14.

5

QUESTIONS OF METHOD
The Philosophy and Practice of Modern Human Genetics

Chitra Kannabiran

The question of scientific method is often dealt with externally as if it were one overarching system by which knowledge is created through a sequence of experimental and conceptual processes that are directed at a specific target, which may be a hypothesis/question or a subject of exploration. Rather than there being one scientific method, a closer look at the evolution of specific sciences shows that historically, there are indeed many scientific methods that have been employed for the systematic creation of scientific knowledge. These methods involve different sequences of steps consisting of observation, inference, and formulation of theories. Apart from the more general logical framework that drives scientific thought and shapes the formation of new knowledge, empirical possibility plays a major role in legitimizing scientific theory. The empirical forces shape the boundaries of scientific knowledge and scientific history indicates that knowledge is created only on the basis of the presence of suitable empirical tests of a theory and through the subsequent process of validating it through repeated experimentation and observation. In fact, the creation of theory goes hand in hand with the design and development of an experimental method to substantiate the theory. Experimental observations generate conclusions which can either prove or refute a theory. A necessary addition to this scheme of routine scientific investigation would be the creative insight of individual discoverers that led to landmark discoveries that changed the course of a field. These creative insights are, by nature, different from the incremental advances that are achieved in most scientific research by the application of the rules of scientific method. Within this framework, I try to sketch the origins and major trends in the field of genetics.

Some Directions of Thought

First, I'd like to state that the development of biology, within which genetics is located, from the 19th century onwards presents us with the threads

DOI: 10.4324/9781003298908-10

of both reductionist and anti-reductionist perspectives. In the former, the structure and function of life forms, such as organisms, organs, and cells, are dissected into the most basic chemical components that are present within them. Life processes such as metabolism, growth, and reproduction are defined in terms of molecules and their structures and interactions. In the anti-reductionist approach, on the other hand, the goal is generally to bring about a synthesis or integration between major realms of biology such as genetics, environment, physiology, and evolution.

Second, I think that the idea that the scientific method is the logical process of arriving at connections between theory and data does not always bear testimony to the practice of science. We know that the inductivist method, pioneered by Francis Bacon in the 16th century, was put forward as an objective method with an unbiased approach to generating knowledge since it involves collecting 'bare' facts, studying them, and then making inferences from the data, leading to a theory. It is a process that is considered to be free of preconceived ideas, hypotheses, or beliefs that exist in the investigators' mind prior to making observations and that may influence the collection of data. The deductive approach, on the other hand, is a framework which begins with a hypothesis that is then tested by suitable experimentation and observation. Baconian principles were an early formulation of a method of science as opposed to the metaphysical approaches that were then in vogue. The philosophy of inductivism was thus held as the model for scientific exploration. Much research in biology superficially at least appears and also claims to be based on inductivist principles. The method used in arriving at a particular discovery may, however, involve some of both inductive and deductive approaches and when examined in retrospect, not present a clear-cut path of the sequence of events that led to it. A prominent example in the history of science that was not an experimental study, but based upon theory and observation, is Darwin's work on the evolution of species and natural selection. Having a deep interest in natural history, Darwin undertook his famed five-year voyage around the world on the ship named HMS Beagle, during which he collected a large number of fossils and specimens. His treatise on the *Origin of Species* elaborated on the theory of evolution, but more significantly, on the role of natural selection in shaping the evolution of organisms and their adaptation to the environment. The question that had to be answered was not *whether* evolution occurred, but *how* so many different species developed and how they adapted to their surroundings. As per records from his dairy, he had formulated a theory of natural selection shortly after his voyage. Thereafter, he reportedly spent the remainder of his life trying to accumulate evidence in favour of his theory (Ayala 1994). Although Darwin himself presented his discovery to the world as if it was based on an inductive process of reasoning, i.e., without any preconceptions (probably due to his perception that this was the most acceptable scientific method at that time), the nature of

128

his work and writings suggest that it was based on a hypothetico-deductive method (Ayala 2009; Blystone and Blodgett 2006). Thus, the very idea that Darwin proceeded to study species without a prior hypothesis has been disputed. During his time, inductivism was the form of scientific enquiry that was upheld as the legitimate and 'unbiased' approach to generating knowledge. Hypotheses were looked down upon as being speculations in the metaphysical realm, and of introducing subjective bias into one's observations. In general, however, the existence of a truly *hypothesis-free approach* has been questioned by epistemologists. In this perspective, every experimental or theoretical investigation in biology starts with a hypothesis that may be stated or unstated. Quite often the framing of a research question has a hypothesis that is implied within it. The term 'theory-ladenness of data' refers to the idea that no data is in fact 'neutral'. Data in biological sciences, even in its raw form, is an entity that is produced by the scientist and is often processed or morphed by instruments that capture and document it, into a form that is devoid of its original context, or any resemblance to the object about which it conveys information. Thus, the way in which data about parts of the genome or the genome sequence is represented has no resemblance to the physical form of the genome itself. As has been pointed out, data is made and not given, and attributing a meaning to it comes from the observer depending on his or her interests and frames of reference (Leonelli 2015).

Creativity and Paradigm Shift

I'd now like to draw upon a well-known event in molecular biology: the discovery of the structure of DNA (the molecule making up the genes) in the mid-20th century as a large illustration of my arguments. This involved the creation of a hypothesis or a model of the structure of DNA by Watson and Crick. The model was based on experimental data from various scientists from X-ray images as well as other chemical analyses of DNA (Pray 2008). The X-ray patterns of crystals of DNA molecules were captured using the technique of X-ray diffraction. This has been a highly effective and widely used method of determining the structure of a large molecule with a regular arrangement of atoms, provided that the substance can be made into crystals. X-ray diffraction is based on the ability of a crystal, in which atoms are regularly spaced, to scatter X-rays that impinge on it. The pattern of X-rays that emerge after impinging on the crystal can be captured on a photographic (or X-ray) film. The images are interpreted based on the patterns obtained to deduce the structure of the molecule. The Watson–Crick model was successful because it fits the basic features suggested by the X-ray patterns of DNA. It is important to note that Watson and Crick worked with a hypothesis that the *DNA molecule is a helix* and tried to fit the existing data into this model. They did not attempt to fit every piece of observed data into the model that they were proposing. In fact, they are said to have built a model that was compatible with major features of DNA – these included the individual atoms and groups in it and their positions in space relative to one another *if* they were part of a helix. In doing so, Watson and Crick sometimes *ignored* empirical data that might not have obviously fit in with this model (Bolinska 2018). This 'model-first' approach starting with the individual components of DNA was an overriding element in their work. Speculation was thus a key ingredient in the process of the discovery of the double helix. It has been noted, however, that this speculation was limited or constrained by empirical data gathered from X-ray images, and other chemical properties of DNA that were already available to them. This form of 'controlled speculation' is, however, within the bounds of rationality. Their approach differed from the method used by other scientists who were also trying to find the structure of DNA. The latter include Rosalind Franklin who looked at every detail of the X-ray patterns of the DNA molecule as a *whole*. Though it was her X-ray image of DNA that was critical to Watson and Crick's discovery, Franklin did not arrive at the correct structure of DNA from her data. The possible reasons for why this connection eluded her, as well as her exact role in the discovery of DNA structure, have been much written about (Gibbons 2012). It is clear that she was a rigid and thorough experimentalist, who made efforts to put in the right type of instrumentation and prepared her DNA crystals with much care and attention to getting a perfect outcome. Yet, she did not make the conceptual leap that was required to arrive at the right structure. This has been attributed to her more conventional approach to the analysis of the data, in fact making a deliberate attempt to avoid speculation and guess work. The latter are looked down upon within a traditional scientific context, since they involve making conclusions that are not supported by evidence. Here, Franklin's effort must be seen as representative of mainstream scientific activity, which concerns itself largely with a systematic process of observation and inference that is rigorously linked to and limited by empirical observations. Radical advancements in knowledge or scientific discoveries (in the sense of landmarks in any field) often represent jumps in perception that are several steps ahead of the current state of knowledge in the field. The intellectual activity involved here may include intuition, insight, and creative thought, which are clearly a break from the more routine epistemic modes in which the scientific process is embedded

Further support for Watson and Crick's model came from its ability to explain for the first time certain important known attributes of genes. One such property is how genes copy themselves faithfully so that an entire set of genes is transmitted from one cell to another when cells divide in the body, and thereby from one generation to another. This discovery encapsulates a hypothesis or speculation, as well as deductive and predictive elements. As mentioned above, the predictive features in the model contributed to the validation of the structure. This came about through future discoveries which uncovered the mechanism by which the genetic material is copied in cells while transmitting the genetic code faithfully from one cell to another. Though other models of the structure of DNA were also proposed, and it is understood that there are other local structures of DNA in specific regions of the genome, which may be different from the Watson–Crick structure, the basic tenets of the latter have stood the test of time.

Multiple Methods in the Route to Discovery

Biological research sometimes has shades of both inductive and hypothetico-deductive processes as part of the method of discovery. Since the formulation of a hypothesis is not always overt and unambiguously presented as a prior hypothesis by the investigator, the distinction between the two types of methods (deductive or inductive) is often blurred when we look retrospectively at the process by which it arose. Quite typically, research begins with a basic question about a molecule, a process, or a phenomenon. While the question may have an implied hypothesis, it is non-directional as compared to a statement. Thus, to ask the question 'Does gene A regulate process X?' may contain an implicit idea about the regulation of process X by gene A but the answer may turn out to be either of the two – that gene A does regulate process X or that it does not regulate it. If on the other hand, one begins with a stated hypothesis 'gene A regulates process X', then the observation of no regulation by gene A falsifies the hypothesis. The prevalent framework of biological research is the former, in which one seeks to answer a question.

An example of the involvement of a complex or combined method in scientific discovery, in which more than one thought process is involved, can be seen in the deciphering of the genetic code carried by DNA. This discovery came from the work of three scientists – Marshall Nirenberg, Har Gobind Khorana, and Severo Ochoa in the 1960s. According to the so-called central dogma of molecular biology, genetic information flows in one direction – from DNA to RNA to protein. Thus, the sequence of letters making up the code on DNA is unique for each gene and consists of a very specific sequence of the four chemicals known as 'bases'. These bases are the letters of the code (referred to as A, G, C, and T). The code for each gene is faithfully copied from DNA onto RNA, and then 'transferred' from RNA to a protein molecule. The difference in the letters in RNA is that it has U instead of T, thus having A, G, C, and U as the four bases. RNA thus acts as a messenger (known as messenger RNA), carrying the code from the nucleus to the cytoplasm of the cell. There it is 'decoded' and 'translated' into a specific protein. In other words, the instructions contained in the code on the messenger RNA molecule are read by the cellular machinery, which acts to make the corresponding protein.

131

Proteins are chains of amino acids strung to one another. They have numerous functions in the cell, and perform a range of tasks such as ferrying other substances into or out of cells, acting as catalysts for all chemical reactions, and forming the building blocks for the various parts of the cell. The decoding of the genetic code present on the messenger RNA involves several other types of molecules and an elaborate apparatus that is essentially a biochemical factory for making proteins within cells. The question to be answered was how can a sequence made of four alphabets in a gene (DNA) code the synthesis of a protein that is made up of 20 different molecules (the amino acids). By looking at various permutations that one can get from the four letters A, G, C, and T, it could be deduced that there are 64 different combinations of three letters (or triplets) that could form a set of unique codes from the four bases. Thus, it was clear from this calculation that one can in fact get more than enough triplets to code for 20 different amino acids. The foundational hypothesis in the search for the genetic code was thus that of a triplet code. The triplet sequences (known as *codons*) each represent or code for an amino acid. The next problem was to find the precise sequence of the three bases in the codons for each of the 20 amino acids. This would lead to the identification of the genetic code in its entirety. Principally, the breakthrough made by Nirenberg and the two other scientists, Har Gobind Khorana and Severo Ochoa, was to carry out experiments that would crack the genetic code. They prepared an extract or a 'cell soup' from bacterial cells that has all the ingredients for making protein. Into this soup, they added short fragments of RNA having a defined sequence of letters. Thus, if we have a cell soup that has all the ingredients in it for making a short protein, Nirenberg postulated that if we put a defined code sequence into the cell soup, we would find that a unique amino acid is recognized in response to the code by the molecular 'machinery' in the soup. Each triplet code was proposed to be read as one amino acid only. To simplify their search, they made RNA molecules having a chain of only one letter repeating itself, for example, they tried a synthetic RNA with the base sequence UUUUUUUUUU (poly-U). This sequence was decoded in the cell soup as an amino acid known as phenylalanine (Phe). The poly-U RNA molecule led to the synthesis of a protein having poly-phenylalanine – in other words, a chain of Phe molecules strung to one another. They deduced from this that the codon for Phe is the triplet UUU. Thereafter, RNAs with all other permutations of bases in triplet combinations, one at a time, were tried in the same fashion (Ling and Soll 2012; Nirenberg 2004). They found that indeed as proposed, each triplet code present in this soup was specific to just one unique amino acid out of 20, which was recognized and added to the protein chain. The codons for all of the 20 known amino acids in nature were thereby decoded in this manner (Caskey and Leder 2014). The method used here first involved a foundational hypothesis of a triplet code, on the basis of the known number of letters in the genetic code, and the number of building blocks in a protein. A series of sub-hypotheses on individual codes for each of the amino acids were then inherent to the process of deciphering the code by means of the experiments with the triplet codes. Finding the correct code for each of the amino acids was then essentially a systematic process of trial and error, since all combinations of three letters were made and tested. However, the initial observations on the genetic code were made in cell extracts from bacteria, leading to the elucidation of the code in bacterial cells. The question that was yet to be addressed was whether the genetic code is the same or different between different species. That is, would the genetic code found in bacteria be true for that of mice and humans as well? This question was answered by testing the codes found in bacterial cells by similar methods using cell soups from other species. It was then confirmed that the code is the same in all organisms. In other words, the genetic code was found to be universal. Other features of the genetic code, such as how many triplets code for a single amino acid, and the punctuation marks on the messenger RNA indicating the 'start' and 'stop' positions for the protein, were deduced by experimentation using the same system mentioned above (Crick 1962).

The Human Genome Project and 'Big Science'

Looking closely at more recent developments in human genetics, the work on the Human Genome Project (HGP) about 20 years ago entailed the collection of the sequence of letters of the entire human genome (the entire set of genes and genetic material in the human cell) (Collins and Fink 1995). Since this involved the identification of the complete readout of billions of letters making up the genome, it was in fact an enormous mining exercise to collect large amounts of data on sequences of our genomes. In terms of scientific method in the HGP and related projects, there was a broad hypothesis of the genome as the key to health and disease. This assumption was the foundation of the HGP effort. Even the repetitive and systematic collection of genome data under this project does not qualify outright as a 'non-directional' accumulation of data, which was carried out in a model-free manner. The idea behind the enterprise as stated explicitly was that the genome sequence will provide answers to human health and disease.

There is an inherent belief here of the paramount importance of the genome sequence as scientific knowledge, which is reinforced by an ideology that genes determine who we are. The emphasis on genome as a 'transformative textbook of medicine' as stated by the proponents of the HGP also attests to the same. However, I would like to emphasize here that the concept that the human genome as a determinant of human health and disease is a broad, overarching idea that might not apply across scale and levels of abstraction. The process of discovering the sequence of the genome is not a single continuous process that opens up to a single or major deduction from the pattern of data therein. We know that data was obtained from multiple players and at several stages. These were then pieced together to derive the complete sequence of the genome. Hence, both 'genome' and health are large, heterogeneous entities – each has many separate terrains that have differences *within* themselves. The genome itself has multiple elements that may define or regulate only some aspect of the body. The sequence of the genome thus provided a huge reference map for looking at how genes influence various diseases. Taking the genome as a whole and the myriad diseases or traits that are within the umbrella of 'human disease/health', there is an enormous and composite arena. Within it, there were some successes in parts of this landscape, in terms of 'finding genes' that provide mechanistic clues to understanding diseases. This group largely constitutes a relatively small set of diseases that are often rare in any population; genes that are associated with them were indeed found. What about the rest of the diseases? They are the common diseases that are therefore generally a major public health concern, but the genome sequence in itself did not provide practical solutions to tackle these diseases in most cases. These diseases include some common forms of blindness in older adults such as cataract and glaucoma, and also conditions affecting various other organs such as heart disease, diabetes, arthritis, and blood pressure. The failure to discover a predominant genetic basis in these latter cases is now understood to be because such diseases are multifactorial, i.e., they are caused by many factors involving one to many genes and the environment. Thus, one cannot easily predict the occurrence of disease in a patient by doing genetic tests even if we know a particular gene that is involved in its development; it is also not manifest that one can find ways of curing such diseases based on the knowledge of just the genes themselves (Epstein 2000). The idea behind the HGP may thus be considered as reductionist in nature, both in trying to pin down the many dimensions of health and disease to one *linear* sequence of the genome and in defining the *normal* genome as (essentially) one reference sequence. This is clearly not the case, since there is a range of normal references due to the extent of natural genetic variation in humans across the world. The effects of the wide range of variations found in the normal populations are still not well understood.

The scientific method here as well has threads of both inductive and deductive approaches if one looks at different levels of the project. There was an all-encompassing concept of the genome sequence as the key to human health as stated above. But the conclusion or deduction from this data was not a uniform or single answer. Since the genetic underpinnings could be identified as significant in certain diseases, but not in many others, we cannot make a generalized and uniform conclusion as to the role of the genome sequence in the causation of disease. If one, however, keeps mining the data from the human genome and looking at specific diseases, there are instances where the genetic causes were found in some diseases, and these attempts appear to follow an inductive paradigm. A somewhat different concept of 'abduction' has been proposed to represent the methods that are inherent to the genome projects. Here, statistics are used as a tool to analyse the data and search for patterns and correlations, replacing the hypotheses made by the scientist. Abduction involves making an inference towards the most *plausible* explanation of the data, although this cannot be proven to be true (Voit 2019). It is statistically the most probable one. The relation between the gene sequence and disease that is being inferred here is not one of causality. Rather, one is looking for correlations between the two entities – correlation hence replaces causation.

Another aspect of the HGP is that the entire effort has been criticised for its ultimate reductionism – that of reducing in which a complex state such as human health or disease to one set of component parts (genes) – and that of reducing the properties of an organism as being dependent solely or predominantly to its genes (Tauber and Sarkar 1993). All human traits, even behavioural and psychological ones, are perceived as genetic in nature under the current scientific world view. Apart from the effects of the environment, different layers of biologic mechanisms that go beyond gene sequences are not part of its conceptual framework. It has been argued that the 'reification' of complex characters such as 'intelligence', 'health', and 'risk' attempts to make them into defined and measurable entities, at the hazard of ignoring the social and political factors that underlie these categories (Kimmelman 2006).

The HGP as a scientific enterprise was unprecedented in scale, and its methodology signalled the advent of 'big data' and 'big science'. These terms essentially imply megaprojects that consume a large chunk of the resources and funding. One of the criticisms of such projects is that diversion of funds to them leads to an exclusion of other types of pursuit in science that are based upon a single investigator or are small-sized research efforts. However, there have been spinoffs from the HGP too, with potential benefits to a larger scientific community (Hood and Rowen 2013). The most significant impact within the scientific community was the impetus provided by the HGP in the creation of new methods and new technologies in the field of genetics that are capable of sequencing and analysing the genome even faster. The billions of bytes of data from a megaproject like the HGP are generated through enormous computational power and high speed. These requirements catalysed the advances in technology that in turn provided the required tools to carry out work on this very large scale. The newer technologies included not only genetic and biochemical instruments that can function at high speed and high throughput, but also advances in computers and information technology to process the data and storage capacity for the exceedingly large datasets. The process of finding the sequence of the human genome also led to the acquisition and analysis of similar big data from various related domains, for example, genomes of different human populations, organisms other than humans, catalogues of messenger RNA and proteins from entire organs and tissues, clinical data of patients from medical records, etc. These 'satellite' projects were also megaprojects that were fashioned on the same patterns of large-scale data collection and subsequent mining of data for patterns and inferences.

Big data was also generated from specific diseases (e.g., breast cancer) in which molecular profiles of all tissues available are maintained in publicly available databases. The transition to big data in these various types of projects is considered to have brought in a new paradigm in science. In this mode, one collects large volumes of data on genes, proteins, diseases, etc., from hundreds to thousands of specimens. Data mining as it is called essentially involves searching for patterns that may emerge from the data. One does not have a prior model or hypothesis in the process of deriving or analysing the data here. Rather, one lets the data speak for itself, *without* the aid of theory (Kitchin 2014). A crucial factor that enabled the push towards big data in biology is that the technological advances needed for this became available, and the problem of *generating* such data was overcome. With a lot of such 'big data' at hand, the more daunting task to be addressed has been the question of how to process and interpret the data and convey its significance to the medical practitioner (Adams 2015). The methodological shift towards a large-scale collection of data which is purportedly hypothesis-free also points to the possibilities created by technology. The age of big data as exemplified by genomics, therefore, is a departure from conventional fields in biological sciences – in the latter, one focuses on specific questions or testable hypotheses regarding a cellular process or component or a molecule. With the emergence of technology that is capable of large-scale profiling and analysis of genomic sequences and other biological data at high speed, it is now cheaper to collect and store huge amounts of data even before thinking about what one will do with it (Goodman 1999).

The Human Genome Diversity Project and the Quest for Origins

An offshoot of the HGP is the Human Genome Diversity project (HGDP). The HGDP was put together with the aim of looking at genetic differences between populations and ethnic groups. Particularly, the focus was on isolated, indigenous populations that had not intermingled with other major ethnic groups. The HGDP has been criticised for attempting to catalogue and establish genetic variations between ethnicities and thereby provide a scientific basis for racial discrimination. Concerns were raised from the outset that the HGDP may be used for racial profiling of people and also end up by attributing health disparities between populations to genetics rather than to socio-economic factors. This criticism has, however, been refuted in a document elaborating on the purpose and design of the HGDP, and prepared under the auspices of the US government (Cavalli-Sforza 1994).

The aims and scope of the project were articulated by the organisations that participated in the HGDP from different countries. It was designed to include samples from individuals belonging to several different populations or ethnic groups, chosen so that they would be representative of the world's populations. In particular, it was proposed that the HGDP would include 'isolated populations or groups' and 'ethnic minorities' so that they could be genetically characterised. The idea of doing genetic studies on these samples using state-of-the-art technology was to provide insight into specific predisposition to disease in a particular population or group, its susceptibility to infections, and so on. Commercial benefits if any were proposed to be shared with the concerned communities out of whom the benefit was obtained. A potential positive outcome that was visualised for the HGDP was that with several countries participating in sequencing their own populations, technology transfer to the Global South could occur under the project, and thus be of use to the laboratories in these countries (Cavalli-Sforza 1994). The above document interestingly contests the idea that the HGDP will encourage racism by genetic discrimination. It clarifies that there is no biological superiority between races, other than the differences in skin colour and certain physical attributes, which are rather superficial and environmentally determined. If anything, the project was expected to disprove the theory that there is any genetic basis for racism and to create a 'bridge between genetics, archaeology, linguistics, and anthropology'. The issue of genetic difference between racial categories was in fact addressed after the first phase of the HGDP. In this phase, genomes of 52 population groups were studied. The data indeed showed a very small extent of difference between populations, amounting to less than 5% of the total genetic variation. On the other hand, differences between individuals *within* the same population account for a majority (almost 95%) of the variation that is present (Cavalli-Sforza 2005). If this was the case, why should one extend the project to further examine the genome sequences of populations? Even with regard to the medical aspirations of finding susceptibility to diseases, there were no records of medical histories of the participants in the project and hence it was not clear as to how this was going to be achieved. The overall goals of the HGDP as put forth were rather fuzzy and did not seem to have enough justification to the public, or even to the proponents themselves. Further, the HGDP stirred controversy among indigenous and aboriginal groups of people due to a clash of cultures. The subjects of the study were not schooled into biomedical patterns of thinking. The Maoris, for example, consider the gene as a 'life spirit' handed down from ancestors; Native Americans hold body parts as sacred and hence view the collection of human tissues including blood as violations of the body (Ilkilic and Paul 2009).

A crucial aspect in the discourse relating to the HGDP is the notion of race itself as a category for scientific study. This has been a subject of controversy even among scientists in the field (Wald 2006). Race and ethnicity are regarded as ever-changing, fluid categories that are socio-cultural rather than biological. Migrations and intermingling of populations throughout history negate the assumption that any ethnic group has clear genetic boundaries from another. This is true of even the indigenous peoples and members of tribes among Native Americans that were being sought for inclusion in the HGDP. Thus, there is no genetically 'pure' or 'untouched' population in the world (Dukepoo 1998). Even within the given boundaries, it has been pointed out that categorizations of ethnicity, including studies of people recruited in research projects in human genetics, are often precarious (Hunt et al. 2008). Despite these uncertainties and questions in the scientific basis for the HGDP, there are two factors that appear to have been the driving force behind it. The first is that technology provided an impetus behind the scientific effort, being employed to create a genetic database that will 'contribute to the world's cultural heritage'. It was evidently a fast and feasible task to create the genome data on multiple races and ethnic groups. Second, the conception and design of this project, even with the problems noted above, are determined by the view that the genome is an arena within which to understand evolutionary history, geography, predisposition to diseases, and biological differences between racial groups.

Science through the Lens of Gender

Despite the claim of objectivity in science, the inevitable 'colouring' of scientific pursuit in terms of its content, objectives, and outcome is a matter that has been intensely debated. Scientific advancement hence often reflects the ideology of the system in which it is rooted and caters to the dominant class, thus often reflecting the inequalities in the system. The supposed 'value-neutrality' of science is contested in various realms that are both within and outside the mainstream of white middle-class male society in which science was originally produced. The concept of theory-ladenness of observation thus applies to a scientist within the bounds of the experiment she is conducting as well as to his or her world view, which encompasses her philosophical and ideological positions in general. This means that no data is truly neutral, but are in fact dependent on the expectations and beliefs of the observer (Tan 2015). In addition, the content and directions of scientific pursuit have been critiqued through the prisms of gender and race. Importantly, feminist studies provided a perspective on the 'biological' divisions based on sexes, genders, races, and classes, attributing them to cultural and historical factors instead (Subramaniam 2009). These criticisms acquire particular significance in realms such as reproductive biology. Here, tools and technologies such as *in vitro* fertilization, surrogacy, and sex determination have greatly increased the prospects for commercialization of women's bodies and reproductive capacity and enabled widespread sex selection against females. These technologies have mushroomed over time to become less invasive and more freely available, giving flesh to the patriarchal value systems in which they were invented. Thus, you not only have sex determination of the foetus and termination of pregnancy but methods for sex selection of the embryo *before* implantation. Though on the one hand, they appear to provide more reproductive choices and freedoms, these methods have in fact promoted the exploitation of women.

Post-reductionism

The reductionist world view in the fields of genetics has been pointed out above, but there was a consequence of such an approach. The search for specific causes of common human diseases in terms of genes revealed that causation of these types of diseases is to a large extent, not explainable by means of the sequence of the genome per se. As stated earlier, this is due to the complex nature of causation in most common diseases. This realization eventually led to changes in method, such that the other factors were added into the equation to better understand the basis of disease. In addition to just the genomes of patients, environmental factors that influence disease (diet, lifestyle, etc.) as well as the interactions of genes with the environment were included as potential causes of a disease. All of these effects were assessed by means of statistical (theoretical) modelling. Thus, there is a turnaround from reductionist to anti-reductionist or synthetic methods as the field advances. The realization of the multidimensional nature of causation in health and disease also led to the evolution of the field of systems biology, which is concerned with the study of systems as a whole. The idea of systems biology is to understand how different properties of the system emerge from a non-linear interaction between its many components. In trying to recreate the whole from a number of entities such as genes, proteins, and other chemicals in the body, as well as health-related parameters, systems biology aims to reconstruct the complexity of the biological system in a more holistic manner (Ahn et al. 2006). The various components of a biological unit (cell, organ or organism) are put together by the use of mathematical models to capture the interrelationships between them. The basic premise of 'systems' approaches is that the system has 'emergent' properties; in other words, the biological system is more than a sum of its parts and hence the field of systems biology can be said to be 'anti-reductionist' or holistic in its perspective (Kesic 2016).

Conclusion

The connections between the empirical, theoretical, and rational aspects of discovery in the biological sciences have been varied and appear to have manifested in different forms. With the progress of time, changes in method were brought on by innovations in technology. Despite the apparent shaping of scientific endeavour by technology, the shift towards a genome-centric perspective in viewing health and disease reflects a world view in which genes are central to who we are. The organisation of this activity itself also shifted from individual scientists or small groups of collaborators to large networks of scientists across institutions and across countries, with private non-state agents playing a major role. The era of genomics also brought in new and unprecedented ways of practising science, moving from a hypothesis-based to a hypothesis-free mode of problem-solving, with a reliance on data-heavy methods of investigation. Here, the data 'speaks for themselves' and patterns within them are often discerned through the use of computer programs and mathematical tools. Thus, theoretical modelling became an important means for understanding and analysing the complexity of life forms. However, the use of models also placed a constraint on the possibilities for 'recreating' the whole from its separate parts. Reconstitution of the whole in this so-called 'anti-reductionist' mode depends on the types of modelling tools available, in terms of the number of features and dimensions that they allow you to use. At this time, a major challenge in the area lies in the synthesis or integration of the molecular data on genes and proteins with the health-related information from clinical records of patients. This is a methodological hurdle that needs to be overcome.

References

Adams, Jill U. 2015. 'Genetics: Big Hopes for Big Data', *Nature*, 527: S108–109.

Ahn, Andrew C et al. 2006. 'The Limits of Reductionism in Medicine: Could Systems Biology Offer an Alternative?', *PLoS Medicine*, 3(6): e208.

Ayala, Francisco J. 1994. 'On the Scientific Method, Its Practice and Pitfalls', *History and Philosophy of the Life Sciences*, 16(2): 205–240.

Ayala, Francisco J. 2009. 'Darwin and the Scientific Method', *Proceedings of the National Academy of Sciences of the United States of America*, 106(Suppl 1): 10033–10039.

Blystone, R.V. and Kevin Blodgett. 2006. 'WWW: The Scientific Method', *CBE Life Sciences Education*, 5(1): 7–11.

Bolinska, Agnes. 2018. 'Synthetic Versus Analytic Approaches to Protein and DNA Structure Determination', *Biology and Philosophy*, 33: 26.

Caskey, C. Thomas and Philip Leder. 2014. 'The RNA Code: Nature's Rosetta Stone', *Proceedings of the National Academy of Sciences of the United States of America*, 111(16): 5758–5759.

Cavalli-Sforza, Luca. 1994. 'The Human Genome Diversity Project', address delivered to a special meeting of UNESCO, Paris, 12 September.

Cavalli-Sforza, Luca. 2005. 'The Human Genome Diversity Project: Past, Present and Future', *Nature Reviews Genetics*, 6: 333–340.

Collins, Francis S, and Leslie Fink. 1995. 'The Human Genome Project', *Alcohol Health and Research World*, 19(3): 190–195.

Crick, Francis. 1962. Nobel Lecture, Dec 11th, 1962. 'On the Genetic Code'. https://www.nobelprize.org/prizes/medicine/1962/crick/lecture/ [accessed 5th August 2021].

Dukepoo, Frank C. 1998. 'The Trouble with the Human Genome Diversity Project', *Molecular Medicine Today*, 4(6): 242–243.

Epstein, Charles J. 2000. 'Some Ethical Implications of the Human Genome Project', *Genetics in Medicine*, 2(3): 193–197.

Gibbons, Michelle G. 2012. Reassessing Discovery: Rosalind Franklin, Scientific Visualization, and the Structure of DNA, *Philosophy of Science Association*, 79(1): 63–80.

Goodman, Laurie. 1999. 'Hypothesis-Limited Research', *Genome Research*, 9: 673–674.

Hood, Leroy and Lee Rowen. 2013. 'The Human Genome Project: Big Science Transforms Biology and Medicine', *Genome Medicine*, 5(9): 79.

Hunt, Linda M., and Mary S. Megyesi. 2008. 'The Ambiguous Meanings of the Racial/Ethnic Categories Routinely Used in Human Genetics Research', *Social Science & Medicine*, 66(2): 349–361.

Ilkilic, Ivan and Norbert W. Paul. 2009. 'Ethical Aspects of Genome Diversity Research: Genome Research into Cultural Diversity or Cultural Diversity in Genome Research?', *Medicine, Health Care and Philosophy*, 12: 25–34.

Kesic, Srdjan. 2016. 'Systems Biology, Emergence and Anti-reductionism', *Saudi Journal of Biological Sciences*, 23: 584–591.

Kimmelman, Jonathan. 2006. 'The Post-Human Genome Project Mindset: Race, Reliability and Healthcare', *Clinical Genetics*, 70(5): 427–432. doi: 10.1111/j.1399-0004.2006.00706.x.

Kitchin, Rob. 2014. 'Big Data, New Epistemologies and Paradigm Shifts', *Big data and Society*, 1–12. doi: 10.1177/2053951714528481

Leonelli, S. 2015. 'What Counts as Scientific Data? A Relational Framework', *Philosophy of Science*, 82(5): 810–821.

Ling, Jiqiang and Dieter Soll. 2012. 'The Genetic Code: Yesterday, Today and Tomorrow', *Resonance*, 17: 1136–1142.

Nirenberg, Marshall. 2004. 'Historical Review: Deciphering the Genetic Code - A Personal Account', *Trends in Biochemical Sciences*, 29(1): 46–54.

Pray, L. 2008. 'Discovery of DNA Structure and Function: Watson and Crick', *Nature Education*, 1(1): 100.

Subramaniam, Banu. 2009. 'Moored Metamorphosis: A Retrospective Essay on Feminist Science Studies', *Journal of Women in Culture and Society*, 34: 951–980.

Tan, Charlene. 2015. 'Investigator Bias and Theory-Ladenness in Cross-Cultural Research: Insights from Wittgenstein', *Current Issues in Comparative Education*, 18(1): 83–95.

Tauber, Alfred I, and Sahotra Sarkar. 1993. 'The Ideology of the Human Genome Project'. *Journal of the Royal Society of Medicine*, 86: 537–540.

Voit, Eberhard O. 2019. 'Perspective: Dimensions of the Scientific Method', *PLoS Computational Biology*, 15(9): e1007279.

Wald, Priscilla. 2006. 'Blood and stories: How Genomics is Rewriting Race, Medicine and Human History', *Patterns of Prejudice*, 40: 4–5.

6

CHEMISTRY, METHOD, SCIENCE, AND SOCIETY

A Conversation

Gita Chadha, Ram Ramaswamy, and Renny Thomas[1]

On Method, Techniques, and Scale

RT: Ram, thank you for agreeing to do this conversation with us. Our volume aims at building a discussion around how the scientific method finds different expressions in different disciplines. We do think that the question of method cannot be divorced from the larger cultures in and of science. Given your vast experience across academic institutions, both as an academic and as an academic administrator, we thought it would be useful to have a broad-based conversation with you.

Can we start by looking back at your early education and training in chemistry? Do you remember any courses on methods that you undertook as a bachelor's or even master's student of science? We find that the teaching of method is not central in disciplinary training, especially in the natural sciences.

RR: First, thank you Renny and Gita. I am glad to be a part of your volume and am happy to do this conversation.

I was an undergraduate student at Loyola College in Madras in the 1970s. Then or now, methodology, as you say, is not actually taught or discussed explicitly in most undergraduate science programmes such as the BSc in Physics, Chemistry, or Maths or Zoology, or whatever. There were things that we were taught as a matter of course, with no particular emphasis on methodology. Even at the master's level, I would say it wasn't as if we were welcomed with a philosophical introduction to the why and how of the methods that would be employed in the discipline. A lot of students in India make career choices when they are very young, and these choices are restricted by the streams of science, arts, and commerce. There are not many options. Things might change now that we have the NEP 2020[2] but at that time in 1969, I went to college and decided that I was going to do a BSc

DOI: 10.4324/9781003298908-11

in chemistry – with not very clear ideas about why I wanted to do it. And then I followed whatever we were taught. On reflection, I would say that we learned a lot of techniques, and in effect the practice – the technique – *is* or *becomes* the method for us.

Chemistry experiments in laboratories, especially in the undergraduate curriculum, were fairly basic. We were taught how to analyse compounds and mixtures, how to do assays by titration – how to identify molecules, how to identify their composition, how to identify the amount, and then, as we learned a little more in the second year and the third years we got a chance to synthesise molecules. And a lot of that was very 'how to' – the method was the instruction. Or the instruction was the method, in the sense that we were told 'add so many grams of this to so much of that, observe what happens' and so on. So we learned a lot of the methodology by doing.

GC: How does theory get taught? We suppose that theory and methods are linked, aren't they? So, is there something like a concept of theory, in the curriculum, at this stage?

RR: I've been talking a lot to a bunch of younger chemists these days and they tell me that increasingly the practical aspect of chemistry is getting lost in our country. In many universities, practical laboratory experiments are not carried out or are not emphasised at any rate. That's why some of these newer, smaller universities, and smaller science universities such as the Indian Institutes of Science Education and Research (IISERs) are important because at least they give students a chance to learn the subject as it is, or as we learned it. In many of our public institutions, the experimental part of chemistry is increasingly de-emphasised.

Hence, in these contexts, the learning of theory happens in two ways. One, in the absence of experiments, all of chemistry is or becomes 'theory'. But that's not the theory we are talking about, right? Second, in most contexts, one takes things in good faith – because we cannot go around repeating and verifying everything. So when a teacher comes in and tells you there are so many elements in nature, or that here is a piece of glass, and this is actually a liquid and this is not solid, one accepts it as the theoretical truth. Many of these things work on a certain kind of faith, right? And sometimes, it is paradoxical because we are supposed to teach our students to question – particularly in the sciences – but there's only a limit up to which the questions are tolerated or can be encouraged.

RT: In many of these places, are 'practicals' – which are basically laboratory experiments – seen as an alternative to the active teaching of methods, and probably also of theory?

RR: In the best places, we do have the 'practicals', many of which are fairly routine. And the routine of it emphasises the method if you like. Even

143

today when you measure the viral load in a COVID patient, to give a current context, one does this via an assay, essentially a titration, and in principle not very different from what you might have done when you were a 16- or 17-year-old, measuring the amount of acid in a solution. The tools have changed, the techniques have changed, but the 'method' is the same. And in a sense, it contains the theory too.

RT: If we look at the history of science, it's clear that the discipline of chemistry has had a strong foundation in the experimental method. Yet we don't find discussions on method in chemistry, unlike, for example, in physics and astronomy. Is there a particular reason for that?

RR: Well, I don't know. As a subject, chemistry is the one with the strongest connection to experiment, because it all begins with fire, in some sense. The first repeatable experiment that we as humans have ever been able to do is to set fire to things and to know what is combustible and what is not. These are the original questions of what is the constitution of matter. An early experiment would have figured out that water doesn't burn and dried leaves do, and so on, so the constituents of matter were all discovered only through early experiments. We wouldn't call these experiments in the modern sense of course, but I see them as being valuable precursors to experiments, because they were always practical and chemistry has, in this sense, a strong experimental tradition.

But in chemistry, there is also the question of scale. There is something about the sheer size of certain numbers that chemistry has to grapple with that makes it very difficult to think in terms of detailed, mechanistic ideas of method.

GC: Could you elaborate?

RR: See, chemists talk about molecules and atoms, and when you ask how many atoms are there in a cube of ice, it is of the order of Avogadro's number, which is about a billion million, such a mind-bogglingly large number. So the basic unit that a chemist works with, the molecule, is much much smaller than the scale of the typical experiment. I believe that this is one of the reasons why the connection to theory or the connection to method becomes difficult. I think that the question of scale is one that chemists have to grapple with all the time.

GC: So, you're trying to explain a very vast field in terms of something very small, is that the problem?

RR: Absolutely. Though this is not the *only* subject with that problem, a lot of subjects have the same issue. But chemistry is also a somewhat 'centrally' located subject because of the kinds of questions it is trying to look at. The basic unit in chemistry is really the molecular scale. And you're doing experiments at 10^{23}, 23 powers above that, right? And that's one of the reasons why some of the methodological concepts can be difficult.

RT: That is interesting. Often one thinks that chemistry as a 'discipline', a body of knowledge about matter, which sets the parameters for the experimental method – but which disappears and loses its identity in the sciences. It becomes associated as the foundation of 'methods' in other scientific disciplines, like biology for instance but loses its place as a theoretical discipline with its own understanding of matter. Is there a reason why it is difficult to distinguish between 'chemistry as a discipline' and 'chemistry as method'? It would be interesting to know if this happens as part of the training of a chemist. Any pedagogic or historical reasons that you might think of? Is it good and advantageous to have such a thin line between discipline and method?

RR: See, actually one aim of the laboratory chemist is to make specific molecules, to engineer substances with desired properties. For them, it's also internal: the practice is the method, as I said. There is a goal and one does not think in terms of whether it is a discipline or whether it is a bunch of methods. What happens as a part of your training is learning the important questions that need to be asked. For the chemist there are certain types of questions that are paramount, one of which can be paraphrased as follows: Given a molecule or given a property that is required (or desired) the aim of the practical scientist is 'Can I achieve that particular property?' Take, for example, Light Emitting Diodes (LEDs) that emit white light, or LEDs that emit red or green light. So one might want, say, a blue LED, and one can do this in many different ways. The chemist would approach the objective thinking 'Can I make a molecule that would emit blue?' Whereas a physicist might turn around and say, 'Can I put enough pressure or can I somehow change the physical conditions of the system so that I can achieve the same thing?' Meaning, the chemist would be led automatically to achieve his/her engineering of the situation by changing the molecule that they are playing around with, whereas the physicist might be more inclined to change the conditions around which the thing is operating. Of course, that distinction is blurred because today you have to ask what the material scientist is doing, and that person might like to do both. Another caricatured kind of thinking that guides, let's say, the practical chemist is the manner in which drugs are designed. If one came up with a molecule, let us say, that could kill the flu virus, and you know that the severe acute respiratory syndrome (SARS) virus is similar in shape, so the chemist might start by changing the molecule around, change its shape, or what have you, and see if it can kill the SARS virus.

In short, the aim is to do what one can by manipulating the things that I know best, namely small- or medium-sized molecules. Molecular biology also deals with molecules, it just happens to be very big molecules.

145

Specialisations, Theory, and Histories

RT: Ram, in chemistry the distinction between inorganic and organic chemistry has emerged historically and has become widespread. We have different departments for these two specialisations in some institutions. Does this divide reflect institutional, historical, methodological, or even ideological/political needs?

RR: The answer is: to varying degrees, all of the above. The way in which the chemistry departments grew in Europe, in the middle 1800s, say, was really along the lines of the kinds of molecules that people looked at. One of the facts of life, as far as chemistry is concerned, is that most molecules that occur naturally, so-called natural products tend to have a preponderance of carbon atoms. It was erroneously thought that life was needed in order to create such molecules, hence the adjective organic. From the early 1800s though it was recognised that there was nothing *organic* about such molecules that contained carbon, although there are special features of carbon that make such compounds interesting from various points of view. It also happens that many organic molecules have industrial applications. All of pharmaceutical chemistry, pharma industry, and medicinal chemistry, eventually also dyes, drugs, petrochemicals, all these dealt with essentially organic molecules, and therefore it was almost natural that when you have so much interest in all this, you have a Department of Organic Chemistry. Inorganic chemistry then got defined by exclusion, more or less, but over the past decades it has been widely realised that such a distinction does not really make much sense.

Most of these distinctions were made in universities and institutions outside India, and by and large we mirrored these here – there was not much by way of turf here in developing a subject. As a matter of fact the only department which was somewhat unusual and new is the Solid State and Structural Chemistry unit at the IISc, and there is probably no such department elsewhere in the world. I suppose one starts defining a new sub-discipline, there is not much that is methodological. There are some instances of entire departments of Physical Chemistry and/or Analytical chemistry, but these are rare, mercifully. By and large the unity of the discipline is emphasised within most departments.

RT: You have already answered parts of our next question which is related to what we asked you in the beginning too. Where do we locate theoretical chemistry in our discussion on methods? The reason we also ask this question is because while theoretical physicists enjoy a dominant, and even higher, epistemological status in institutes of fundamental research – such as the Indian Institute of Science (IISc) or Tata Institute of Fundamental Research (TIFR) – in India, theoretical chemists don't. Does this distinction between theoretical vs applied/or experimental have to do with different methods in the two disciplines? Or is it something else?

RR: The big distinction between theoretical physics and theoretical chemistry (and I know that a lot of theoretical physicists will disagree with me on this) is that a theoretical chemist has to work very closely with an experimental chemist. If you don't, the theory is not worth all that much. This is not as crucial in theoretical physics.

To give an example, I was a postdoc with a well-known theoretical chemist, Rudolph Marcus who got the Nobel Prize for his very important theory of electron transfer. This theory of electron transfer applies to how materials corrode, how oxidation happens, some of the crucial steps in photosynthesis – electron transfer is everywhere. Marcus came up with his idea in the late 1950s and 1960s, and there was an assumption that he had to make. The theory was applied from the beginning and was widely used, but he only got the Nobel Prize in 1992 because it was only in the middle 1980s that a very crucial result of his assumption – the so-called Marcus inverted regime – was experimentally verified. There is some similarity to Higgs, who was given the Prize only after the discovery of the Higgs particle, but nobody had any doubt that there *was* a Higgs particle. Even if it had somewhat different properties, most physicists would concur that the Higgs boson was a very real entity. As also gravitational waves. Maybe I don't understand these issues very well, but my impression is that fundamentally people did not disbelieve in the existence of the Higgs particle. But this is not the same thing when it comes to belief or disbelief of theory in chemistry.

In a way, there are few conceptual frontiers that remain in chemistry – at least not in the same manner as there are in physics – so even the theory is to a large extent 'applied'.

Intuition, Innovations, and the Silos

RT: We are also keen to explore the nature of the creative process in the method of science. If we look at the history of innovations in chemistry, some of the innovations were products of what we may call 'serendipity'. But then serendipity is not *seen* as a crucial element in the production of scientific knowledge. As a practising chemist, how do you view this debate?

RR: See, many accomplished chemists that I have met or interacted with are great masters of intuition. And that comes from a lot of experience, from repeatedly looking at a problem. Many chemists think in pictures, so there's a lot of very strong intuition that gets built up through this practice. Very fine chemists build up their repertoire by just seeing example after example after example, and then that serendipitous discovery comes about. It could not happen in the absence of a wealth of experience. A young graduate student walking into his or her lab and discovering something absolutely wonderful – that does not happen easily in this field.

GC: The word intuition is used very differently by scientist across different fields and disciplines, isn't it? The mathematicians understand intuition very differently from, say the physicists. Within physics, the theoretical physicist understands it differently from the experimentalist. So if we really want to look at the question of intuition, it would actually make sense to see how the scientists themselves are looking at intuition. Anyway, Ram, also, how do we contest – or nuance – this popular idea of the 'eureka moment' in scientific discoveries where insights are believed to come out of nowhere?

RR: No, it does not come out of anywhere! Regardless of discipline, it has to be grounded in some experience. The people whom I would class as being intuitive are really people who have spent a fair amount of intellectual time thinking about very specific kinds of issues. For example, the question of what is the shape that a molecule might take, or what shape a group of molecules might take? Large molecules for which the answer is not immediately obvious. This is related to problems like protein folding, which in turn is related to things like how the ribosome acts and similar questions. Even a casual conversation with someone who thinks deeply about such matters would tell you that the intuition was based on (a) a *lot* of examples that they have seen, and (b) obsession. That is, you've got to keep on thinking about a problem over and over again, in different contexts and so on, and then you may get your flash of insight. But of course, that moment took it's time coming. That moment was actually built up of a lot of other moments.

GC: Right. The trouble comes in the pedagogy of science, or what we call the meta-narrative of science in society, where on the one hand science is presented as a rational and linear method, on the other hand there is a drama around how genius discoveries are made, and how they rely on exceptional talent often associated with the great intuition of individual scientists.

RR: I think as academics, particularly as practising academics, we try to convey that our subject is interesting. I used to frequently have discussions with my mother and she would be asking me questions like, 'What do you actually do?', 'Is it satisfying to just sit in your office?' She was not really able to understand the *how* of it – her job was a practical one, taking things from one place to another, doing specific tasks, going somewhere, and she couldn't understand what it was that I did. This kind of incomprehension is quite common – most people just don't get it. So in an effort to make our work interesting, we talk mainly about the high points, the grand ideas. But very frequently when I mentor young students I try to explain to them that a lot of science is boring. It's

148

boring in the sense in which you think certain things are boring – repetitive without very spectacular occurrences every day. As we grow older, we do realise that a lot of what we do just takes a lot of time, is repetitive, and is not particularly insightful. But, we gloss over these facts.

GC: Yes, and when we want to present it, we want to present it in this very hyperglamorised way, very dramatic, and almost mystified manner.

RR: Exactly! I mean, one can't say 'I went to the lab, I worked from morning to evening. Not sure I made any progress.' What's there to say about your typical working day, right?

GC: Yes. It's interesting that you put it in this way. Because often we sociologists are asked by our friends in natural sciences: 'What do you actually *do* in sociology?' The fact that we do 'field work' – or formal and abstract work sometimes – and not laboratory work is never good enough!

RR: I taught at JNU for 30 odd years and many colleagues would show a naive surprise as to why there was a need for a School of International Studies? 'What do they do except read the newspaper, right?' Colleagues in the sciences would be very upset that their social science colleagues would be away from their workplace in the afternoon. 'Gone to the archives, gone for field work? *Yeh kya hota hai field work? Kya archives? Ganga dhaba mein baith ke baat kar rahe hai, that's all!*'[3] I exaggerate, but this was not an uncommon reaction!

Language, Grammar, and Culture

RT: The next set of questions are about language, Ram. It would be interesting to hear your views on the question of language in chemistry – its methods and its larger pedagogic context?

RR: Well, I haven't thought so much about it because I've grown up in the English language tradition of Indian education. I did learn German because I thought it would be useful for me as a chemist and that I can read some of the texts in the language they get written in. I actually did, I studied German in (undergraduate) college and had to pass an exam to qualify for my PhD. Many US universities did have a foreign language requirement even in the 1970s. But it has not been useful anymore because even the Germans write in English these days! But, more in our context in India, especially given the huge difficulty that many of us have with English, I think that many science subjects are, in fact, somewhat intimidating to non-native speakers of English. The 'natural' language of a science or of a discipline can sometimes be very off-putting, because the nuances are lost in the learning and non-native speakers don't pick up as much.

I've seen this happen in JNU and have been concerned about this in the language of examinations, for instance. When setting a national examination that is given to let's say, 200,000 students or 200,000 aspirants – all of whom are thinking in different languages – the manner in which a question is framed can itself be very intimidating.

RT: And alienating.

RR: Yes, alienating and off-putting. As it happens I'm grappling with this issue now, and in formulating questions, I try to imagine how a person who thinks in Hindi or in Tamil is going to see the way in which the question is phrased.

GC: We would like you to dwell a little more on this question, especially in the way it pans across the silos. Some of us in the social sciences tend to think that this question of language, this problem of language faced by many a student in India is graver in the social sciences and the humanities. And that in the sciences it's probably less so because science is mathematical, uses symbolic conventions. So, in that sense, a functional knowledge of English is sufficient. What do you think of this?

RR: I think that is a bit of an illusion. Thanks to JNU, I have had several students who are first-generation learners. First generation, but coming from remote districts of India. Being a first-generation learner in Dharavi is very different from being a first-generation learner who's coming out of Champaran or someplace like that. In one case, you may not have education at home, but many of your neighbours are educated. But what do you do when your entire village is uneducated? And somehow you have aspirations and you manage to come to university?

I have students whose lack of English has been such a major handicap: in spite of their ability to work phenomenally hard, sometimes they just don't understand. Science is also built up on jargon; science is also built up on tradition. So it's not as if one can just get away with a rudimentary knowledge of English. Sometimes, the best contribution I am able to do to a collaboration is to reformulate an idea. In a sense, just by changing the language, one can give something to the concept. Frequently when you give a thing the right name and the right description, it is halfway to explaining it. And that is true in science as well as in any other discipline.

GC: Do you think that the language of modern chemistry is Eurocentric? Can we say that? Would you say that? This is a provocative and polemical question, but we cannot escape asking it

RR: It is Eurocentric, but as is much of the language of all of science.

GC: It's almost a grammar that we've inherited, right, a culture of that grammar and also a cultural grammar?

RR: We've inherited a grammar but we've not been able to steal that grammar, to make it our own. We've not even been able to create a creole!

GC: Or reinvent it? Hybridise it?

RR: We have created a tradition of sorts in mathematics, in the way in which a lot of the inventive mathematics was done in the subcontinent, so it can be done. But most of Indian science, and I will be massacred for saying it, is really quite derivative. It is derivative either by mimicry, in the sense that you look at what has been published in the latest journal and then you ask related questions in your own context. Or it is derivative by tradition, what you did for your postdoctoral training in a research group elsewhere, you continue to do for the rest of your career. I don't think that we've created a new indigenous tradition – what I am calling a creole – through this process.

GC: It's possible though, you think?

RR: Of course, it is possible. Japan for example has its own idiom, as does China. I lived in Japan for a while and I know that it is possible to do science in Japanese. It is not really possible to do much science in Hindi or Bangla or any other Indian language. Eventually you'll just have to come back to English.

GC: Yes, while on the one hand we have so many dialects and diverse languages on the other the dominant languages, often the official ones, are deeply embedded in caste and other social structures of power. For instance, the kind of regional translations that are available in the social sciences is extremely sanskritised and brahmanical. They reproduce another hegemonic order and culture, don't they?

RR: There used to be this joke that in the Tata Institute of Fundamental Research (TIFR)'s Mathematics department, the faculty meetings could be conducted in Tamil.

RT: Upper caste Tamil! But let's link this to the question of access. The question of language is central in understanding caste and gender-based exclusions in India. Recent attempts to teach engineering in Telugu, or Kannada, in some engineering colleges in the South, have met with resistance from the Dalit and Bahujan communities. While many local politicians have welcomed these moves to 'indigenise', there is a large resistance from Dalit and Bahujan intellectuals because they see English as the language of emancipation for their people. We must also see this in the context of the larger patterns that are emerging with the privatisation of higher education in India. While the privileged upper caste students go to elite colleges – and will study in English – in the city, students from the rural areas will be forced to study in their 'mother tongue'. Consequently, the upper caste elites will continue to dominate these professions because of their English skills.

RR: Exactly! Who's going to employ them? This is a serious concern. The discussion on language is very difficult in our country and it's fraught with all sorts of problems, but, if we are to internalise it to make an idiom which is truly somehow non-western, truly somehow 'ours', it has to be done. Otherwise, you are just a derivative. If your only source of information is (the magazine) *Nature*, or *Science*, or another of the fancy journals, then you've already lost. In a lot of our institutions, the main

151

focus of publications is really not to address a problem of relevance, but to do something that will use the correct jargon and get into *Nature*.

One of the brightest students at JNU who was admitted to virtually every PhD programme he applied for in the country (he could not afford the cost of going abroad) nevertheless chose to join the Department of Atomic Energy (DAE) for the assured job it gave him. When making this choice, he basically told me that his family could not afford having him not contributing for another five or six years.

GC: Let us examine the language – academic language across the silos – question further. You've been the Vice Chancellor of one of the leading central universities of the country, the University of Hyderabad, and you have had experiences with JNU another Central University with a rich culture of dissent. You have this vast experience of being an academic administrator and also of being a public intellectual; all of that. Typically, our experience has been that in a conversation between the scientist and the social scientist, the latter is often accused of using jargon. The scientist tends to think that everything that a social scientist says should be accessible to them, simply because they understand English, and because it is all common sense in any case. How do you see this? Isn't this epistemic intolerance a very serious problem when we are attempting to go beyond the silos? How do we educate ourselves? And challenge Snow's assumptions of the two cultures that shall never meet

RR: This happens even in a pluralistic university like JNU, one which is very sensitive to democratic ideals.

RT: And progressive.

RR: And progressive, but not in all ways. There are political positions that are actually very strange, for example, many of my colleagues from the social sciences would say that the 'science-types' are all right wing, no question: if you are from a science school, then you must be right wing and today of course it goes along with a particular kind of ideology. Many of them actually seriously believe it. I used to find that offensive.

But this is very much a case of the two cultures. What was true of Cambridge in Snow's time was true of JNU too. Scientists on the faculty had a very poor understanding of what scholarship in the social sciences actually meant. And sadly, most of the good scholars in the humanities and social sciences did not engage with their peers in the sciences as equals. Maybe it was the fact that most of the science faculty were not very sophisticated on political and cultural issues, or maybe because as far as they were concerned the science-wallahs were interested in some mundane issues and small questions, not in something grand in scope or importance.

I was aware that JNU was well known as a social sciences university because of the stellar contributions that had been made in those areas, and the people who had been there. But most of these contributions of JNU could not be quantified in the same way as the contribution from the sciences could. The science contributions were straightforward to measure: how many papers have you written, where have they appeared, how many people have cited them, and so on. Social science contributions would focus on the number of books, the impact on policy, the seminars attended, and so on. Similar, but different.

So when I went to Hyderabad, one of the things that I was conscious of was these very different yardsticks by which to evaluate disciplines. I believe the JNU experience made me quite sensitive to the fact that contributions from different disciplines come in different packages. And, at some point, one has to go beyond the numbers, and also to realise that peer validation and peer recognition can be a very real way of evaluating contribution. If a whole group of people whom I believe are serious academics, tell me that someone is a fine young sociologist, then I don't have to sit in further judgment and ask for the *h*-index data as well.

GC: In fact, quite a few of the social scientists in Hyderabad University say that this is something that really helped them in your tenure as Vice Chancellor, you strove to delink measures that were used for the social sciences from those that were used for the natural sciences.

RR: I think that it is useful to have some recognition that people can aspire for, especially at the early stages of one's academic career. You must have noticed that in India there are a lot of awards for 'young' scientists and very few for social scientists. So one of the things I was able to introduce at the University of Hyderabad was a set of early career awards that were actually open to all disciplines, with different modes of evaluation for different areas.

RT: As we conclude, a very generic question but a question that brings us back to *the question of method*. What is going to be the future of chemistry as a discipline? More particularly, do you think chemistry as a discipline has to rethink its methods? If yes, what are some of the methodological challenges the discipline of chemistry will have to deal with in the coming future?

RR: I'm probably not the best person to answer this. But if I were to essay a response, I'd say that one of the things that will increase in importance in chemistry is the notion of complexity. The reductionist approach that is characteristic of much of modern science is changing. What we've realised at least in the last couple of decades is that systems are complex and the reductionist approach is not always possible. When dealing with large (very large!) numbers of molecules, it can be very profitable to factor in features such as cooperativity and to look at the system as a whole.

Chemists have been doing this for some time already – the realisation that one can use the idea of groups of molecules as a fundamental unit rather than just using atoms and molecules is dawning on many chemists. If I may make an analogy with your discipline, if instead of the individual being the important constituent, one might think about a family, or a clan or a tribe – as the basic unit. This 'supra' molecular approach gives a very different approach to thinking about how to carry out chemical processes. Similarly, one increasingly sees that instead of carrying out a set of reactions sequentially, it is possible to carefully manipulate the chemical environment in a reaction vessel so as to make the system effectively control the outcome. This is where intuition comes in, knowing how to manipulate the circumstances to get a desired chemical outcome. Let me just say in a catch-all terminology that a recognition of complexity and factoring it in is an important way in which this subject is evolving.

Institutions, Excellence, and Exclusions

RT: Let us, in this last section, shift somewhat to questions of scientific communities and cultures. And examine how objective and inclusive these are. In the Indian context, for instance, rarely does one see scientists publicly talk about caste and gender, leave alone an easy entry into the domain. As a scientist who is interested in sociological questions in science, what are your observations on the discipline of chemistry itself? Would you say it is dominated by upper-caste, Hindu, Brahmins? At least in the elite institutions. What are the possibilities of democratising the discipline of chemistry in India?

GC: Also, in the hierarchy of sciences and if you examine them on the axes of purity–impurity, what we see is the hierarchy of sciences. The 'pure sciences' are at the top. And what are the pure sciences? They are abstract, they are theoretical, the queen being mathematics. Disciplines like chemistry which are much messier are placed lower down in this hierarchy of abstraction and purity. You have the hierarchy between the pure and the applied and experimental, even within disciplines. This phenomenon recurs in the social sciences too. How do you respond to this?

RR: These are some of the issues that have come to the forefront recently, the democratisation of science. We have to learn how to do it and learn how to do it fast.

Actually, a proper answer to this would need much more thinking on my part, but let me tell you my immediate thoughts. The fact that most faculty members in most departments are savarna is just a consequence of the fact that only the savarnas have had access to higher education by large. Dalits have a much more difficult time coming into education and being able to do science is even more of a challenge.

That said, the establishment, namely us, has not taken this as a solvable challenge, at least not in the way that gender disparity is seen as a solvable challenge. Namely that it is possible to get equal numbers of women and men into science, it is possible to keep the percentage of women working in science to something like 40%, and so on. By and large, people see this as something that can be achieved.

Some years ago, I tried to get information from the Indian Academy of Sciences about the composition of the summer research fellows that they support each year. Out of something like 20,000 students who applied for a summer fellowship, what are the different categories that students fall into? Most often one can figure out gender from the name, but caste and religion are much more difficult. And there's a backlash against even trying to collect this information: Although there have not been many studies on this matter, when asked, many Indian scientists would claim to be atheist and non-casteist, and even caste blind. In this context, asking about religion or caste can be a challenge.

The fact of the matter is that we have not really given access. It is not enough just to assert that one's doors are open, one has also to make sure that others can enter. From the word go, we have excluded entire categories by imposing impossibly high standards. By the time someone from less privileged backgrounds is even able to contribute, they have to climb against a huge gradient. What I guess I'm trying to say is that unless we try hard to give proper access at the entry stages to different groups, we will not see any change in the composition of departments. We need to work hard in order to make our institutions truly representative of our social diversity.

Another thing that works against inclusivity is the small numbers that we admit into higher education. Especially in our elite institutions, there is the attitude of exclusivity that communicates 'What I do is very difficult – you can't work with me, you won't be able to keep up'. And this leads to small numbers of students that these faculty are able to mentor. In a career of 30 or so years (which is fairly typical), if I happen to have one student, or two or three, chances are they'll also be *savarna*, and the consequence is that the composition of the academic pool doesn't change as fast as it needs to. On the other side of the spectrum, at some universities it is possible to mentor a larger number of students (I'm talking about PhD mentorship) – something like one per year, so in a career of 35 years one could have had about that many students. Were they all brilliant? No. But over the years, it doesn't matter – they all have positions, and the effect of the training tells. And in those 35, there were students from reserved categories, there were students from different religions, from a large number of states, there were women. Even so, this was not as representative as it should have been. When the numbers are large enough, this level of diversity almost comes for free.

GC: This despite the original discouragement of being told and made to feel that this science stuff, particularly the abstract and theoretical stuff is too difficult, and complicated. 'And just too hard for someone like you'.

RR: Yes, 'It's too hard, you won't be able to do anything significant ... and you won't get a job after that. You know you have to support your family'. This kind of apparently realistic mentoring is in effect very discouraging.

Moving from TIFR to JNU saved me in a sense, because at TIFR I could have lived out my working years quite peacefully. I may not have had more than five or ten students, if at all, because TIFR works that way. That was actually the major motivation for me to get out of TIFR because I realised that the longer I stayed there, the more difficult it was going to be for me to have students of my own. Having been educated in the United States, the US University was the model that in my mind was worth emulating. That's how you did things: you had your own research group, own grants, own research problems.

The best thing that happened to me was that by coming to JNU, I got this independence. I have also very consciously never turned away a student and have quite consciously tried to help them reach whatever level they could, telling them in effect that the quality of their PhD would be as good as they could make it. Just because I worked with a Nobel laureate, am I going to get a Nobel Prize? It's not going to happen! The point of mentorship is to help the student achieve their goals to whatever extent, and this is something that I really internalised at JNU. Many of the leading social scientists were really exemplary in this regard.

GC: Could this also be because of the divide between the research institutes, the so-called centres of excellence, and the universities, the former set-up exclusively for reassert and the latter for teaching, once again a problematic divide that sets up hierarchies?

RR: Yes! Absolutely.

RT: So, do you think that the university could be the site for a reimagination and reinvention of quality education and research? And not these centres of excellence? We seem to be putting so much of a premium on these centres of excellence.

RR: What is happening also is that these so-called centres of excellence will realise that they have to become more university-like.

GC: You think so?

RR: I mean, look at what is happening already in TIFR. And the new national laboratories. They will realise that they have to get out of their exclusivity which is so much at variance with the rest of the ethos of the country. And I don't mean it in a good way. They are somehow upholding a standard that is truly not representative. Much of what they do is an elitist pursuit of elitist ideals, which are in the long run, unsustainable.

I believe that we will see many of these institutions – if they are not able to become more like universities, that is to have more students and more turnaround, with the vitality that comes from the flow – we will see them wither away. When I came back from the USA, that was a higher education model that I knew, it was also a model that I admired. I did not find myself admiring the actual conditions in TIFR, although to be honest, what's not to like? I mean the exclusive West Canteen, the magnificent art collection, the dramatic setting by the sea, the manicured lawns and gardens ... Yet I left TIFR in 1986, very clear that this level of exclusivity was something I did not want.

GC: Ram, let us look at your own social location more reflexively. You were born in an upper-caste, upper-class family. Was your choice to do science influenced by these social locations?

RR: In the late 1960s when I had to make choices, there were few disciplinary choices available. I didn't have biology in school, so medicine was out, most other fields I had little interest in, but I did have the freedom to do so, which came from the fact that I belonged to an upper-middle-class family.

GC: Being male and heterosexual gives you a further advantage. Were you conscious of these advantages in your early life, maybe even as child?

RR: I don't think I realised all this at the time, but I was aware that I had the luxury of pursuing an academic career because there was no pressure on me, from my parents or others in the family, to take up a professional career (or a 'well-paying' job).

GC: Also, do you think these structural privileges helped you gain the power and success you did? Recognising privilege is a step in the direction of making scientific cultures more inclusive. How did you learn to do that?

RR: At some point, I did articulate to myself (and a few others) that the biggest gift that I got from my parents was this freedom to do what I wanted. Several of my contemporaries did not have this luxury – they had to start earning earlier, had to have more well-defined career paths, they had a responsibility to their families. I see that now with several of my students, and the choices that they have had to make. A first-generation learner who has the freedom to pursue a PhD has a very different sense of what he or she 'owes' to those that in a very real sense made it possible. Being from a privileged background makes us much less sensitive to this matter.

GC: Ram, at the very end we want you to address a rather sticky debate in the relationship between science and society. What do you think is the relationship between *scientific method* and the *scientific temper*? We think that the former is a very specific academic tool used by the sciences and that it varies so much in practice. We think that it cannot be applied as a corrective to all social and cultural belief patterns. Science, or its method, cannot always serve as an antidote to the unreason of various kinds, can it? In fact, science often reproduces social hierarchies, like when it justifies racial, caste, or gender inequalities. We'd suggest that a more powerful epistemic and moral tool is required to counter unreason in our world, science is not enough, and neither is scientific temper. A critical temper of enquiry, wherever it comes from, needs to be promoted. What do you think?

RR: Like many of us, I am not overly fond of the term (scientific temper) nor do I think that years of scientific practice makes it even possible to clearly define what this temper is. The use of the word temper in this phrase is both subtle and unfortunate. Subtle because it is uncommonly used in this context, and therefore minute distinctions need to be made between temper and temperament. It is unfortunate because of the implied arrogation of all wisdom or knowledge of a truth to science and its methods.

There is much wisdom in the adage that the more one knows, the more one needs to know, and that is more zen than the scientific method.

GC: Ram, thank you very much for this conversation. Your vast experience of being an administrator and academic across several constituencies and domains has helped us discuss several important issues not only about the method of science as it travels in and from chemistry, but also about how this scientific method does or does not find its way into larger scientific cultures and practices. It was also important to speak about the problems of going beyond methodological silos. Thank you.

Notes

1 Ram Ramaswamy is a well-known scientist trained in chemistry. He has held faculty positions in Jawaharlal Nehru University, Delhi, and has also been the Vice Chancellor of the University of Hyderabad. Currently, he is a Visiting Professor at IIT, Delhi. Gita Chadha and Renny Thomas are editors of the volume. The listing of names is in alphabetical order. The transcript of the interview was done by Oskar Kinny and ran into 22,000 words. We reproduce here an edited version of the interview.
2 The National Education Policy of India, 2020, emphasises the integration of the liberal arts with the science-centric streams in Higher Education Institutions.
3 Roughly translated from Hindi: 'What is all this field work? What archives? These people are just sitting at the Ganga *dhaba* and chatting away!' Ganga *dhaba* is an outdoor canteen on the campus of the Jawaharlal Nehru University in Delhi where students, research scholars, and faculty members often 'hang out'.

7

'BETWEEN CLEARING AND CONCEALMENT'

Knowledge-making in Physics

K. Sridhar

Introduction

The question of method in physics is an important one because it is of concern not just to the narrow group of physicists but also to much wider sections of society that go beyond the boundaries of academia. This is due to a host of reasons – largely historical but some systemic – as the method of physics is paradigmatic for method in all sciences, including the human and social sciences. And science, through its relation to rationality, has also become the face of modernity. In a sense, the physics sets many paradigms of and for modernity.

It was in the 18th century, the period referred to as the Age of Enlightenment,[1] that the scientific world view took the form of an overarching philosophy of the world going much beyond the strict confines of science. Alongside science, the project of Enlightenment brought other values to the fore such as rationality, democracy, liberalism, and individualism. In an essay called '*What Is Enlightenment?*' (Kant 1995), Immanuel Kant sought to equate Enlightenment with an ability to speak up for oneself not simply with an ability to reason. Kant argued that it was a lack of courage, and not of intellect, that made people unable to voice their thoughts. It was this latter condition that Kant identified with a lack of Enlightenment. Kant defines Enlightenment as 'man's emergence from his self-incurred immaturity' and he claims '*Sapereaude*' (Dare to think for yourself!) as the motto of the Enlightenment. This influential essay went a long way in establishing this period as the age of expression and not just the age of reason. Enlightenment philosophy further made an identification of rationality with science, thereby establishing the triad of reason, expression, and science. Reason, especially scientific reason, then gave the individual the right to expression. This had far-reaching implications: on the one hand, it gave voice to the common individual who needed to appeal to no authority to

DOI: 10.4324/9781003298908-12

be heard but, on the other hand, it privileged a scientific rationality leaving little room for other discourses.

In establishing science as the voice of reason, the Enlightenment project also portrayed religion as allowing no scope for reason or for a freedom of expression (Sobel 1999). There was an attempt to see the history of modern science as a struggle against religious dogma. In fact, what marks the departure of modern science from the science that preceded it was its opposition to a religious culture with this point of departure usually identified with the advent of the heliocentric planetary model of Copernicus. Copernicus was certainly not the first to conceive of a heliocentric model of the planetary system and purely as a model of this sort, it may not have exerted the influence on human thought that it did. It was revolutionary because its scope went far beyond the strict confines of scientific academia – it was setting the stage for a confrontation with the authority of the Church by asserting that the order of the heavens is to be understood by using scientific reason and not by ecclesiastical prescription. Copernicus' *De revolutionibusorbiumcoelestium* (On the Revolutions of the Celestial Spheres) and Galileo's *Dialogue Concerning the Two Chief World Systems* were eventually named in the Catholic Church's *Index of Prohibited Books*. Copernicus had refused to publish his work in his lifetime for fear that his revolutionary ideas may provoke criticism. For his advocacy of Copernicus' heliocentrism, Galileo had to face an inquisition by the Church and was placed under a lifelong house arrest and the publication of his books was forbidden. But even Galileo did not suffer as much at the hands of the Church as Giordano Bruno, the Italian philosopher-mathematician, did – Bruno was burned at the stake.

In spite of their heroic defence of science against the tyranny of the Church, the relation of these and other scientists to the Church and its authority was ambivalent: Copernicus dedicated his book to Pope Paul III, and Galileo sought papal permission for publishing his book. Galileo also strove to reconcile heliocentric ideas with the Scriptures and, as for Bruno, it is not entirely clear whether his being condemned as a heretic had anything to do with his defence of the Copernican system; it may well have been due to his theological heresies. The position of the Church on the Copernican doctrine also kept changing and, even in this period, there were several high-ranking church functionaries who showed great intellectual interest in the new science of Copernicus and Galileo. The relationship between religion and science cannot be cast in the simple binaries as many readings of this period are wont to do, but recent literature on this subject does bring justice to this complexity (Sobel 1999, 2011; Rowland 2008).

In the latter half of the 17th century, Newton published his new mechanics which made a break with the mechanics of Aristotle that had held sway for two millennia. Newton's laws of motion and his theory of gravity went a long way in establishing the world view of modern science. In particular, Newton, by propounding universal laws of mechanics, did away with the difference between the terrestrial and the celestial that Aristotle's system had upheld. The 'fall of the celestial sphere' had a visual quality to it that made it an overarching metaphor for all contestations of power: divine, papal, royal, and political. Implicit in this wide-ranging acceptance of the truths of the new science was the all-round endorsement of its method. Indeed, philosophers and scientists of the time went about the task of abstracting a 'method of science' because of the popular appeal that science had generated, especially because of its ability to answer questions ranging from the terrestrial to the cosmological. In abstracting this method, the focus was on Newton and the new physics that he had discovered. This is how method questions in science got to be focused on physics and that legacy has lasted a good three centuries.

The Enlightenment of the 18th and 19th centuries in Europe was a period of consolidation of science. On the one hand, science was providing a new method of addressing those questions that had traditionally been the domain of theology and philosophy. But with its insistence on both mathematical rigour and quantitative, empirical verification science provided a way of arriving at an understanding of the natural world that was demonstrably true. The precision in understanding phenomena also meant control so that research breakthroughs in science often led to the development of new technologies. This was a period of rapid development of mechanical technologies that had immense implications for industry, navigation and military ammunition. These developments fuelled the imperialist ambitions of the European countries of the time so that science managed to grab the state favour traditionally reserved for the Church.

This period was also one of several path-breaking scientific discoveries: the development of modern chemistry, the ideas of atomism, thermodynamics, and evolution were all great achievements that went beyond the mechanics of Newton. But, in an attempt to consolidate the power of science, the issue of method was left untouched by these discoveries so as to not perturb the universal and immutable nature of scientific method. It is my belief that if these discoveries were allowed to converse with accepted notions of scientific method in this period, we would not have had to wait till the 20th century to address these issues.

Following this general discussion, in the next three sections, I will discuss the understanding of method culled out of classical physics and then go on to discuss the challenges that have come up when this method had to confront new developments in the last hundred years. In the final section, I will share some unconventional speculations about approaching the method question in physics.

Method in Classical Physics

The classical view of science starts by clearly dividing the domain of science into two parts: the formal and the empirical. The formal system consists of a set of propositions that are deductively linked to each other. Some of these propositions have epistemic connections with another set of propositions that are inductive and represent the extant empirical information about the physical system or phenomenon that is being studied.

In the simplest version of the classical view of physics (science), it is with the observation of physical phenomena that science begins. The spatial or temporal regularities of phenomena or causal correlations between them are usually considered the triggers for new work in science. The observation of an interesting phenomenon then leads to the next phase which is experimentation. Experimentation is distinguished from observation in that it is the study of the phenomenon in controlled, often isolated, circumstances. As much as can be done, the system is prepared for the experiment. Experimentation also involves the measuring apparatus and devices specially built for making the measurements possible. The results of the experiment are quantitative and presented after a statistical analysis of the data is carried out. So while commonplace observation may suggest a connection between two events, an experiment will seek to find out the statistical significance of the correlation between the two events.

In this simple version of the method of science, the next step that gets listed is the making of a theoretical model or a hypothesis. This step involves making a model of the system studied in the experiment. Such a model in physics is almost always mathematical and so the mathematical model is analysed and new predictions from the model are made which can then be tested and verified or falsified in subsequent experimental measurements. If the theoretical model is successful, then it is generalised to bring other experimental observations within its purview.

In physics, especially as it is practised today, quite a bit of this simple description gets modified. In the first place, and that is a fairly obvious thing to state, both the measuring apparatus and the statistical analysis that goes into extracting data from experiments are highly sophisticated as compared to the time when much of these early ideas of the method of science were worked out. Moreover, the building of measuring instruments and apparatus is also considered as part of routine experimental work. The fact that the apparatus plays such an important role in experimental work implies that even the kind of research questions that can be explored get determined with the kind and accuracy of the apparatus that are available or can be developed. Experiments are done in collaborations which could involve anything from three or four members to thousands of members. So an important part of experimental work involves standardisation of procedures and techniques so that different parts of the collaboration, usually spread over different locations, can work consistently. While this may seem like a trivial thing to do in a small group, it is a challenging task when large collaborations have to function and this forms a major part of the experimental activity. This challenge is exacerbated when it involves

162

the fabrication of equipment required to conduct large experiments, which is also done in parts over several locations and then assembled and put together at the experimental site. Particularly, when precision is important, the achievement of similar levels of precision at the different locations where the components are fabricated is a daunting challenge. Large experimental collaborations are not just spread out over space but also over time. Often two or sometimes three generations of experimentalists may end up working on the same experiment and much effort at standardisation needs to be achieved in order that information across these different generations is transmitted smoothly and efficiently.

But, more significantly, almost no experiment originates in an act of simple observation. Every sub-area of physics has, at any given time, a set of interesting questions that provide impetus to new experimental projects. Even the most novel and unexpected research questions that physicists may arrive at rely on the extant knowledge in that area, so that what is chosen for study is very largely determined by what has already been studied. It is within this frame that creative breakthroughs, intuitive leaps, and path-breaking discoveries are made alongside what one may consider the quotidian, 'bread-and-butter' research.

In most other areas of science, theoretical work involves making a model of an existing experimental finding, in accordance with the simplest understanding of method in science. If the model is successful in explaining the features of the experimental data, then the attempt to generalise it to other experiments is made. But theoretical research in physics is of an entirely different nature and there is a very clear separation between experiments and theory in physics. There are several reasons for this separation: of all natural sciences, physics is the one that is most amenable to mathematical reasoning given that the observations of the physical world are measurable and quantifiable. Physics has also had the advantage that there is always a large theoretical system at its base: classical mechanics, electromagnetic theory, statistical mechanics, relativity, or quantum mechanics and theoretical calculations are done in the context of this theoretical framework. It is not an ad hoc modelling of an experimental situation that the theoretician attempts, but rather the calculation is done independently, staying true to the consistency and rigour of the chosen theoretical framework and then finally the comparison is made of the results of the theoretical computation to the relevant experimental information. What is required then is a pool of physicists who have a mastery of the theoretical foundations of the subject and not necessarily the technical aspects of experiments, their design, instrumentation, and data-taking. Consequently, at a rather early stage in physics education, a division between theory and experiment is made and postgraduate training in physics, in particular, is aimed at making a student either an experimentalist or a theorist. With the growing sophistication of the knowledge systems in experiments and theory, this separation has almost become watertight from around the early part of the 20th century. There are several detours that method takes because of this separation. In the standard under-standing of scientific method, when a theoretical model disagrees with the data from experiment, the model is discarded. In the way physics is actually practised, this does not quite happen. That is because theoretical physicists fall back on a much larger theoretical system of which the calculation they are doing is only a small part. A disagreement with experiment is not an indication of the failure of the grand theory but rather a cue that the specific calculation needs to be redone with some tweaking of inputs. Even when the experimental result stubbornly refuses to fall in line with the theoretical calculation after several modifications, the entire edifice of the grand sys-tem is not given up. For that to happen, a much larger breakdown of the grand theory is required as happened when Newton's mechanics gave way to quantum mechanics, for example. Also, because theoretical calculations are done independent of the experimental situation, it is often the case that a theoretical calculation is done not in response to a particular experiment but before the experiment has actually measured a particular phenomenon. In other words, a theoretical prediction is made for what may be observed in experiments. Again, if the prediction is not confirmed by the experiment, it is not a complete rejection of the theory that is called for but a mere tweaking of the specifics of the model. In fact, there are several examples in the history of physics where theoretical predictions have been made and the experiment was planned and executed in response to the prediction. In some of these cases, the experiments could not be executed or had to be shelved for financial or other reasons leaving behind a trail of theoretical predictions which have not been verified or falsified. The independence of theory from experiment also results in the creation of new theories which are so far ahead of their times that experiments to test these theories are not even conceivable because the required technologies do not exist. In the last 50 years, much of the finest work in theoretical physics is of this kind in areas like high-energy physics or cosmology – belonging to the realm of being not testable even in principle.

Of course, theoretical research in physics is also largely determined by the extant knowledge in the field and on the research questions that the community has singled out as important, though it may not be quite as constrained in this respect as experimental physics is because it is not quite as dependent on external inputs like apparatus and devices; however, physi-cists doing theoretical work using massive computers may find themselves constrained in ways similar to experimentalists.

Challenges to the Classical Perspective

Having discussed how the actual practice of physics, especially in the con-text of the present, poses challenges to the simple standard view of method,

I will now discuss the challenges that have been posed to the classical view from developments within physics – especially those due to the radical departures from the Newtonian world view that took place in the 20th century. These issues have been discussed extensively in the literature (Weyl 1949; Margenau 1950; Heisenberg 1958).

To do this, I will first present a more formal description of the classical world view, in terms of a set of fundamental guiding principles:

1. The physical system to be studied can be separated from the rest of the universe and, in particular, its local environment. This is a very important axiom that guarantees that the physical system under study can be isolated and that its features can be studied without reference to the rest of the world so that its properties are not affected by the surroundings. If the surroundings do have an effect on the system, they are so small that they can be neglected.

2. The physical system to be studied can be separated from the observer (the physicist studying it) or, more precisely, the system is not affected by the act of observation. The act of observation involves the interaction of the physical system with not just the observer but also the measuring apparatus.

3. The next principle is one that refers to the preparation of the system for observation. The act of preparing the system for observation does not change the features of the system so that the experimental set-up, contrived though it may be, does not interfere with the observable properties of the system.

4. Then there is the conception of space in classical physics. From the early days of Greek philosophy, two conceptions of space have existed in the Western world. There was the substantivalist view of the Greek atomists like Lucretius and Democritus for whom space was fundamental and had an existence independent of matter. The void, which was space devoid of matter, was an important part of their philosophy. For the relationalists like Aristotle, space was a positional quality of the phenomenal world. In this view, space is intimately linked to matter and there is no space that is devoid of matter. The conception of space in classical physics was very much a substantivalist position. Absolute space, as Newton called it, was very much like the void of the atomists and existed independent of matter.

5. Time too was conceived of in absolute terms in the classical view. Time flowed independent of all matter and was independent of the nature or state of matter. There was one obvious problem with the conception of time in the Newtonian framework. Newton's laws of motion were symmetric with respect to time: the past and the future stood symmetrically with respect to the present. The arrow of time, i.e. the flow of time from the present to the future, which is part of our everyday experience, is

not something that emerges naturally out of Newton's laws of motion. It is a problem that has not been addressed very satisfactorily even now, though it is believed that somehow in dealing with large, complex systems the unidirectional flow of time is recovered.

6. Finally, there is the issue of causality. It is to Descartes to whom we owe the clearest conception of mechanical causality that is the foundation stone for much of classical science. The conception of causality that had dominated philosophical thought before Descartes was teleological and was due to Aristotle. In a teleological understanding of causality, there is a notion of an end or a final cause that yields the causal connection and an explicit reference is made to *telos* or the end of the causal process. Aristotle argued that the effect was latent in the cause (potentia) and it was an unexpected and magical transformation that changed cause to effect. The discontinuity of cause and effect was crucial in Aristotle's conception. Descartes argued, however, that cause and effect look totally unlike each other only when we consider large temporal separations. Descartes would have us focus on tiny steps in time and then it is possible to map the entire evolution of the system in a gradual and continuous fashion. The system at any time feels the forces acting on it and gets displaced to a state very slightly displaced from the previous one. Through such a series of infinitesimal displacements, the system moves continuously from one state to another. Discontinuity is replaced by continuity and magic by continuous mechanical evolution. Due to the fact that the evolution can be continuously mapped, prediction becomes possible and it is this predictive power that is the hallmark of classical physics. In fact, mechanical causality ensures that if we have complete knowledge of a system at an initial time and the forces acting on the system at that time, then the entire time evolution of the system can be predicted. With the invention of calculus, the mathematical machinery needed to do this was in place and classical Newtonian science was firmly established.

The above six principles are guiding principles that form the skeletal framework of classical physics. Such guiding principles/assumptions form the basis of any scientific theory. Most of these seem to very obviously accord with our everyday experience of the world and so we take these to be true. Typical empirical tests do not validate or falsify these guiding principles, but they are challenged when there is a Kuhnian paradigm shift. Indeed, such an exigency materialised in physics in the early part of the 20th century when the new theoretical frameworks of relativity and quantum mechanics were proposed. I will describe the changes wrought by these theories in the remainder of this section.

As mentioned earlier, the guiding principles seem very obviously true because they match very neatly the world as we experience it: the world of familiar length scales and timescales and, in this neatly constructed world a metre is a metre and a minute is a minute, no matter what. The theory of relativity, or more correctly the special theory of relativity, showed that this is not true and that distances and time intervals are relative to the state of the observer. More to the point, two observers moving at different speeds will report different length and time measurements for the same system that they observe. But the reason we do not see this in our usual experience is that the speeds we deal with usually are too small compared to the speed of light. But close to the speed of light, space and time go awry and the solidity of the classical world evanesces, moving from a substantivalist to a relationalist account of space. In fact, in relativity, it makes little sense to talk of space and time separately – instead the concept of space-time is introduced because intervals in space-time are what appear the same to all observers. When we advance from here to what is called the general theory of relativity, then space-time is curved by the matter that inhabits it and is not flat as we always assumed space was – which means that a straight line is not the shortest distance between any two points and the sum of the angles of a triangle do not add up to 180°, for instance. Most importantly, space and time are not a skeletal framework like they were in the Newtonian framework but part of the dynamics of the new theory.

The first big challenge to this conception of causality came in the early 20th century with the advent of quantum mechanics. Quantum mechanics was the new mechanics that governed the dynamics of the microcosm – the world of atoms – where classical Newtonian physics could no longer be applied. In the classical framework, physical systems were modelled either in terms of Newtonian particles, i.e. entities that were localised or in terms of waves which were not localised but spread out in space. These are to be thought of as two mathematical models that classical physics uses to understand the physical world. The expectation was that in the subatomic world, one or the other of these pictures would hold except that this expectation was too naive. The stuff of the microcosm demanded an altogether different mathematical picture that does not correspond to anything that we know of in the macroscopic world given to our senses. This new mathematical model corresponds in some ways to particles and in other ways to waves, but it is, in fact, neither one nor the other. The upshot of this is that the state of a quantum mechanical system cannot be completely specified: there is a trade-off between the particle-like and wave-like manifestations of this subatomic entity and the state of this entity can be specified only with some uncertainty – not as sharply as can be done in Newtonian physics. So Cartesian causality gets rewritten somewhat in the quantum world because the initial state of the system cannot be completely specified, as Descartes would want us to do. However, the time evolution of the system is still specified uniquely so that the final state of the system can still be predicted, albeit with a level of uncertainty that comes from that in specifying the initial state.

The second major challenge to Cartesian mechanical causality, and a much more serious one at that, came from a different direction. It was realised that much of our understanding of causality in classical Newtonian physics came from systems that were experiencing forces which were linear, which is essentially a mathematical way of saying that small causes would lead to small effects. In the middle of the 20th century, some mathematicians revisited Newton's physics but by considering non-linear forces, i.e. with no guarantee that causes and effects will be commensurate. These studies have led to the development of completely new paradigms in classical physics, none of which could have been anticipated by studying the simple systems that were classically investigated. The key point again is that this is a departure from Cartesian causality. The initial state is classical and so precisely determined but because of the non-linearities in the system, the evolution of the system is no longer the simple evolution anticipated by Descartes. Instead, starting from an initial state one would, depending on a choice of parameters, end up in a whole range of disparate final states. These new paradigms go under the name of chaotic dynamics and are now understood to be the models for complex systems.

Quantum mechanics goes further in challenging every one of the other guiding principles of the Newtonian system. In particular, the principles that guarantee the separation of the physical system from the environment and from the observer are no longer valid in the quantum framework. This shift occurs because in quantum theory, the information about the system is encoded in a probabilistic manner in an object called the wave function. Rather than the definite values that we associate for observables in the classical case, the quantum object only reveals itself through propensities. For example, we would need to observe a classical coin after it has been tossed to know if it is showing heads or tails, but we would know for sure that it is either a head or a tail even before we observe it. The quantum analogue of this coin does not quite behave in this manner: a quantum coin exists in a state which is a mixture or a superposition of the two possibilities before it is observed. It is the act of observation that collapses the system into one of the two possible states and we observe either one or the other. The subtler aspect of the theory is that it is not just the system that is described by the probabilistic wave function but the system and the measuring apparatus are all implicated in this probabilistic quantum description. It is clear then that the principles of separability of the physical system, that were assumed to be true in the classical case, are no longer valid.

If relativity and quantum mechanics indicate a complete breakdown of Newtonian mechanics, then what is the status of Newtonian or classical mechanics, post these 20th-century revolutions? Or why were Newton's laws so successful in an older historical epoch if we have found them to be wrong now? The standard understanding is to invoke what is called the correspondence principle. So take relativity, for example. So while relativity and not Newton's theory is the correct description of nature, the physical effects predicted by relativity are discernible in very special situations, i.e. when the systems being studied are moving at incredibly high speeds close to that of light. At speeds which we normally deal with, it can be shown that the mathematics of relativity neatly reduces to that of Newton's mechanics. This is the correspondence principle by which the theory that is correct over a larger range of phenomenal experience reduces in some situations to match with the theory that works in a more limited domain. The correspondence principle also works in the case of quantum mechanics; i.e. quantum effects are only important at microscopic length scales, but when we get to macroscopic length scales, Newton's mechanics take over. The matching is not quite as neat in quantum mechanics as it is in relativity and that has been a persistent problem in the foundational understanding of quantum mechanics. I would like to argue, however, that if we take theoretical systems to mean not just their mathematical structure but also their fundamental guiding principles, then there is no limit in which we can recover Newton's mechanics from relativistic mechanics or quantum mechanics. The demonstration that in some limit the equations of relativity or quantum mechanics reduce to that of classical mechanics is not sufficient because these are fundamentally different world views and they are truly incommensurable in the Kuhnian sense.[2]

Before I end this section, I will discuss one binary which becomes an important way to frame method question in physics: subjective/objective. I will do this briefly by alluding to a discussion by Carl von Weizsacker where he problematises this binary (von Weizsacker 1952). Weizsacker describes himself seated at his desk working on the manuscript of his book and his attention is drawn to a block of crystal that is on his desk which he uses as a paperweight. He says if he is asked to describe the crystal, then, as a physicist, he will talk about its physical properties, its optical properties, its lattice structure, and so on and so forth. But, as Weizsacker says, this is true of every other crystal of the same kind in this world. Then what is it that singles out this particular crystal, its uniqueness, the property that philosophers call haecceity. Weizsacker says if he is asked why this particular crystal was on his table, he will give his account of how he was given this crystal as a gift by his university professor at the time of his graduation. This information, of course, never enters a description of this crystal as physics would have it because it is considered subjective and physics concerns itself with the objective description of the crystal. But the problem is precisely that with quantum mechanics, it has become somewhat unclear where this line of delineation needs to be drawn.

Realism/Anti-Realism, Atoms, and Beyond

The 18th century was the pinnacle of the industrial age – the age of machines and engines. The study of machines became one of the applications of

Newton's mechanics in engineering and while it may have lacked the elegance of the conventional physics applications, it was considerably more useful. It was utility that also motivated the study of engines, but the more mathematical studies of engines evolved into a new branch of physics called thermodynamics. The question that these 18th-century physicists asked themselves was how one can translate gross, macroscopic properties like temperature and pressure of a gas, for example, into concepts in classical mechanics like force or energy. This translation to Newtonian terms was the way to make their study a respectable science! They found a way out that turned out to be remarkable. They invoked the idea of atoms, which philosophers had talked about 2,000 years before them, and imagined that the gas they were dealing with was made of several millions of tiny 'atoms' or corpuscles. Each individual atom was governed by Newtonian mechanics: one could think of the energy of each atom or the force that each atom exerted on the walls of the container. Through this act of translation, it was possible to identify the temperature of the gas with the average energy of all the atoms in the gas or the pressure as the average force exerted by the atom, i.e. it was possible to explain thermodynamic concepts like pressure or temperature using purely mechanical concepts like force or energy. The price to be paid for this was the introduction of a whole new substratum of matter which was not observable. Atomistic ideas gained currency because of their introduction in chemistry where the introduction of molecules and atoms was necessitated as a way of explaining the systematics and regularities of chemical processes. A great synthesis of several strands of this story was achieved when quantum mechanics was used in the early 20th century to explain the structure of the atom in terms of its constituents: protons, neutrons, and electrons. Of course, the story of atomism did not end here: it went deeper first with nuclear physics and then with particle physics in the search for the ultimate constituents of matter.

The important methodological question that arises in the context of atomism is: Are atoms real? Are electrons real or, in the context of particle physics, are quarks real? The question of the reality of the external world has been discussed extensively in philosophy with idealists like Berkeley taking the extreme position that the sensations of the external world do not guarantee its existence. But I will discuss this question more in the context of the philosophy of science. The basic question that one starts from here is: Are the entities of science real? A related question that arises is whether that reality is independent of us human observers (Rothbard 1998). The reality question has more charge, of course, with entities like atoms that are not observable. The realists' answer to the reality question is in the affirmative arguing that if there is a chain of links, even theoretical, that connects any entity that science posits to a set of observable phenomena that are verifiable in experiment, then that is reason enough to say that these entities exist and they are not merely explanatory devices. Rather, the fact that these entities have to be invoked for explaining phenomena is to be taken as a guarantee of their existence. The realist accepts the scientific theory literally *and* they believe in the truth of this theory. Giving up either the literal construal or the truth of scientific theories is the movement towards an anti-realist position.

Anti-realists refuse to equate the validity of a scientific theory with the existence of the entities it postulates. There are several different anti-realist positions, but largely they say that the entities that are posited are shorthand devices that we invoke in the process of comparing a scientific theory to experiment and the success of the theory in explaining experimental results cannot be taken to be literally true.[3]

I am of the opinion that a recent theoretical development within physics can actually shed light on the realism/anti-realism question. This new idea is called duality and I will attempt to explain it very briefly. In its simplest form, duality relates one theoretical model to another in a mathematically specified manner. Usually it would mean that a parameter specifying theory A is related to a theory B which is described not by the parameter but by its reciprocal. A and B are called dual theories. The mathematical analysis of theory A may become intractable for some values of the parameter in which case it would be easier to work with theory B and the reciprocal parameter. Duality is seen as a mathematical device – it gives ease of computation. But at some point one is forced to reckon with the realism/anti-realism question. Does the system described by A exist or the one described by B exist? Recent versions of duality relate seemingly disparate theories – what would be a relatively normal state of matter in the laboratory could have a dual interpretation in terms of black holes. This makes the realist position untenable and the only way out is to adopt an anti-realist position.

Reflections

In the last two sections, I put forth the standard view of method in physics and its 'standard' reworkings. In this section, I want to put down some reflections that are tentative but suggestive of a new way of approaching the method question in physics.

I start by quoting Ludwig Boltzmann, the great 19th-century physicist, who said, 'Once we examine the simplest elements, where would be the boundary between science and philosophy at which we could stop?' (Boltzmann 1974). Almost as though he is responding to Boltzmann, Heidegger in *The Origin of the Work of Art* makes a characteristically insightful statement:

> [S]cience is no original happening of truth, but rather in every case the cultivation of an already open domain of truth, namely by conceiving and grounding what shows itself to be possibly and necessarily correct within its circle. When and insofar as a science comes

171

over and beyond the correct to a truth, i.e. to an essential unveiling of being as such, it is philosophy.

(Heidegger 1993)

The activity of scientists seems to Heidegger like that of explorers who have set out deep into the forests with compasses and sextants and charts and go about the business of exploring and mapping the terrain without doing the self-reflexive turn of asking themselves the question: Why are we doing this exploration? This is why question may have both an epistemic and an ethical charge, but the scientist does not seek a standard of judging his enterprise from anything without but assumes it has already been justified within his realm of exploration. Truth, Heidegger says, is 'the strife between clearing and concealment' – it is through the conflict between clarity and obscurity that truth emerges where it may allow science, as Latour says, to divert our attention from 'matters of fact to matters of concern' (Latour 2004).

Physics, more than any other science, has pursued the fact-driven path of clarity to the extent that not just considerations of ethics but even considerations of empiricism seem to no longer matter. The research goal is arrived at by the consensus of a sub-community of researchers, a race to a known end begins, results are published and citations are collected and very often without the worry whether any experiment will either verify or falsify their explorations. Instead, physicists seem eager to complete papers ending with 'So we conclude' so much so that physics becomes a discourse of conclusions. This is the reason fundamental physics suffers from what I like to call the final-theory anxiety – the search for a theory that will answer everything – without even bothering to stop it in its tracks and ask itself the question: 'What is it?'

Is there a way out of feeding into this anxiety-ridden culture of physics and move towards a different understanding of its method so that the practice may also evolve in a different way? There are no definitive answers here, but even tentative suggestions may be of value. If with Heidegger we believe that truth is not something that the sciences are after and the pretence that a science like physics has towards truth should be shed, then what should be the new direction towards which physics should head?

Deleuze and Guattari (1994) suggest that one should deflect one's attention away from truth to look for what is interesting: 'Thought as such produces something interesting when it accedes to the infinite movement that frees it from truth as supposed paradigm and reconquers an immanent power of creation'. This is probably a much-needed gestalt shift to perform within physics. In practice, the research area that a young physicist may choose to work on is very often determined by interest rather than grand claims that the project may have towards knowledge and truth. Interest also seems much less forbidding a goal: too many sacrifices have been made on the road, rather highway, to truth, whereas the path mapped out by interest has several small intersections and branches, by-lanes and diversions, and much more accepting of particularities and contingencies. This will also allow moving away from the unitary conception of physics as a thrust towards that one truth, one knowledge, one theory of everything to a physics which is more amorphous and accepting of multiple viewpoints, something that will allow us to experience the roughness of the texture of the domain and not keep sliding along its smooth surface. For too long we have looked at research projects that are being simultaneously undertaken as competitors insisting that of the multiple studies that are being done, only one will be true and that will get selected, as though through a process of natural selection – the first, correct paper wins the race of evolution in physics and the rest are simply forgotten. Why would the third paper or the fifth, or even the 'wrong' paper not have a perspective, an understanding that the community should not gain from, that researchers should not acquaint themselves with or students not be made to study? Is it not possible to read physics as one would read literary fiction – worry less about its 'truth-claims' and engage with what is interesting? Why, instead, do we have in place this ahistorical monstrosity of a system where only the 'first, correct' publication remains in the collective memory? In fact, this system is so firmly in place that the second, third, and nth papers very often offer no perspective seriously different from the first because there are really not too many ways that you can essay a 100-metre run. If one wants a genuine multiplicity of standpoints to emerge, a complete break with the old model has to be made and the new one allowed to seep into the physics community long enough for a new perception to emerge. Would we not exult then over a paper which is 'so deliciously wrong' – can we even begin to dare think of this alternative universe?

Notes

1 Though this is well known, it is important to state here that science did not emerge in the Europe of modern times but has a long history in several cultural locations. Science, reason, and the relation between the two has been a part of this history and the closely associated understandings in philosophy like materialism have also been developed to a great degree of sophistication in many cultures like that of ancient China, India, and Greece. The age of Enlightenment granted science the centre stage and it became the overarching discourse across disciplines and that is essentially different from what existed as science prior to that.

2 In his famous book, *The Structure of Scientific Revolutions* (Kuhn 1962), Thomas Kuhn argued that in the course of a scientific revolution when a new paradigm replaces an older one, the terms of the old paradigm are not explicable in those of the new one. This lack of a common measure with which to assess competing scientific paradigm is known as incommensurability.

3 It is worth pointing out that the logical positivists, with their insistence on empirical verification of scientific propositions, take a strongly anti-realist position. For them, atoms and subatomic particles are only convenient ways of describing the observable phenomenon and these explanatory devices should be treated as such and not elevated to the status of existents.

References

Boltzmann, L. 1974. 'On Statistical Mechanics' in B. McGuinness (ed.), *L. Boltzmann: Theoretical Physics and Philosophical Problems*, p. 163. Dordrecht: Reidel.

Deleuze, G. and F. Guttari. 1994. *What is Philosophy?* New York: Columbia University Press.

Heidegger, M. 1993. 'The Origin of the Work of Art' in D.F. Krell (ed.), *Heidegger: Basic Writings*, pp. 143–212. New York: Harper Collins.

Heisenberg, W. 1958. *Physics and Philosophy: The Revolution in Modern Science*. New York: Harper & Row.

Kant, I. 1995. 'What is Enlightenment?' in I. Kramnick (ed.), *The Portable Enlightenment Reader*, pp. 1–6. New York: Penguin Books.

Kuhn, T. 1962. *The Structure of Scientific Revolutions*. Chicago: University of Chicago Press.

Latour, B. 2004. 'Why Has Critique Run out of Steam? From Matters of Fact to Matters of Concern', *Critical Inquiry*, 30: 225–248.

Margenau, H. 1950. *The Nature of Physical Reality: The Philosophy of Modern Science*. New York: McGraw Hill.

Rothbard, D. 1998. *Science, Reason and Reality: Issues in the Philosophy of Science*. San Antonio, TX: Harcourt Brace and Company.

Rowland, I. D. 2008. *Giordano Bruno: Philosopher, Heretic*. New York: Farrar, Strauss & Giroux.

Sobel, D. 1999. *Galileo's Daughter: A Historical Memoir of Science, Faith and Love*. New York: Walker and Company.

Sobel, D. 2011. *A More Perfect Heaven: How Copernicus Revolutionized the Cosmos*. New York: Walker and Company.

Von Weizsacker, C.F. 1952. *The World-View of Physics*. Chicago: University of Chicago Press.

Weyl, H. 1949. *Philosophy of Mathematics and Natural Science*. Princeton, NJ: Princeton University Press.

Part III

SHIFTS WITHIN THE SILO: SOCIAL SCIENCES

INTRODUCTION TO PART III

The social sciences silo is an interesting one. Like the natural science silo, it places the disciplines on a positivist hierarchy, with economics at the top of the pecking order. The more quantitative a discipline is, the more value it gains. Economics is like physics, sociology like biology. Yet, unlike in the natural science silo, in the social science silo one encounters relatively more discussion on methods. There seems to be more reflexivity produced in this silo, obviously due to the nature of the subject matter. Almost all the disciplines within this silo have grappled, and continue to do so, with the question of what it means to be objective about subjective things. And probably how to make subjective things into objective ones, in order to study them. Interestingly, the question of method, central to this debate, marks many of the origin stories of these disciplines. Questions of method bring in many contestations and ruptures around 'scientificity'. Most of these debates arrive at the proposition that knowledge is often a function of methodological orientations, which are influenced, if not determined, by the historical, social, political and cultural contexts in which they are articulated. Almost all the chapters suggest the eurocentrism of method – and theory – in these disciplines needs to be continually seen on the axes of power. Almost all the chapters further suggest the need to free ourselves from the hegemony of the positivist/scientific paradigm of knowledge production. Almost all the chapters suggest ways of opening the disciplines to multiple methods and multiple ways of knowledge-making, particularly foregrounding marginal perspectives and epistemes. The theory-ladenness of knowledge and the grounded nature of theory become evident in almost all the chapters, in different measures and on different scales.

DOI: 10.4324/9781003298908-14

8

DECOLONISING METHOD

Where Do We Stand in Political Studies?

Aditya Nigam

Introduction: The Present Moment

For the purposes of this chapter, I will use the term 'political studies' to include both the empirical discipline of 'political science' and the field of 'political theory'. Unlike most other disciplines, political science and political theory have existed in a very tense and uneasy relationship, deriving from the conflicts that were played out over decades in the American academy – though political theorists continued to be housed in departments of political science.[1] The term 'political studies' has occasionally been used in a broad and generic sense, but I use it here mainly to sidestep the baggage that the 'political science'–'political theory' divide has come to connote in general. It is interesting that in India, in the early 1970s, when Jawaharlal Nehru University was set up, there was a larger concern with interdisciplinary teaching that led to the institution of the Centre for Political Studies, alongside ones like the Centre for Economic Studies and Planning (economics), Centre for Historical Studies (history), or Centre for the Study of Social Systems (sociology). The idea was to delineate the field of study rather than the discipline – the assumption being that one needed to draw from different disciplines, whatever one's object of study.[2] My use of the term here does not really follow that logic but is more immediately concerned with moving beyond the debilitating divide mentioned above.[3]

If debates on method in an earlier era focused on and emerged out of a concern with 'scientificity' and the positivist insistence on producing a 'political science' that would be akin to, and modelled on, the natural sciences, we are in an entirely different moment today. The 'scientificity' issue was particularly an obsession, as Ball (2007) has underlined, of the post–Second World War tendency that is known as the 'behavioural revolution', which ended up exacerbating the fateful divide between the empirical discipline of 'political science' and 'political theory' (Ball 2007: 1). Debates between the proponents of 'scientificity', who favoured quantitative methods and 'empirical' evidence geared to 'discovering' invariant laws of

DOI: 10.4324/9781003298908-15

political behaviour, and political theorists who either moved towards normative philosophy or in rarer cases, to critical theory, tended to make divide insurmountable. Then there were theorists of another kind who took microeconomic theory, based on methodological individualism, as their ideal and worked towards building rational choice and game theory models (Petracca 1991; Satz and Ferejohn 1994). They were equally invested in making the discipline of political studies a 'science' and were extremely influential in the 1970s, 1980s, and 1990s.

In contrast, the debates are very different today, in what I will call the moment of decolonisation. This is especially so in the field of political theory, which has been less concerned with questions of scientificity and of late, more seriously engaged in the business of 'de-parochialising' it by opening it out to texts, thinkers, concepts, and modes of thinking in non-Western cultures, by instituting a new subfield called 'comparative political theory', or 'comparative political thought' which would work through 'dialogue'-as-method (Dallmayr 1997; Tully 2016). What this means, above all, is that the assumptions of universalism that undergird all quests of scientificity are no longer seen to hold. It is now not even a matter of debate that 'political theory', by definition Western, is, like most other bodies and forms of knowledge, a highly 'parochial' or 'provincial' entity and can only be rescued by opening it out to other intellectual traditions. Interestingly, the earlier poststructuralist moment, its critiques of universalism, the linguistic or cultural turn, and the concern with 'difference', all seem to have bypassed political studies altogether. Even a cursory look at recent works on the history of the discipline is enough to bring out this strangely insular nature of political studies, especially in the United States (Farr 1988; Dryzek 2006; Leca 2010; McBride 2016).[4] However, one of the indirect ways in which critiques of universalism have impacted the discipline in more recent times has been by forcing it to look beyond its 'scientific' and 'universalistic' pretensions and try to hear 'other voices' – that is to say, when the 'other voices' became more vocal. This is something that had not quite happened earlier, despite decades of serious feminist and critical race scholarship in the United States and elsewhere (Philipose 2007; Hawkesworth 2010).[5] The criticism has been made by feminist scholars in particular, that feminist and critical race theory have not only challenged key distinctions in political theory like those between the 'public' and the 'private', but have also done something that traditional political theory has failed to do, which is to 'engage with the political world' (Hawkesworth 2010: 690–691). It may not be incorrect, however, to say that these voices acquired greater resonance as political theory started facing bigger challenges with the overall move towards decolonisation of the social sciences gaining momentum. Today things have changed so much that, according to Frances Widdowson, the Political Theory Section of the 2012 meeting of the Canadian Political Science Association, hosted a number of workshops and panels that sought to 'critically examine the colonial impulses and decolonising potential of political theory'. It was even recognised that 'western political thought' has 'served, either implicitly or explicitly, to justify the dispossession of indigenous and/or other non-European peoples' lands and self-determining authority', though it was also simultaneously underlined that colonised peoples have been able to 'selectively appropriate and critically transform these theoretical frameworks to support their own discourses and struggles over land and freedom' (Widdowson 2013: 1).

It is worth underlining that one of the reasons Mary Hawkesworth, a political scientist who works with a feminist perspective, thinks might lie behind political theory's elision of both feminist and critical race theory is their 'context-specific mode of theorising' and 'a particular philosophical impatience with the messiness of complexity' (Hawkesworth 2010: 691). I bring this up here because this question of 'context-specific mode of theorising' is a concern that will be with us throughout this chapter as we grapple with other traditions of intellection in other parts of the world.

This chapter is divided into five sections. In the first section, I deal with the empirical political science versus political theory division that has, despite its very narrow US origins, replicated and reproduced itself in universities and political science departments across the world, thanks to the power of the institutional–disciplinary apparatus. It underlines the point made by Hawkesworth from our very specific postcolonial vantage point: theory in the postcolonial/non-Western world cannot but be at once engaged, 'empirical' and context-specific, though a context-transcending ambition remains crucial to its enterprise. The second section deals with the problem of the 'field' as it were, where the central question will be of 'theory' produced in one time-space/context being usually 'applied' to data or material from the 'field' to either 'interpret' the field or, at best, 'test' the theory – which as we know is never ever either falsified or invalidated. As opposed to this, some recent work – work that is at once empirical and theoretical – underlines the need to treat these spaces not merely as 'field' in the above sense but as the ground for theorising by paying more attention to the categories of thought and specific practices that structure the lives of social agents. As an instance of this, the second section briefly discusses the ways in which the Indian democratic experience has been studied and theorised by some scholars. The third section deals with some very important developments within the discipline (or political theory referred to earlier) emerging largely within the US academy which make some attempt to actively engage with other thought traditions. In the fourth section, I enter the relatively new – and vibrant – debate on decolonisation where I will contrast these trends in the American academy with other voices that go well beyond the attempts to 'open out' the discipline, for I suggest, that the historical moment for this may be long past. In the fifth section, the conclusion, I return to theoretical forays made by some Indian scholars which have a direct bearing on the issues at hand.

Since a lot of story of the birth of political theory and its battles with empirical political science are primarily an American story, the first part will be largely an account of that story where I draw heavily from the Introduction to a volume that I edited with two other scholars in the discipline (Menon et al. 2014). Thereafter, in the second, third, and fourth sections, I will move into a series of questions that arise from the business of de-provincialising and de-parochialising political theory, which also have implications for the reconstitution of the very idea of political science–political theory as it exists now. The fifth section will look at the articulations that are taking place outside the Euro-American world, which do not directly address themselves to 'political theory' but to theory as such – which cannot but have an implication for political theory as well.

Political Science versus Political Theory

The theory–'empirics' divide is actually quite pervasive across most disciplines, but it has a special resonance in the field of political studies where the two (political theory and political science) have been at daggers drawn for a very long time, perhaps from the time of the very birth of the discipline of political science as we know it today, which is to say in the late nineteenth and early twentieth centuries. It is by now a widely accepted fact that as an academic practice and discipline, political science was an American affair (Far and Seidelman 1997; Adcock and Bevir 2005) and that was where the extremely fraught nature of the relationship between these two entities was shaped (Wolin 1969; Gunnell 1988; Kettler 2006).

Ball (2007) sees the initial period as one of a decent partnership between the two that was all but destroyed with the emergence and influence of the 'behavioural revolution' in the 1950s and 1960s. Before that, says Ball, there was 'no strong and irreconcilable split between political "science" and political "theory"' and all major political science departments offered courses in political theory as well (Ball 2007: 2–3). Indeed, Ball points out, all the major protagonists of the behavioural revolution like Robert Dahl, David Easton, Heinz Eulau, and David Truman had started off as students of political theory and had published writings on Western political thought. However, as the behavioural revolution took off, there emerged a disdain for the 'backward-looking "historicism" of their elders' and David Easton complained in the early 1950s that political theory's primary interest was in the history of ideas, whereas if 'political science is to be a science, this preoccupation with the pre-scientific political thought of Plato and his progeny must cease' (Ball 2007: 3). 'Traditional political theory' seemed to be more concerned with the past than with the present, more interested in the study of 'classical texts' and political scientists needed to study the 'real world' rather than keep reading these old texts (*Ibid*: 3). It wasn't that the behaviouralists shunned theory as such, but they were interested in strictly scientific, positivist theories, separating 'facts' from values and 'is' from 'ought' (*Ibid*: 3).

This battle became more and more fraught as time went by and there came a point in the 1950s when the 'death of political theory' was famously pronounced. The only theory that was recognised as legitimate was 'empirical' theory that tried to 'explain' the 'real world' of politics by looking for regularities and correlations among different kinds of facts. Normative political theory or political thought were no longer thought to be relevant within the political science establishment.

However, there was something else that had started happening right from the 1940s, as more and more European intellectuals, mostly Jewish, started migrating to the United States to escape persecution in Nazi Germany and elsewhere in Europe. Whatever may have happened in the established departments of political science, in the intellectual life of the United States, their arrival had another kind of impact in terms of defining the place of theory and philosophy in relation to the world. The émigré intellectuals came from very different thought traditions and did not all see politics in the same way. So, on the one hand, there were political philosophers like Leo Strauss, Hannah Arendt, and Eric Voegelin and, on the other hand, there were critical theorists of the Frankfurt School like Franz Neumann, Max Horkheimer, Theodor Adorno, and Herbert Marcuse. It has been argued that in some ways, put together, the work of these scholars amounted to a sharp critique of both liberalism and scientism, for many of them saw these as having been complicit with the rise of totalitarianism (Gunnell 1988; Dryzek 2006).

It should be clear from this brief account that the battles that were being fought within the political science departments in the United States were all very specific battles that had to do with the influence of positivism, the rise of behaviouralism, and their relationship to the specific formations of 'political theory' that therefore began to be seen in the way mentioned earlier. Neither the so-called 'death of political theory' nor its grand return with John Rawls later in the 1970s were anything more than very 'provincial' episodes in the history of the discipline. But the fact of the matter is that such is the power of the disciplinary–institutional apparatus that for decades those battles continued to be replayed in political science departments in India.

All this of course has to do with the specific formation of the academic disciplines of political science and political theory within the American academy, which is not co-terminus with political thought, political reflection, and analysis as such, which are very much present in other cultures and societies as well. In societies like ours in India, it seems quite clear that the sharp divide between the 'empirical' and the 'theoretical' did not and does not have the same resonance that it did in the United States and other parts of the Western world. This may have to do with the fact that political reflection has been primarily a matter of existential engagement rather than a professional academic practice and, to recall the point made earlier by Hawkesworth, has therefore tended to be more context-specific rather than universalist and abstract in its mode of theoretical reflection.

A brief examination of the practice of political reflection and analysis in India in relation to Indian political science might be of some help in making the point clearer. In the 1950s and 1960s, the influence of American political science on Indian universities was extremely powerful and all the theories – substantive and methodological – that were read and taught here came from the US academy, ranging from modernisation theories to behaviouralism and structural functionalism. The most important studies in Indian political science on Indian politics of the time, however, were those undertaken by American political scientists. We can immediately think of Robert Hardgrave, Myron Weiner, Paul Brass, Morris-Jones, Harold Gould, Rudolph and Rudolph, Francine Frankel, and such other outstanding names. Even though trained in the same schools of political science, however, we can see a very distinctive flavour emerge in the work of a scholar like Rajni Kothari – a flavour that came out of a desire to evolve a specifically Indian-experience-based understanding of democracy, which Kothari never failed to relate to larger social and cultural contexts. What is more, in the later Kothari, we also see the empirical political scientist engaged in theorising about caste, 'non-party political process', Dalit discourse, and such other matters that do not rely on quantifiable 'data' in the way early election (and democracy) studies did.

It is interesting that there were, in this same period, other scholars too, who wrote on politics and political economy in non-specialist journals like *Economic and Political Weekly*, *Social Scientist* and *Seminar*, most of whom were not professional political scientists but simply politically engaged intellectuals, often economists, historians, or sociologists by training.

It is from the 1970s that we can identify a more clearly defined body of academic research and reflection coming from political science scholars. Scholars like Partha Chatterjee and Sudipta Kaviraj started charting out a distinctly different, theoretically engaged, historically and empirically grounded study of politics. Already in their work we can see the difference between 'theory' and 'empirics' being transcended in interesting ways – to some extent a legacy of their earlier Marxist entanglement.

Politically engaged theorising received further impetus since the 1980s, first with the emergence of a powerful current of feminist scholarship and later with the rise of Dalit studies, which together had three significant consequences. Firstly, the 'subject–object' of feminist or Dalit scholarship was no longer the 'object' of any of the established disciplines but cut across them. Since a large part of the endeavour inevitably had to be devoted to producing histories where none existed, existing disciplinary methodologies, especially of a scientistic nature, made no sense. Secondly, therefore, new archives had to be constructed alongside evolving new methods. Memoirs, oral narratives, literary artefacts, deconstruction of cultural artefacts, and study of discourses – all these naturally became crucially important. Thirdly, even though housed within different social science departments, research and therefore teaching, too, saw the increasing collapse of disciplinary boundaries. Political scientists too were increasingly forced to look at archival sources, oral narratives, literature and discourse, and political speeches, aside from their usual surveys and ethnographic field studies (borrowed from anthropologists). The development of cinema and visual studies in the past few decades in India has also opened up possibilities of looking at other archives that had so far been neglected. This then brings us to the next part of our discussion.

The 'Field' Talks Back

The important – as also the very obvious – thing about these different kinds of archives mentioned above is that they are culturally situated and unlike quantified tables of specific sets of abstract data that are comparable across cultures, none of these archives allows us the possibility of 'scientific' explanation. Methods through which these archives may be approached, material or 'data' collected, and the methods through which they may be analysed or read presuppose a very different relation to the field. It is no longer possible to simply gather 'data' in some neutral fashion and interpret it by 'applying' to it theory produced in other contexts. As I mentioned earlier, the only way in which 'theory' related to the 'field' was through *application*, to either interpret its data or use it to 'test' the theory which of course would, at best, be tweaked a bit but never falsified. Though there have been some honourable exceptions, especially among anthropologists studying alien cultures, particularly in more recent decades, this hasn't quite been the case with political studies which sees itself as studying 'universal' structures and phenomena like 'state', 'government', 'elections', 'democracy', or concepts like 'equality', 'rights', 'sovereignty', 'justice', and so on. Its primary mode therefore, despite the great emphasis of the positivists and behaviouralists on 'facts', 'empirical reality', and the 'real world', remained one of application. In a strange way, this is a problem with all purely 'empirical' studies for they must take all their concepts around which 'data' has to be generated as given. They cannot subject those concepts to any theoretical or philosophical scrutiny.

The theory–'empirics' divide actually acquires a very different and troubling resonance when the field is the non-West, and all theory is produced in the Western academies. As it is, in the best of situations, there is something very troubling about the assumption that we can simply 'apply' theory produced in one context to understand practice/s in another. For behind this assumption lie two other possible assumptions: (a) that 'political practice' is 'non-theoretical', completely bereft of any discursive-theoretical content so that any theory from one context can be used to make sense of practice in another context; (b) that all thought and meaning-making is actually *universal* and it makes no difference where the theory is produced for it helps 'understand' the field and its constitutive practices regardless. The second assumption has few takers today. So, essentially, it boils down to an unstated reliance on the former. The fact of the matter is that things are not quite so simple, because there are levels and levels of conceptuality and there is no practice that is completely devoid of any kind of meaning and concept-making. All political practice is always constituted by some form of reflection and thought, which may or may not be theoretical in the academic sense of the term, but is never devoid of meaning-making and conceptuality and at least one part of theorisation must be about making sense of 'practice' through the subject's own world and categories of thought. This should not be understood to mean that academic theorisation must therefore always remain caught within the agent's thought-world but it does mean that the additional theoretical labour of the academic political theorist must not end up imposing his or her meanings on the practices of the agent. It is not the argument here that this thought-world of the agent is always and only 'cultural' or 'religious', for there are any number of ways in which agents think of their practice/s.

Increasingly, with the opening up of the newer archives, a different challenge came to the fore. Scholars had to start recognising that 'the field' was no passive entity from where 'raw material' could be gathered to be processed to produce the final product in academies; that the field and the material it generated resisted easy fit into the moulds of the already-given theory. In a manner of speaking, it talked back.

To take one instance, Indian democracy has been studied endlessly by Indian political scientists, and the entire field of election studies, starting from the 1970s, has generated huge amounts of data – some of which allowed the scholars concerned to understand the 'party system' specific to India and something like changing caste and community dynamics in the context of elections. Though these exercises helped texture the bland understanding of the democratic process as it existed earlier, cases like India's always remained 'special cases', making no impact whatsoever on democratic theory. It was a bit like the great debates on capitalism in the 'peripheries' during the 1960s and 1970s: it was not as if the protagonists did not recognise that the much anticipated universal development of capitalism was simply not materialising. They certainly did but the entire theoretical effort was to 'explain' the non-development of full-blown capitalism outside the West in terms of 'dependent development', 'imperialism', 'world-system', on the one hand, and the 'backwardness' and a 'weak bourgeoisie' in those societies, leaving the core of the theory totally untouched, on the other hand. In the instance of democracy, it wasn't even the larger ex-colonial 'peripheries' that were being theorised for the focus was on Indian democracy alone.

In the late 1970s and early 1980s, some scholars at the Centre for the Study of Developing Societies (CSDS), Delhi, started theorising about the 'non-party political process' which did not emerge from the accumulated data of election studies but from a keen engagement with the world of political activism and intense study of Indian politics. They saw in the rise of the 'non-party political process' a symptom of the limits of representation and of formal electoral politics and a redefinition of what constituted the political in some sense (Kothari 1984, 1986; Sethi 1985, 1993; Sheth 1982, 1983). This theorisation had the potential of making larger forays into postcolonial democracies, their levels of institutionalisation, the domains of the popular, the limits of the electoral-representative system, and the rise of social movements – but that step was never taken. Nevertheless, in a manner of speaking, the 'field' was certainly resisting being simply incorporated into the theory of democracy, forcing other questions to come into view.

It is in the late 1990s that we see an attempt to talk of the specifically postcolonial experience of democracy, with Partha Chatterjee's studies of what he called 'political society'. In his later development of the idea, he even referred to these studies under the rubric of 'popular politics in most of the world' (Chatterjee 2005). 'Political society', in the way Chatterjee deployed the term, meant something completely different from what it means in Western political theory, for it now designated a space separate from both 'state' and 'civil society', the latter being the high ground of modernity.

'Political society', in Chatterjee's hands, was a way of recuperating a sphere of politics that had been a permanent source of anxiety for theorists of Indian (and postcolonial) modernity and democracy – the vast domain that existed *outside* the designated spheres of modern politics, where the untutored masses made claims on the state and formed their own associations and organisations, unmindful of the formal grammar of rights and citizenship. A crucial part of what defines activities in this domain is 'illegality', or, at any rate, non-legality, where the state itself places the law in suspension in order to recognise the claims of the governed. Thus, for instance, squatting by the poor on government land, that is, strictly speaking, encroachment in legal terms and can never acquire the status of a 'right', is nevertheless allowed by governments to continue through the recognition of some kind of moral claim of the poor on governments and society at large.

Despite some problems with the way Chatterjee formulates the concept, linking it very closely to welfare, the fact that it immediately found acceptance among a large number of scholars from different parts of the postcolonial world showed that the concept spoke to a range of experiences of popular politics in such democracies. I will actually go so far as to say that once the concept was enunciated, it illuminated an entire area of politics even in Western democracies and has the potential to say something about democratic politics as such.

Comparative Political Theory and the De-parochialising Project

It was primarily as a consequence of globalisation in the 1990s and the fact that it brought different societies of the world in closer contact that a fresh round of thinking began on how to open out political theory to other 'previously (more or less) segregated civilizations and cultural zones' (Dallmayr 1997: 421). In the opening statement, Fred Dallmayr, the initiator of the enterprise, stated:

> As practiced in most Western universities the study of political theory or political philosophy revolves basically around the canon of Western political thought from Plato to Marx or Nietzsche – with occasional recent concessions to strands of feminism and multiculturalism as found in Western societies.
>
> (Dallmayr 1997: 421)

Given this situation, he went on to add that the proposed disciplinary subfield of 'comparative political theory complements, but in many ways also transgresses and unsettles the established field of "comparative politics" which to a large extent is either empirical-descriptive in character or else governed by stylised or formal models of analysis' (Dallmayr 1997: 421). What Dallmayr and his colleagues found 'most dubious' about these models was 'their unabashed derivation from key features of modern Western politics', including the structures of a secular nation-state with its accent on proceduralism, separated powers, and public–private bifurcations (*Ibid:* 421).

However, Dallmayr's introductory essay to the special number of *Review of Politics* that in a sense initiated comparative political theory (henceforth CPT) also recognised the circumstance that

> political theorising in the 'non-Western' world today involves in large measure the (avowed or unavowed) effort to respond to or come to terms with Western political ideas, especially with the whole welter of ideas bunched together in the label 'Western modernity'
>
> (Dallmayr 1997: 422).

From this recognition followed Dallmayr's conclusion that to that extent, 'the dichotomy between the West and non-West is deeply problematized [by the papers in the special issue], making room instead for reciprocal questioning and critique' (Dallmayr 1997: 422).

How one reads this important 'empirical fact' (that political theorising in the non-West is largely about coming to terms with Western political ideas) will decide which direction any project like CPT, desirous of opening out to other traditions, can take. If one takes this simply as a given, then the direction that CPT took in later years was perhaps inevitable. However, there is at least another way of reading this 'fact' – something that has now become far clearer, especially in the post-11 September 2001 world, as that moment inaugurated a more and more serious excavation of, and engagement with, these non-Western traditions by scholars from that world. Anticipating my argument in the next section, let me suggest that this is also a moment when many critiques of Western knowledge-hegemony in different parts of Africa, Asia, and the Americas also come to the fore very strongly – and they begin to define a new relationship to Western knowledge that is no longer simply concerned with 'coming to terms' with Western political ideas. They are now keen to draw critically on their own traditions of intellection as well (Dallmayr and Tingyang 2013; Hui 2014; Nigam 2020).

Perhaps in the end of the 1990s, a project like the CPT originating in the US academy could only have taken this relationship of non-Western scholarship and Western theory as a given. The other difficulty with CPT, as one of its practitioners, Farah Godrej, put it more than a decade later, was that while it rightly wanted to push political theory to open out to other traditions of thought, it had not actually faced up to the immense methodological challenges that the enterprise posed. She identified the challenge of interpreting texts from radically different traditions as one of the key difficulties: 'is there a hermeneutic approach that would allow us to understand effectively the concepts and ideas contained therein?' (Godrej 2009a: 137)

Godrej proposes 'three hermeneutic moments' in the encountering of a text or an idea from beyond the Western canon, each of which 'involves confronting a particular methodological issue, as well as a proposed solution to the challenge'. The three moments according to her are of the hermeneutic of existential understanding', where the encounter must take the form of an immersion in the world of the text, the idea, or the thinker; the moment of the 'reconstructed cultural account' which involves the more fraught and complex task of bringing that world into academic discourse; and finally, the moment of cosmopolitan political thought that requires the theorist to 'struggle to reconcile the conflicting imperatives' of the first two moments (Godrej 2009a: 138, 156).

Jenco (2011) too addresses the methodological consequences of engaging such 'foreign sources' but in a slightly different mode – 'not by constructing a "third space" of dialogue or contrast but by taking seriously the broader ambitions of their claims to wider-than-local significance' (Jenco 2011: 28). Jenco pits her argument vis-à-vis positions like that of Godrej whose 'plea for including non-Western perspectives within a cosmopolitan political theory ... does not expect to advance political theory along non-Western lines so much as enhance the discipline's capacity for self-reflection' (Jenco 2011: 32). As opposed to this, she argues that 'by accepting that its research findings may put its very self-identity at risk', her proposal of a *recentred political theory* is not concerned merely with understanding 'how historically excluded others can remind "us" of our own specificity, or trouble the finitude of categories implied by secular, rationalist social scientific approaches'. Rather, she argues, 'its knowledge becomes increasingly disciplined by resources, audiences, and concerns sited in other, globally diffuse communities that discourage return to a parochial starting point' (Jenco 2011: 29–30).

It is not possible within the space of this chapter to map in any detail of the debates and methodological concerns that have emerged from the move towards CPT and the concern with de-parochialising. I refer to some of the positions here to indicate how the field of political theory is in the process of being gradually reconstituted as a result of these newer concerns now being debated in mainstream political theory journals.

Another development is worth a brief mention in this regard – and this has to do with political theorist James Tully's call for 'de-parochialising political theory' (Tully 2016). Tully's call for de-parochialising political theory bases itself on an acknowledgement of the fact that abstract 'moral principles can literally mean anything the user wishes them to mean, unless they are grounded and articulated in relation to the experiential self-understanding of those to whom they are applied' (Tully 2016: 52). He argues that moral principles like 'treat each other as free and equal' and 'as ends in themselves', however important they may be, 'they have been and continue to be used to justify the greatest inequalities in human history; modern wars of intervention, conquest, subjugation and modernization; environmental destruction and climate change' (Tully 2016: 52). Tully's sharp indictment of these high moral principles of Western political philosophy is not unconnected to his being from Canada, which has set up a Truth and Reconciliation Commission in an attempt to recognise the 'cultural genocide' that was meted out to is First Nations. Tully's call for the de-parochialisation of political theory therefore calls for a 'dialogue' based on 'deep listening' – an exercise where the obvious burden of listening is on the white Western theorist and who must learn and cultivate an ethics of dialogue.

The first step in de-parochialising political theories, according to Tully, is to 'reparochialise' them, 'to recontextualise their presumptively general or universal terms back into the parochial contexts in which they make sense'. This is a step, says Tully, that is common to the decolonisation and provincialisation projects as well (Tully 2016: 56). Since other traditions of political thought are forced to change in the face of the spread of such parochial institutions and practices of Western power; since they are 'rendered marginal, lower, particular, primitive, exotic, assimilated and subordinated', the second step in de-parochialising must be 'the hard work of studying the complex relationships between political theories and forms of power' (Tully 2016: 56). As a third step, Tully urges us to recognise that political theory as we know it is a specific genre as it emerged in the West and proposes that we use the expression 'political thought' in its stead. In contrast to political theory that claims universal validity, 'political thought' he says, 'is practical' and can be used in different contexts without causing much problem. Political theory, he suggests, can continue to be a subset of this larger rubric of 'political thought' (Tully 2016: 56). Political thought, Tully underlines, takes place *within* traditions and so the task of 'comparative political thought' should be to 'always place political texts in their background traditions in order to make sense of them'. Then alone can a genuine dialogue 'across and among the world's traditions' become possible (*Ibid*: 57). Finally, because political thought is always located in practices and places, Tully proposes 'locating dialogues of comparative political thought in the places where they are practised' for two reasons: apart from being an issue of *epistemic justice*, 'it also gives oppressed minorities confidence and courage to speak truthfully to the powerful in their own languages and ways' (Tully 2016: 59). Scholars have responded to Tully's proposals in a variety of ways (Cooper et al. 2017). While generally welcoming the overall move, Mills, for instance, points out that among the 'obstacles' to genuine dialogue that Tully lists, 'race' and 'racism' are absent and that this is a specificity that has to be recognised. 'In the United States, for example', he argues, 'the white narrative within which the oppositional black narrative speaks, sanitizes the history of indigenous expropriation and genocide, African slavery and post-bellum subordination'. He therefore underlines that 'A genuine dialogue would thus require an admission of the truth of this past, and an end to the claimed white "innocence"' (Cooper et al. 2017: 161).

This is obviously a debate that continues and as will be evident from the discussion above, methodological issues are being raised and debated – though they are still at a very early stage of elaboration.

The Moment of Decolonisation

As we can easily see, the projects discussed in the previous section have all emerged from within the Western, particularly American, academy and were initiated by American scholars. The agency of the transformation therefore still lies in their hands. So, even the generous 'deep listening' that Tully proposes is ultimately to be practised in dialogues set up by such scholars as himself. The problem however is twofold. Firstly, the moment for such an initiative from Western scholars may have already passed as scholars actually based in the Global South take the initiative of establishing direct South–South conversations in different parts of the world. I am aware of such initiatives emerging from South Africa (Wits University, Johannesburg, and Cape Town University, Cape Town) and China (Tsinghua University but also other institutions).

It might be appropriate to begin this section with an account of one such instance that will give a sense of the widely differing concerns and priorities that shape this moment. In his account, 'Reading Ibn Khaldun in Kampala', Mahmood Mamdani (2013) discusses his attempts to restructure research and teaching at the Makerere Institute of Social Research (MISR) in Kampala, Uganda, in this phase of MISR's transformation from a former colonial think tank-turned-consultancy-unit to a centre for African and Ugandan social science researchers. After a year-long brainstorming exercise among scholars in 2010, says Mamdani, 'we agreed that nothing less than the development of a process of *endogenous knowledge creation,* including a full-time, course-work based, *inter-disciplinary* PhD programme would do'. These two elements need to be noted right away: the emphasis on endogenous knowledge creation and the stress on interdisciplinarity. Apart from more contemporary concerns, this latter question (interdisciplinarity) also relates to the point I have raised earlier, where I have indicated that in many of the countries like India, the field of social and political reflection has always cut across disciplinary boundaries precisely because they were never so powerfully entrenched in them.

The interesting thing, says Mamdani, is that

> [t]hough we started with this ambition, the tendency was to borrow the curriculum from the Western academy – wherever each of us had just taught or graduated from – as a turnkey project. So students in the MISR doctoral program were supposed to take two courses in theory, *Western Political Thought, Plato to Marx* in their first year and another titled *Contemporary Western Political Thought* in their second year. At the same time, *The Muqaddimah* [of Ibn Khaldun] was to be read in a third course titled *Major Debates in the Study of Africa. It is the students who began to ask whether we could redesign the theory courses so they are less West-centric and more a response to the needs of this time and this place.*
>
> (Mamdani 2013, emphasis added)

I have argued elsewhere that this has been the most fundamental problem with higher education in India (and perhaps most postcolonial countries), as they went about setting up their institutions of higher learning and research. The entire bodies of knowledge that social sciences made available to them, that then became the bases of our new curricula and syllabi, were all produced within the universe of Western social and political thought. The needs of the new nation-state to rapidly develop 'their own' expertise meant, ironically, that we simply produced our own experts in Western knowledge (Nigam 2019, 2020).

Mamdani's essay then goes on to actually write about the experience of reading Ibn Khaldun's *Muqaddimah* in Kampala and the kinds of questions that raised in terms of the classical heritage of knowledge – long held to be Greece, then challenged by revisionists who countered it by claiming Pharaonic Egypt as the source. Ibn Khaldun presents Persia as a parallel source of classical knowledge alongside Greece. Similarly, questions arose about what 'Africa' might have been or meant in Ibn Khaldun's time. The point of all such discussions, says Mamdani, is not really to trace an alternative source or origin but to put into question the very notion of a single point of origin or source of knowledge.

The larger point that I want to make by bringing in Mamdani's reflections is that in this moment of 'decolonization of thought', which as I have indicated earlier owes a lot to the ruptural moment of 11 September 2001, it is no longer possible to simply go on reading and teaching Western political thought and conducting research on that basis. It goes without saying of course that different societies have had their own different histories of the move for intellectual decolonisation and '9/11' does not have the same resonance everywhere. Undoubtedly, an earlier moment goes back, for instance, to Said's (1978/1994) powerful intervention by way of his critique of 'Orientalism' that unleashed a wholesale re-examination of Western knowledge in all its dimensions and birthed the whole new field of postcolonial studies. In Africa, we can trace a similar turning point with Ngugi wa Thiong'o's intervention *Decolonising the Mind* (1981), starting once again like Said, from within literature and literary studies. I have traced the complex genealogy of a similar tendency within India which, at the academic level, can be traced back to Ashis Nandy's *Intimate Enemy* (1983) (Nigam 2020).

All these different currents of thinking in different places came together in the post-Saidian moment of postcolonial studies. While postcolonial studies made significant contributions in terms of the critique of Western knowledge, it seems to me that the present conjuncture of decolonisation is different precisely in the sense pointed out by Mamdani – of moving beyond critique towards endogenous knowledge creation.

It may also be worth mentioning in passing that it is the last two decades that the move for decolonisation of the human sciences or the 'decolonial option' has acquired immense traction with the works of Quijano (2000, 2007) and Walter Mignolo becoming extremely influential (Mignolo 2002, 2007, 2009, 2011). An entire body of work has emerged around the work of these scholars, which cuts across disciplines and which has contributed to the growing realignment of the global knowledge formations – by which I mean the relationship between the hegemonic Western and the emergent non-Western. In this new constellation, it is not the rejection of Western knowledge that is crucial but rather the emphasis on 'epistemic reconstitution' via 'epistemic delinking' and 'epistemic disobedience' (Mignolo 2007, 2009). To put it in Mignolo's words,

> The struggle is for changing *the terms in addition to the content* of the conversation ...
>
> If delinking means to change the terms of the conversation, and above all, of the hegemonic ideas of what knowledge and understanding are and, consequently, what economy and politics, ethics and philosophy, technology and the organization of society are and should be, it is necessary to fracture the hegemony of knowledge and understanding that have ruled, since the fifteenth century and through the modern/colonial world by what I conceive here as the theo-logical and the ego-logical politics of knowledge and understanding... That is, *one* strategy of de-linking *is to de-naturalize concepts and conceptual fields that totalize a reality.*
>
> (Mignolo 2007: 459, emphasis added)

The two points that I take away from this quote from Mignolo are, firstly, that delinking aims to change not just the content of the conversation but the very terms on which it should or can be conducted; and secondly, one key strategy for doing so is to move beyond the familiar categories and the conceptual fields that structure – and therefore totalise – 'reality'. In that sense, it is about setting aside totalising conceptual frameworks in order to allow other viewpoints and possible conceptual frameworks to emerge.

A detailed discussion of the decolonial position is beyond the purview of this chapter, but it should be underlined that the emphasis here is as much on a suspension of the constricting hold of disciplinary formations as it is about totalising knowledge, in order to arrive at new understandings, new concepts, and categories, as other knowledge systems, hitherto marginalised, gain entry into the world of new knowledge creation. To that extent, it has implications for the discipline of political theory as well.

To conclude this section then, I want to briefly discuss the tasks of the 'postcolonial political theorist' (reflecting on their epistemic methods) as proposed by Chatterjee (2011) before discussing an instance from my own work. In the 'Introduction', Chatterjee spells out his larger methodological concerns by subjecting the practice of 'political theory', which is largely dominated in the Anglo-American world by 'normative political theory' that we know works with concepts and their normative justifications. 'Normative justifications' are usually made on the basis of certain assumed 'universal principles' but, more importantly, the assumption about their universality is made on the basis of a virtually total abstraction of those principles from any kind of actual empirical–historical context. Chatterjee underlines that it is not as if there is no empirical–historical referent here, but that these 'normative debates take place in a time-space of epic proportions which emerged fully formed only after the victorious conclusion of an epochal struggle against an old order of absolutist, despotic or tyrannical power', which is to say that it posits a *definite historical past* against which it positions itself (Chatterjee 2011: 1). Even though this past is a very specific past, says Chatterjee, it is also 'limitlessly elastic in its capacity to include virtually any geographical space and historical period', which then makes it possible for this 'modern political theory' to assume the stance of universality. Strikingly a personal note, he contends that 'as someone from the postcolonial world' who has been exposed to this mode of theorising very early on, it was always quite baffling that 'all the bitter and bloody struggles over colonial exploitation, racial discrimination, class conflict, suppression of women, marginalisation of minority cultures', have had practically no impact on the 'stable location of political theory within the abstract discursive space of normative reasoning' (Chatterjee 2011: 3).

Much of this liberal normative theory actually also emerged in tandem with and became quite influential in the 19th-century practices of government – including colonial governmental practices. However, argues Chatterjee, in the context of colonial government and the great debates that took place in the British parliament as to how the colonial subjects should be governed – on the basis of which laws and the position with respect to local laws and customs – there emerged disjuncture. Long debates led to the formulation of what he calls the 'norm-deviation' structure, where the universal norms embodied in the modern secular law, once established as the unquestioned norms, would then become the standard against which all other societies, their own 'moral and religious' biases and customary practices, could be mapped. The instrument of 'policy' would become the means by which colonial societies would be administered in a way that would aim at reducing the distance between the empirical state and the desired norm (Chatterjee 2011: 7–8). From here emerged, says Chatterjee, the second sense in which the norm could be understood as often placed in abeyance – to 'make an exception' as Chatterjee puts it, by governments – since they had to rule societies that were not yet ready to be governed by universal laws. 'The universally valid norm would be withheld in favour of a colonial exception' (Chatterjee 2011: 9).

It is this same structure of norm-deviation and norm-exception that carries on in postcolonial governmental practices. The universal norm, embodied in the law, continues to make or had to keep making exceptions in order to accommodate 'deviant' practices from its point of view. Such exceptions, says Chatterjee, keep piling up because the expected result of reaching the desired norm never materialises. 'The relation between norms and practices has resulted in a series of improvisations. *It is the theorisation of these improvisations that has become the task of postcolonial political theory*' (Chatterjee 2011: 19, emphasis added).

From this then emerge two theoretical – and methodological – challenges, according to Chatterjee. The first is the challenge of breaking 'the abstract homogeneity of the mythical time-space of Western normative theory by emphasising the real history of its formation through violent conflict and imposition of hegemonic power'. The second is 'the even greater challenge' of redefining the norms themselves in the light of the considerable accumulation of new practices 'that may at present be described only in the language of exceptions but which in fact contain the core of a richer, more diverse, and inclusive set of norms'(Chatterjee 2011: 22).

In Lieu of a Conclusion

In this concluding section, I offer an instance from my own recent work (Nigam 2020, Chapter 4, 'Theorizing the Political') in order to illustrate as well as delineate some of my own methodological moves in this direction. Since 'political science' or 'political studies' is centrally about the 'object' called 'politics' (or 'the political'), how we understand these expressions has a direct bearing on our understanding of the discipline itself. As opposed to the behaviouralist revolution's rejection of history, for the purpose of putting political science on a scientific footing, I have found it necessary to engage with both actual empirical history and texts from earlier times – *but with one crucial difference*. Recall David Easton's declaration cited earlier that if 'political science is to be a science, this preoccupation with the pre-scientific political thought of Plato and his progeny must cease'. As opposed to this reduction of political thought to that of 'Plato and his progeny', in order to understand the constitution of the political across different contexts and different parts of the world, it has become necessary to engage with political thought across the world – from an Al Farabi or an Ibn Khaldun in the West Asian and North African context to texts like the *Arthashastra* or the *Nitisara* in India or non-religious state-like formations as in China.

To focus here on the Indian context, it becomes equally necessary to understand the political or state structures that actually existed – say during the Sultanate and Mughal rule in the North and the Vijayanagara Empire in the South. What are the forms that preceded it? Did the Sultanate or Mughal rule involve any radical rupture with the past? How did the imposition of the colonial state affect the formations that existed before it came along, considering that it involved an entirely different conception of the law and statecraft? In what relation did the emergent nationalist movement and its takeover of the state, the deliberations of the Constituent Assembly, and the formation of a new constitution change things?

In order to address such questions, it is not necessary that political scientists become historians but it is necessary that they engage deeply with the work of and debates among historians. The point here is not to trace the roots of the present through some linear historicist development of the past but rather to begin by looking at the different histories and sequences that come together in, say the time of colonial rule, to uncover different imaginations of and excavate histories of the actual political formations that existed.

In doing so, I found the repeated reference to the *mandala* form everywhere – one did not actually have to go looking for it. And its prevalence extended from South-East Asian regimes studied by Tambiah (2013/1973) to Balinese kingdoms studied by Geertz (1980) to the 'Zomia' studied by Scott (2010). The mandala form of what Tambiah called the 'galactic polity' with a number of smaller kingdoms linked to a bigger one led in effect to a whole network of small kingdoms. Historians of medieval India have referred to this phenomenon as that of the 'segmentary state' or the 'dispersed foci of power'. This form is fundamentally different from the unitary state that existed in Europe, around which notions of 'sovereignty' have emerged. The fusion of the Church with such a unitary form of 'temporal power' and their eventual separation, which we understand as the beginning of 'secular' statecraft, are all irrelevant to the formation of political power in India, South and South-East Asia. The important thing is that even the period of the Sultanate or Mughal rule did nothing to change this form and institute in its place a unitary state.

There are important ways then in which political theory concepts need to be thought differently once we understand the significance of these different forms. No less important is the fact that for scholars studying contemporary Indian democracy, for example, the continuation in different ways, of this dispersed form of power becomes evident in the pervasive figure of the regional satrap who exists not only in state institutions but also in different political parties. How we pose questions pertaining to parties – and why most of them remain regional or differ regionally even within the same party – or how we understand the impulse towards centrifugal tendencies in Indian politics are all likely to appear very differently, once embark on this exercise.

I want to end then by underlining that such a move towards decolonising method is not about 'return' to some pristine source of precolonial past. On the contrary, it is about broadening the base of our disciplinary understanding by going beyond the narrow confines of the European experience. The move towards decolonisation is not about an 'Indian' or even 'Asian' or 'African' theory but rather about opening out the discipline towards a far richer and wider range of experiences on the basis of which to develop its understanding.

Notes

1 This is a specifically American story and I will not dwell on it any further here. There are literally volumes written on this conflict that has played out not only in terms of disciplinary issues relating to the method and empirical research

but also in institutional matters. The interested reader may usefully see issues of *American Political Science Review* for greater details of the disciplinary and institutional history of this conflict. (See for instance, Smith 1957; Wolin 1969; Farr 1988; Gunnell 2010; Leca 2010; Rehfeld 2010; Corbett 2011.) However, the impact of the divide, it is often recognised, has not been a happy one. In the words of a historian of the discipline: 'The problem with this estrangement between political theory and political science was that political science lost its reflective dimension, and political theory tended to lose contact with politics as it embraced various external intellectual domains. Political science, intent on maintaining its empirical, scientific credentials, sublimated its normative heritage, and although political theory was initially committed to redressing the failure of political science to be politically relevant, it tended to increasingly become an anomalous, dislocated rhetoric.' (Gunnell 2006: 22)

2 It was as part of this imagination that students were encouraged to take optional courses in other centres than their own, to develop a sense of interdisciplinarity. Gradually, that enthusiasm seems to have waned, though students still can opt for courses elsewhere.

3 In that sense, it should be noted, the expression 'studies' in political studies has a different resonance from the way it appears in say women's studies, environmental studies, or Dalit studies – where the term refers to a new field being carved out in the wake of movements and struggles outside the academy.

4 It is beyond the scope of this chapter to explore why this may have been so, but it is interesting to compare the situation with other disciplines like anthropology or sociology where such concerns seem to have come up quite significantly.

5 In the context of political studies, this is equally true of India as well. This however requires a much bigger discussion than we can go into here.

References

Adcock, Robert, and Mark Bevir. 2005. 'The History of Political Science.' *Political Studies Review*, 3: 1–16.

Ball, T. 2007. 'Political Theory and Political Science: Can This Marriage Be Saved?' *Theoria: A Journal of Social and Political Theory*, 113: 1–22.

Chatterjee, Partha. 2005. *Politics of the Governed: Reflections on Popular Politics in Most of the World*. Ranikhet: Permanent Black.

Chatterjee, Partha. 2011. *Lineages of Political Society: Studies in Postcolonial Democracy*. Ranikhet: Permanent Black.

Cooper, Garrick, Sudipta Kaviraj, Charles W. Mills, and Sor-hoon Tan. 2017. 'Responses to James Tully's "Deparochializing Political Theory and Beyond".' *Journal of World Philosophies*, 2: 156–173.

Corbett, Ross J. 2011. 'Political Theory Within Political Science.' *Political Science and Politics*, 44 (3): 565–570.

Dallmayr, Fred. 1997. 'Toward a Comparative Political Theory.' *The Review of Politics*, 59 (3): 421–427.

Dallmayr, Fred, and Zhao Tingyang (eds). 2013. *Contemporary Chinese Political Thought: Debates and Perspectives*. New Delhi: KW Publishers Pvt Ltd.

Dryzek, J. 2006. 'Revolutions Without Enemies: Key Transformations in Political Science.' *The American Political Science Review*, 100 (4):487–492.

Farr, James. 1988. 'History of Political Science.' *American Political Science Review*, 38 (4): 1175–1195.

Farr, James, and Raymond Seidelman (eds). 1997. *Discipline and History: Political Science in the United States*. Ann Arbor: University of Michigan Press.

Geertz, Clifford. 1980. *Negara: The Theatre State in Nineteenth Century Bali*. Princeton: Princeton University Press.

Godrej, Farah. 2009a. 'Towards a Cosmopolitan Political Thought: The Hermeneutics of Interpreting the Other.' *Polity*, 41 (2): 135–165.

Gunnell, J. 1988. 'American Political Science, Liberalism, and the Invention of Political Theory.' *The American Political Science Review*, 82 (1): 71–87.

Gunnell, John G. 2006. 'Political Science: Orthodoxy and Heterodoxy.' *The Good Society*, 15 (1): 21–25.

Gunnell, John G. 2010. 'Professing Political Theory.' *Political Research Quarterly*, 63 (3): 674–679.

Hawkesworth, M. 2010. 'From Constitutive Outside to the Politics of Extinction: Critical Race Theory, Feminist Theory, and Political Theory.' *Political Research Quarterly*, 63 (3): 686–696.

Hui, Wang. 2014. *China From Empire to Nation-State*. Cambridge, MA: Harvard University Press.

Jenco, Leigh K. 2011. 'Recentering Political Theory: The Promise of Mobile Locality.' *Cultural Critique*, 79: 27–59.

Kaviraj, Sudipta. 2005. 'An Outline of a Revisionist Theory of Modernity.' *European Journal of Sociology*, 46 (3): 497–526.

Kettler, D. 2006. 'The Political Theory Question in Political Science, 1956–1967.' *The American Political Science Review*, 100 (4): 531–537.

Kothari, Rajni. 1984. 'The Non-Party Political Process.' *Economic and Political Weekly*, 19 (5): 216–224.

Kothari, Rajni. 1986. 'On the Non-Party Political Process: The NGOs, the State, and World Capitalism.' *Lokayan Bulletin*, 4 (5): 6–26.

Leca, Jean. 2010. 'Political Philosophy in Political Science.' *International Political Science Review*, 31 (5): 525–538.

Mamdani, Mahmood. 2013. 'Reading Ibn Khaldun in Kampala.' *Critical Encounters*. https://criticalencounters.net/2013/07/05/reading-ibn-khaldun-in-kampala-mahmood-mamdani/ (accessed on 28 August 2020).

March, A. 2009. 'What is Comparative Political Theory?' *The Review of Politics*, 71 (4): 531–565.

McBride, Keally. 2016. 'Radical History as Political Theory.' *Political Theory*, 44 (2): 178–186.

Menon, Nivedita, Aditya Nigam, and Sanjay Palshikar (eds). 2014. *Critical Studies in Politics: Exploring Sites, Selves, Power*. New Delhi: Orient Blackswan.

Mignolo, Walter. 2002. 'Geopolitics of Knowledge and the Colonial Difference.' *South Atlantic Quarterly*, 101 (1): 57–96.

Mignolo, Walter. 2007. 'Delinking.' *Cultural Studies*, 21 (2): 449–514.

Mignolo, Walter. 2009. 'Epistemic Disobedience, Independent Thought and De-Colonial Freedom.' *Theory, Culture and Society*, 26 (7–8): 1–23.

Mignolo, Walter. 2011. *The Darker Side of Western Modernity: Global Futures, Decolonial Options*. Durham and London: Duke University Press.

Nandy, Ashis. 1983. *The Intimate Enemy: Loss and Recovery of Self under Colonialism*. New Delhi: Oxford University Press.

Ngugi wa Thiong'o. 1981/1986. *Decolonising the Mind: The Politics of Language in African Literature*. London: James Currey.

Nigam, Aditya. 2019. 'Decolonising the University.' In Debaditya Bhattacharya (ed.) *The University Unthought: Notes for a Future*, pp. 62–73. London and New York: Routledge.

Nigam, Aditya. 2020. *Decolonizing Theory: Thinking Across Traditions*. New Delhi: Bloomsbury.

Petracca, M. 1991. 'The Rational Choice Approach to Politics: A Challenge to Democratic Theory.' *The Review of Politics*, 53 (2): 289–319.

Philipose, Elizabeth. 2007. 'Decolonising Political Theory.' *Radical Pedagogy*. https://radicalpedagogy.icaap.org/content/issue9_1/philipose.html (accessed 26 August 2020).

Quijano, Anibal. 2000. 'Coloniality of Power, Eurocentrism, and Latin America.' *Nepantla: Views from the South*, 1 (3): 533–580.

Quijano, Anibal. 2007. 'Coloniality and Modernity/Rationality.' *Cultural Studies*, 21 (2–3): 168–178.

Rehfeld, Andrew. 2010. 'Offensive Political Theory.' *Perspectives on Politics*, 8 (2): 465–486.

Said, Edward. 1994. *Orientalism*. New Delhi: Penguin India.

Satz, D., and J. Ferejohn. 1994. 'Rational Choice and Social Theory.' *The Journal of Philosophy*, 91 (2): 71–87.

Scott, James C. 2010. *The Art of Not Being Governed: An Anarchist History of Upland Southeast Asia*. New Delhi: Orient Blackswan.

Sethi, Harsh. 1985. 'The Citizen's Response: A Glimmer of Possibilities.' *Lokayan Bulletin*, 3 (1): 59–73.

Sethi, Harsh. 1993. 'Survival and Democracy: Ecological Struggles in India.' In Ponna Wignaraja (ed.) *New Social Movements in the South: Empowering the People*, pp. 121–148. New Delhi: Vistaar Publications.

Sheth, D. L. 1982. 'Movements.' Seminar, 278 (October): 42–52.

Sheth, D. L. 1983. 'Grassroots Stirrings and the Future of Politics.' *Alternatives*, IX: 1–24.

Smith, David G. 1957. 'Political Science and Political Theory.' *American Political Science Review*, 51 (3): 743–746.

Tambiah, Stanley J. 2013/1973. 'The Galactic Polity in Southeast Asia.' (Reprint From *Culture, Thought and Social Action*). *Journal of Ethnographic Theory*, 3 (3): 503–534.

Tully, James. 2016. 'Deparochializing Political Theory and Beyond: A Dialogue Approach to Comparative Political Thought.' *Journal of World Philosophies*,1 (Winter 2016): 51–74.

Widdowson, Frances. 2013. 'Decolonizing Political Theory: Exploring the Implications of Advocacy for Political Science.' Paper Presented at the Annual Meeting of the Canadian Political Science Association, University of Victoria, 4–6 June 2013. https://www.cpsa-acsp.ca/papers-2013/Widdowson.pdf (accessed on 26 August 2020).

Wolin, Sheldon. 1969. 'Political Theory as a Vocation.' *The American Political Science Review*, 63 (4): 1062–1082.

9

BETWIXT AND BETWEEN?

Anthropology's Engagement with the Sciences and Humanities

Kamala Ganesh

Introduction

Crisis in Triple A

A crisis of sorts broke out in the world of anthropology when, in 2010, the American Anthropological Association (*Triple A*) decided to cut out the word 'science' in its self-definition. For long, a key debate within anthropology on whether it belongs to the sciences or humanities had had a 'neither-fully-one-thing-nor-the-other' tenor. Then at its 2010 annual meeting, the American Anthropological Association created a stir by dropping the word 'science' from the purpose statement of its long-range plans. The 2009 statement was 'The purposes of the Association shall be to advance anthropology as the science that studies humankind in all its aspects'.[1] The commitment to science had been there right from its inception. The next year, it was changed to 'The purposes of the Association shall be to advance public understanding of humankind in all its aspects'. The wording, Association officials said, was meant 'to address the changing composition of the profession and the needs of the AAA membership' (Hirst 2019), meaning one section of members was uncomfortable with the label science, when so much had happened in the previous two decades to question this.

Founded in 1902 by pioneering anthropologist Franz Boas, *Triple A* is the world's largest scholarly and professional organisation of anthropologists and arguably the most influential. The furore that ensued among anthropologists in the Western world following the dropping of the word 'science' brought out into the open latent and deep tensions between the two major factions – those who espoused the discipline as a science and those who saw it as fundamentally a humanistic discipline. As Peter Peregrine, president of the Society for Anthropological Sciences colourfully put it, the word 'science' being dropped 'just blows the top off the tensions … even if the Board

DOI: 10.4324/9781003298908-16

goes back to the old wording, the cat's out of the bag and is running around clawing up the furniture' (Wade 2010). *Triple A* had to recognise the widespread indignation its action had triggered.[2] Within a few weeks, it issued a conciliatory statement that 'anthropology is a holistic and expansive discipline that covers the full breadth of human history and culture ... and draws on the methods of both the humanities and the science' (Ibid). By the end of 2011, it officially brought 'science' back with the statement, 'The strength of Anthropology lies in its distinctive position at the nexus of the sciences and humanities, its global perspective, its attention to the past and the present, and its commitment to both research and practice' (Hirst 2019). The explicit inclusion of 'science' has continued since then; the current purposes of the Association as per its website 'shall be to advance anthropology as the science that studies humankind in all its aspects, through archaeological, biological, ethnological, and linguistic research'.

Triple A's backtracking was not about going back to the status quo. Catalysed by the rich history of anthropological discussions on its self-definition, the issue had now got transposed onto a different key. The bitter polarisation somewhat softened into a compromise that it cannot be just one or the other, nor betwixt and between, neither this nor that. Rather the discipline must be inclusive of both. A powerful argument for bridging the sciences and humanities was being made.

Science versus Humanities Debate

The science versus humanities debate is a larger and older one, precipitated by the emergence of the social sciences as a category in the mid-19th century in Europe. In this initial stage, anthropology, sociology, and other social sciences unambiguously defined themselves as heirs to the natural sciences and as votaries of positivism. Science was seen as the great redeemer and contrasted with religion, belief, and superstition. By the early 20th century, the sciences became contrasted with the humanities and the creative arts as irreconcilable opposites, each with fixed attributes, creating concern among philosophers and historians of science in the West. Since C.P. Snow's warning on the perils of this polarisation and Thomas Kuhn's analysis that dented the idea of exclusively objective criteria to establish scientific truths, ways of understanding and doing science have been changing. This is true at least at its margins and in its more reflective domains. But these arguments have not touched all the social science disciplines equally. Anthropology, which, to begin with, espoused the notion of scientific rigour with gusto, was also quick to reflect on the difficulties of sustaining the polarisation between sciences and humanities. Way back in 1950, in his Marrett lecture, Evans-Pritchard made a powerful critique of anthropology's pursuit of science and suggested that it identify itself with the humanities. In the decades that followed, beneath the overarching positivist structure-functionalist framework in anthropology, different currents flowed expressing greater ease in identifying with the humanities.

There was a backlash. By the late 1980s and early 1990s, what are called 'science wars' had erupted as several scientists struck back at what they read as antirealist and politically motivated attacks on science from the humanities and social sciences (Candea 2016). The contending factions in the *Triple A* controversy reflected aspects of this major debate in the intellectual currents of the West.

Anthropology in the West: From Science to Non-science

A central argument of this chapter is that anthropology in the West tackled the divide between the sciences and humanities in a distinctive way, due to its subject matter of studying 'other cultures' and due to the methodology that evolved for this, namely ethnography with participant observation. However, writing, as I do, from an Indian location where there is a strong tradition of treating sociology and social anthropology as a unified field, questions on scientific methodology have to take into account both the disciplines.[3] In the institutional architecture of Western academia, the two disciplines were kept distinct and separate. Anthropology was born and developed in the context of colonialism and became the study by Western scholars of cultures other than their own, in far-flung geographical regions and cultures that were small-scale, and considered to be in a primitive stage of evolution. Its scope included all aspects of being human. Sociology, on the other hand, was about Europe looking at its own society, as it was transforming and taking a leap into modernity. Initially, both aimed to be part of the sciences rather than the humanities. Debates on scientific methodology were part of both disciplines. Anthropology must be among the rare instances in which the geographical location of the 'object of study' and the 'origins' of the subject/researcher defined the identity of a discipline. Due to this foundational character, it became quickly self-aware. The inherent problems of a definition based on location precipitated, over the decades, internal cogitation over its own claims of scientificity.[4] Sociology had, through its hermeneutical, phenomenological, and ethnomethodological traditions, evolved its own critiques of positivism, but anthropology's fieldwork methods provided a unique epistemic vantage point for its critique.

Indian Context

In India, sociology and anthropology developed as formal disciplines in the early 20th century, under British rule, producing a cadre of trained professionals. Initially, the two disciplines were separate, but the scope of both, mostly, was within India itself, i.e. it was the study of one's own country, society, and culture by Indian scholars trained in social sciences. This struck at the very heart of the definition of anthropology as the study of 'other cultures'. It rendered one of the central distinctions between anthropology and sociology invalid.

There was a concerted effort among the generation of sociologists who were engaged in professionalising it to treat sociology and social anthropology as a unified field, under the overall label of sociology. This position continues to be an influential one, but it is not without caveats and opposition. Where separate departments of anthropology had existed historically, they continue to do so, including archaeology, linguistics, physical, and social anthropology. However, social anthropology is for the most part delinked from the other three and situated in sociology departments. The debates on scientific methodology in sociology and social anthropology partake of each other's specific lineages of thought. There has thus been a considerable cross-fertilisation of ideas. The initial phase of the development of sociology (by which henceforth I mean social anthropology as well in the Indian context) was under colonialism and a burgeoning nationalism, followed by a nation-building thrust post-independence. The debate on scientific method was mediated by all the above.

My chapter starts with a discussion on how anthropology in the Western academia has debated the issue of its own identity as hovering between the sciences and humanities, and what contribution it has made to the issue of scientific method. In doing so, it also reflects on the distinctness of the anthropological critique of positivist empirical science. It then looks at the Indian scenario, where there is limited direct writing on theory and methodology per se. The positions on scientific method have to be inferred through the larger and persistent cogitations on what a sociology for India should be. The recurrent themes here are the unity of sociology and social anthropology, the search for indigenous categories, and the notion of value neutrality.[5] By revisiting these themes, we get some glimpses of the approach to science and scientific method. Finally, this chapter discusses the specific case of the oldest Department of Sociology in India at Bombay University, where the effort to bring together sociology and social anthropology began.

Anthropology as Science and Science as Anthropology's Object of Study

Both anthropology and sociology started off by explicitly defining themselves as sciences. They enthusiastically espoused the new positivistic methodology, aiming to generate scientific knowledge, characterised by certainty, based on radical empiricism that confined knowledge to observable and verifiable data.

An important point is that both disciplines also treated 'science' itself as an object of study. It gave them a handle to historicise and critique science as a cultural and social system, instead of uncritically adopting its identity and methodology. This was a point of inflexion and creative contradiction that changed the tone of methodology and cast a shadow on positivism. According to Candea (2016), sociologists and historians were somewhat ahead of anthropologists in taking up science as an object of scrutiny, but they largely used ethnographic and anthropological materials for substantiation.

Even before the formal advent of a sociology of science, sociologists were already reflecting upon it. In the 1950s, Robert Merton co-authored a fascinating and entertaining book titled *The Travels and Adventures of Serendipity: A Study in Sociological Semantics and the Sociology of Science*. Merton recognises that scientific discovery is often a chaotic blend of failures and accidental successes. 'Some scientists seem to have been aware of the fact that the elegance and parsimony prescribed for the presentation of the results of scientific work tend to falsify retrospectively the actual process by which the results were obtained' (Merton and Barber 2004: 159 cited in Campa 2008: 5).

Yet, in his 1973 work *Sociology of Science*, considered to be a standard early text, there was no exploration of the actual content of science. The question 'how does science define knowledge and how is scientific knowledge created'? was not asked. Rather, he investigated science as a functionally integrated social institution whose role was the extension of certified knowledge. How does one explain this gap between Merton's two works? Perhaps it bears out the point that the 'New Ethnography' school was to make in the 1990s – that the writing of ethnography had never been a conventionally scientific enterprise, it was only made to seem so to serve the demands of the profession. In contrast, sociology of scientific knowledge which overtook the Mertonian genre provided sociological explanations of scientific ideas themselves. Its allied genres drew from anthropology and made detailed ethnographic studies of particular laboratories seeking to demonstrate the social construction of scientific knowledge in concrete settings.[6]

Scientific Rigour in 'Notes and Queries'

Early anthropology's aspiration to scientific rigour in its methodology had also to do with the subfields included in its scope like archaeology and physical anthropology that were directly using methods of the natural sciences and were amenable to structured observations, measurements, and calculations. This aspiration is palpable in a publication titled *Notes and Queries on Anthropology* that became a go-to for anthropologists of the time.

Ethnographic fieldwork with participants' observation is widely seen as an innovation perfected by Malinowski in the early 20th century. Urry (1972: 45) points out that from the mid-19th century itself, there were numerous efforts in Britain to collect ethnographic material and formulate methodological propositions systematising collection of field data on a scientific basis.

In the late 19th century, during the time when ethnographic protocols were nascent, a field manual called '*Notes and Queries on Anthropology*'(N & Q) was published by the British Association for the Advancement of Science, with several organisations and committees and 'a roster of the most distinguished British anthropologists' (Vogt 1954: 1154) collaborating in producing the various editions. N & Q represented the currents and concerns of anthropology of that time not only in Britain but also wherever it was an established discipline.

The preface to the first edition (1874) states that it was aimed at travellers and other non-anthropologists in far-flung colonies to help them collect information systematically and supply it for scientific study by anthropologists back home in Britain: a precise description of how the discipline of anthropology came into being. Five more editions were published, in 1891, 1897, 1912, 1929, and 1951.

I browsed through the 1899 edition. It was divided into two parts. The first comprised of Anatomical Observations, Physiological Observations, and a Medical Section. The second part was titled Ethnography. Clearly, the instructions given were meant for observations and measurements of human subjects of non-European descent. The range of topics in Part 1 included instructions and tools for measuring parts of the human body, its physiological features, physiognomy, illnesses, and diseases. In a similar vein, Part 2 was about the culture and lifestyle of specific groups of people, mostly tribes, material culture, dwellings, implements, basketry, jewellery, religious and ritual life, food, occupation, and so on: what we are till today familiar with as ethnography. The comprehensive scope, minute details, and elaborate instructions to ensure accuracy and precision in observation, collection, collation, tabulation, and classification give us a panoramic view of positivist empirical science of that time. By treating human biological and cultural data together in one sweep, N & Q illustrates how the science of anthropology furnished a base to produce stereotypes and create biases.

More pertinently to my argument on the tradition of internal debate in anthropology, the N & Q editions, cumulatively in their shifting emphasis in consecutive editions, throw light on ethnography as a self-consciously envisaged work-in-progress.

Firstly, they lay bare the deeply racial stereotypes that underlay attempts to classify and comprehend people of non-European descent.

Secondly, the same broad format continues, but there are important changes in every edition – a result of intensive reviews. There is a growing concern to enhance scientific techniques, a need to hone the tools of the questionnaire to elicit 'facts' with greater accuracy and fidelity to the source.

Thirdly, the exigencies of the field situation throw up nascent issues of cultural translation and ethics that decades later led to a fundamental questioning of the conventional ethnographic mode itself.

To give a few examples, in 1898/1899, folklorist Andrew Lang criticises the authors of N & Q 'for selecting scraps of information relevant to their case' – the data collectors are often 'ignorant of the language of the people they talk about or who are themselves prejudiced by one or the other theory or bias. How can one pretend to raise a science on such foundations?' (Urry 1972: 47). Read who edited the ethnography sections of the 1899 edition suggested the need for long-time residence among the people whom one was studying, which later became a *sine qua non* of ethnography (Garson and Read 1899). The difficulties of 'transference of meaning from one culture to another, problems in questioning informants and avoidance of bias in reporting' were addressed after the very first edition (Garson and Read 1899: 50).

The idea that ethnographic information may come in useful for colonial rule surfaced only by 1892. Subsequent committees started asking for trained anthropologists to directly collect information from the field. By the time of the 1951 edition, the space for physical anthropology and archaeology was much reduced, and the contents were heavily weighted on the social anthropological and ethnological side. The emphasis on observation and description rather than theory and interpretation continued, but the value of theory as a stimulus to relevant observation was recognised (Vogt 1954: 1154). In 1913, the influential anthropologist W.H.R. Rivers was engaged in reviewing earlier editions and was critical of the 'survey'. Facts themselves were not enough; the manner of gathering them was important. He recommended intensive and immersive fieldwork studying every detail of the life and culture, speaking the language and building close personal relations. 'It is only by such work that it is possible to discover the incomplete and even misleading character of much of the vast mass of survey work which forms the existing basis of anthropology' (Rivers 1913 cited in Urry 1972: 50). The next two editions reflect Rivers' approach.

Critique from Within

Early anthropologists tended to assume that they were representatives of science which belonged to the West, and that their objects of research were non-Western peoples whose ways of thinking and being were non-scientific or not-quite scientific. This is palpable in the N & Q approach. In a departure, Malinowski had argued in 1925 that, in fact, scientific and non-scientific ways of thinking existed alongside each other in all human cultures, 'primitive' as well as 'modern'. Despite his patronising aura, his line of enquiry was to eventually culminate in the emergence of ethnoscience, and even to the position that all science, including modern Western science, is ethnoscience (Candea 2016).

The preoccupation with empirical data, its collection, and classification in pursuit of scientific rigour, that began with *Notes and Queries*, continued in the first half of the 20th century but a critical stream was appearing in parallel within social anthropology itself. A.R. Radcliffe Brown, who dominated social anthropology well beyond the mid-20th century, defined it as the 'science of comparative sociology which seeks universal laws governing human social behavior' (Srinivas 1958: x). Srinivas who was his PhD student avers that Radcliffe Brown saw social anthropology as a science like physics, chemistry, and biology, and one of his aims was to apply the logic of the natural sciences rigorously to social anthropology (*Ibid*:xiv).

There were challenges from his contemporaries. Edmund Leach rued Radcliffe Brown's approach to data and classification by skimming the visible surface. He famously disparaged Radcliffe Brown's followers as 'butterfly collectors' (*Jabal-al-Lughat*). Evans-Pritchard, in his Marrett Lecture, criticised the modelling of social anthropology on the constructs of the natural sciences

> which had led to a false scholasticism and one rigid and ambitious formulation after another … (whereas) regarded as one of the humanities, social anthropology is released from dogmas and could become truly empirical and in the true sense of the word, scientific.
> (Evans-Pritchard 1950: 123)

The internal division in social anthropology on this issue was being discussed in detail already in the late 1950s (Pepper 1961).

The succeeding giants of anthropology like Claude Levi Strauss, who gave new meaning to the concept of structure as that which lay beneath the visible level of social relations, and Clifford Geertz who treated anthropology as an interpretive exercise, laid the ground for a radical dismantling of the ethnographic enterprise as classically preached and practised.

The sparks from this dismantling process sometimes flew into the public and popular arenas, as, for instance, in the much publicised attack on anthropologist Margaret Mead that her ethnography on Samoan society had many flaws of observation and interpretation, and had been written to support academic theory rather than as an unbiased account.

So there has been within anthropology in the West a considerable internal churning since the mid-20th century onwards on the issue of its identity (does it belong to the sciences or humanities) and on methodology (is it scientific or only scientistic?). Questions were being asked about its claims to comprehensiveness and objectivity. From the late 1980s, these questions got consolidated into a sustained critique which came to be known as 'New Ethnography'. Clifford and Marcus's 1986 work *Writing Cultures* is considered to be an important early intervention. Conventional ethnography had been the locus of a distinct encounter in which the researcher and the subjects from vastly different cultures, and unequal economic and social status, were locked in a long-drawn relationship of intimate interaction. The 'New Ethnography', as a genre, draws from this fertile anthropological lineage but goes beyond, with its threadbare analysis of knowledge creation in conventional ethnography, anthropology's colonial history, and its neocolonial contexts; the hierarchy of relationships between researcher and researched, the subjective nature and hidden biases of the knowledge emanating from this method, the critical role of social location, and the mode presentation of the research. Apart from critically revisiting personal accounts of anthropologists of the past, there are innovative experiments through which anthropologists are attempting to expose their own locational, personal, and social biases, to reduce the unequal power dynamics between researcher and researched, to intersperse their own voices with those of their informants, and to share and vet their findings with those who were co-creators of this knowledge. They challenge the scientific idea of objectivity in the creation of knowledge. How far and in what ways these innovations have mitigated the failings of the old ethnography is a question I am not taking up here.

Critique from Without

It is not my case that the transformation of ethnography as a critical methodology central to anthropology has been exclusively fuelled from within. The overall thrust of the postcolonial, postmodern critique of knowledge creation in recent decades has had an impact on several fields. Each new field's experiential engagement with issues specific to it has also gone into the critique of science in the social sciences, with its own ramifications in anthropology. Studies on environment, race, ethnicity, and gender are a few examples of this.

In some cases, the critique is not completely from outside, like feminist anthropology, which draws from anthropological traditions in methodology and combines it with impulses and insights from feminism. From the 1980s onwards, feminist anthropologists have formed a well-knit group, across countries, engaging in a variety of measures to highlight the gender biases in anthropology. By revisiting older ethnographies, they have shown how a supposedly scientific activity was deeply gendered. Unlike some social sciences, women were present as part of the field being researched but were not taken into account or rendered invisible or objectified. The concept of androcentrism or the cognitive ignoring or sidelining of women as subjects was backed with ethnographic material. One of the strong contributions of feminist anthropology has been to retrieve from the fieldwork accounts of some women anthropologists of earlier generations and their sensitivity to the intersubjective nature of knowledge creation. Such material speaks for the prescience of women anthropologists to issues that got raised only much later in mainstream anthropology.[7] Others have done fresh ethnographies drawing from feminist studies as well as some of the ideas of new ethnography.[8]

The end of colonialism created an entire genre of ethnographies by anthropologists from the former colonies working in their own societies and cultures. Their critical analyses of the mode of studying 'other cultures' by Western anthropologists are also part of this self-reflection from both within and without. It has led to a critique of scientific methodology inflected with issues of nationalism and nation-building and their corollaries. We shall take this up in the next section.

Sociology and Social Anthropology in India

In India, the inauguration of anthropology and sociology as formal disciplines happened more than 50 years later than in the West. That they were established in India under the aegis of colonial rule and inherited British disciplinary lineages and that their early development was in an environment of nationalism, independence, and nation-building are much-discussed topics. The issues and debates on methodology are inevitably mediated through this lens.

There is a paucity of writing on sociological theory in India. It has to be traced from research in substantive areas like caste, kinship, religion, urbanisation, and so on. Method appears to be the defining feature of the discipline, but again there is limited scholarly work on it as also its link with theory.[9] There are also certain themes that come up repeatedly for debate and discussion. There are periodic reviews and taking stock of performance and growth of the discipline. From these, in an indirect way, we can recognise or infer the positions on scientific method.

A recurrent theme is the inadequacy of borrowed categories and the need for indigenous categories. Another is dissatisfaction with value neutrality and a desire for committed research. Lastly, and importantly for this chapter, the debate on sociology and social anthropology as a unified field and the place of Indology in understanding Indian society. Discussions on some of these have passed their prime. The last decade and a half has seen major shifts in the global scenario and in national priorities. Responding to them, sociology in India has somewhat muted its preoccupation with the old themes. Yet they continue to play out, by proxy, through other themes. It is widely recognised that an overarching feature of sociological debates in India is a coexistence of a variety of – even contradictory – paradigms without one being totally replaced by another. I will briefly touch on these themes, drawing inferences on methodology, before elaborating on the relation between sociology and social anthropology at Mumbai University.

Until 1970, the two most influential University Departments of Sociology in India were in Bombay University and Delhi University. The latter was established in 1959, with M.N. Srinivas as the first head. The Centre for Study of Social Systems at Jawaharlal Nehru University was established in 1971 and became another powerful centre for sociology, but its thrust areas and approaches have been more accommodative of diversities (Chaudhury 2011). Institutionally, the strongest proponent of the unity of sociology and social anthropology has been the Sociology Department of Delhi University.[10] The Bombay Department is much older, set up in 1919. Due to the influence of G.S. Ghurye who was helming the department for many years, the idea of an integrated sociology and social anthropology was first put in practice here – and this has not been explored in any depth so far.

Search for Indigenous Categories

I have argued earlier that self-reflexivity is part of the constitution of anthropology in the West and sociology in India. In part, it is also the outcome of an independent nation ruminating over colonial memories. The angst that Uberoi et al. (2008: 2–3), for example, see as endemic to sociology in India, the preoccupation with originality versus mimicking the West, and the search in India's traditions for categories – these anxieties were often cast in the idiom of modernity versus tradition. The study of tradition in itself, as a separate topic, and the engagement with Indology, somewhat unique to sociology in India, were a reflection of this anxiety. The debate on modernity was in a way a proxy for scientificity. Except for a few fundamental critiques of scientific modernity, for example by Uberoi (1968) and A.K. Saran (Pandey 2021), the overall thrust has been towards accommodating a modified, chastened, positivist empiricism combined with some insights from the field on objectivity, subjectivity, and intersubjectivity.

Value Neutrality

One of the powerful corollaries of positivist scientific empiricism is the idea of value neutrality. In India, it was stoutly espoused by mainstream sociologists for several decades. But with increasing attention to the multiple dimensions of social inequality and deprivation, a debate ensued on taking sides and on acting on behalf of the people or group being researched. The idea of public sociology and the academic-activist is now not merely acceptable but in fact validated in a range of fora. The side-lining of value neutrality stems from a keen consciousness of the relevance and role of sociology in a country of overwhelming structural discriminations. But it can also be read as an attack on the notion of scientific objectivity. In contrast, in the West, especially in social anthropology, it has had a somewhat different trajectory of siding with interpretivist methodologies.

Quantitative survey-based research and ethnography with participant observation were the two main methods in sociology in India. Generally, sociology did not attempt advanced quantitative methods. Ethnography overtook survey research as the more favoured research method in mainstream sociology for decades (Sundar et al. 2000: 2000). Later, historical analyses, both Marxist and non-Marxist, became more prominent.

In so far as the question of scientific method in sociology is discussed in India, quantitative methods follow the positivist ideas of science and try to adapt them by honing techniques towards greater accuracy. A need to combine statistical data with qualitative methods like depth interview, group discussion, and open-ended narrative is acknowledged. Historical analysis in sociology does not directly engage with the scientific method. A frontal critique of science and scientism in methodological terms from within the discipline of sociology is found in a distinct form in the mature phase of ethnography. Of course, as I argued in the earlier sections, this critique has also come from outside the discipline through multiple sources. And usually, this initial impulse is absorbed within a sociological framework to create subfields within the discipline that are not reducible to either parent.[11]

Sociology and Social Anthropology: A Unified Field

When they entered the Indian system, sociology and anthropology were distinct and separate from each other. Sociology was part of the colonial project of bringing modernity to India through higher education in scientific subjects. Anthropology was closely associated with the requirements of the colonial state for gathering accurate and relevant information for effective rule. As social sciences, both had allegiance to positivist empirical scientific methodology.

Within a few decades, through the consensus of a group of pioneering and influential scholars, social anthropology largely disengaged from physical anthropology and started being considered together with sociology as a unified field. The reasons for and implications of this have been debated extensively in the sociological literature. I will only deal here with how this unification impacted methodology.

The fundamental definition of anthropology at its point of origin in the West was a paradox and a contradiction. If its claim to science had to be validated, how could the identity of the subject/researcher and object/other cultures be fixed and circumscribed?

The overwhelming majority of sociologists in India were studying their own society. G.S. Ghurye, considered to be one of the founders of sociology in India, who initiated the project of bringing the two together, pointed out, 'in India with its huge numbers of communities and castes and ethnic groups in all stages of culture, there is no room for distinguishing and clearly separating social anthropology from sociology' (Upadhya 2007: 223–224).

Demarcating boundaries by saying that sociology dealt with complex, large-scale, industrial society with quantitative methods and anthropology with small-scale primitive tribal cultures with qualitative methods like ethnography have been found untenable. Today in the West, in the era of interdisciplinarity, both sociologists and anthropologists do both kinds of studies. The distinction is more in the historical lineage.[12]

Beginning with G.S. Ghurye, the next generation of pioneers and institution-builders like M.N. Srinivas, Andre Beteille, T.N. Madan, and others in subsequent generations pushed for sociology and social anthropology to be treated as a unified field. There have been strong dissenters but the criticism has not always been on principle. Rather, it is the perception of the dominance of social anthropology in the partnership at the expense of understanding Indian society through macro-frameworks and of the settings, networks, and interrelationships with the meso and the macro that have always characterised micro-communities like village, caste, and tribe.[13]

In practice, the relationship of sociology vis-à-vis anthropology in India is not homogeneous.

There is a coexistence of a few separate anthropology departments where physical and social anthropology both are taught with many sociology departments teaching predominantly sociological courses with a paper or two in social anthropology as optional. There are journals that are named as anthropological but the journals of sociology include social anthropology in their scope. The ethnographic method is used by both sociologists and social anthropologists (Sundar et al. 2000: 1999).

As a concluding section, I will elaborate briefly on the dynamics between sociology and anthropology in the University Department of Sociology in Bombay.

Anthropology in Bombay

The first postgraduate department of sociology in India was established in 1919 at Bombay University with Sir Patrick Geddes, sociologist and town planner, as its head. In his brief tenure, no courses in anthropology were taught. G.S. Ghurye, who succeeded him, initiated and developed the project of integrating sociology with social/cultural anthropology in his long tenure. Ghurye was a Sanskrit scholar who in 1920 had been selected by Bombay University to study sociology abroad. Geddes, who had interviewed him for the scholarship, advised him to study under L.T. Hobhouse at the London School of Economics. Ghurye did so, but within a month moved to Cambridge to study anthropology under W.H.R. Rivers. The reasons for this shift are unclear (Upadhya 2007: 204) but proved fateful for the future of sociology in India. In return, he joined the University Department of Sociology (UDS) at Bombay University in 1924 and led it till his retirement in 1959.

213

Ghurye was supported in his project of an integrated sociology by N.A. Thoothi, an anthropologist trained in Oxford, who joined the department in 1925. Ghurye implemented his project at three levels: teaching, research, and outreach through journals and institutional collaborations. He introduced courses in anthropology and created two permanent faculty positions in cultural anthropology.[14] The topics of his more than 80 MA and PhD students were distributed between sociology and social/cultural anthropology.[15] He was one of the founders of the Indian Sociological Society with its flagship journal *Sociological Bulletin*. The membership of the former and content of the latter reflected the objective of integration. He nurtured collaborative links with two important anthropological institutions in Bombay: the Anthropological Society of Bombay (ASB), and the Institute of Indian Culture (IIC). The history of sociology's integration with anthropology in Bombay at this early time with the UDS at its hub deserves to be reconstructed more elaborately that I am able to do here. I merely gesture to the key institutions, their activities, and their interrelations.

The ASB was set up in 1886 by a group of the city's educated elite from academic and professional fields, led by Edward Tyler Leith, Professor of Law at Bombay University. The majority of the members and office bearers were British and European, but a significant minority was Indian. It aimed to promote research, especially of the various races inhabiting the Indian empire. Its major activities were organising scholarly lectures and publishing the *Journal of the Anthropological Society of Bombay* (*JASB*) from 1886 to 1936 and again after a hiatus from 1946 to 1973 when they were both closed down. Leith in his inaugural speech bemoaned the sparsity of anthropological investigations in India – probably no other country in the world offered so interesting a field for anthropological research (*JASB* 1886 1-1:1).

JASB, the earliest journal of anthropology in India, was a lively forum, featuring an array of topics beyond strict anthropological concerns: folklore, Indology, art and antiquities, caste histories, contemporary social and social reform issues.[16] The geographical canvas included the whole of India, and also other regions in Asia. Most contributors were Indians, especially in the later issues.

Shah (2014: 356) describes the ASB as a voluntary initiative of individuals with scholarly interests unlike the Ethnographic Survey of India set up in 1901, which was directly sponsored by the Government. Without minimising the colonial imprint on institutions of knowledge-making, one can see differences in its extent and quality and the involvement of Indians over time. The first few decades were the pre-professional stages of the discipline, with some leeway for different types of scholarship. In the ASB proceedings, we can see the beginnings of the conversation between anthropology, sociology, and Indology as also the nascent scholarship by Indians strengthening over the decades. The ASB became from 1896 onwards a part of the

Bombay Branch of the Royal Asiatic Society, which had a similar approach to knowledge-making in India.[17]

Several faculty members of the UDS were active members of ASB. Ghurye became a member in 1924. In 1942, he became President of ASB and from 1946 he revived the *JASB* in a New Series. N.A. Thoothi was also a member and was secretary when Ghurye was President. Irawati Karve, M.N. Srinivas, and I.P. Desai from the UDS had presented papers at ASB's scholarly meetings. J.V. Ferreira, A.R. Desai, and other faculty were members.

The other strand of anthropology in Bombay – the Institute of Indian Culture – was associated with the SVD[18] Catholic mission in India. Its founder Stephen Fuchs, an Austrian missionary and an anthropologist, was trained by Wilhelm Schmidt, a pioneer of the culture historical approach who had set up the Anthropos Institute in Germany. Fuchs worked for the SVD mission in India at Indore from 1934. He went back to do his PhD in the University of Vienna on a topic straddling Ethnology and Indology. Back in India in 1950, he helped introduce anthropology courses at St. Xavier's College Bombay with the blessings of Ghurye.[19] In 1950, he established the Indian branch of the Anthropos Institute to pursue anthropological research. It was renamed as the Institute of Indian Culture in 1976. Fuchs has done decades of fieldwork as well as missionary work in Central India among Dalit and tribal communities.[20] IIC had a close link with the UDS. J.V. Ferreira, Professor of Cultural Anthropology at UDS, was close to Fuchs who was instrumental in enabling him to go to the University of Vienna for his PhD.[21]

The UDS has always had fewer anthropologists than sociologists on its faculty. Ferreira and A.R. Momin had, in their time, held both the anthropology positions and were part of the senior faculty for many years. Both contributed to conceptual and theoretical areas. After his early work on Totemism, Ferreira was engrossed in building bridges with philosophy and Momin's attention was largely on the civilisational dimension. Neither of them had engaged in any substantial ethnographic fieldwork.[22]

Ghurye's student A.R. Desai, who was Professor and Head of UDS from 1967 to 1976, was a Marxist who introduced historical materialism and political economy to sociology in Bombay and in fact, in India. He was also an activist who supported radical political and social movements. As a charismatic teacher, his ideas attracted several students and young faculty. An active group coalesced around Desai, engaging with Marxist perspectives in general. The two anthropologists Ferreira and Momin and a few other faculty were not part of this group. The culture of the department during the 1960s–1980s was one of cordial interaction, but there was seldom real dialogue on sociological issues across the two groups.[23] Indra Munshi who was Desai's PhD student in the 1970s and early 1980s remembers that in Desai's group, no one had time for anthropology. There was a vague unease about its colonial origins, but nobody had any expertise. There was no academic confrontation with the other group, they remained in their own bubbles, but each was privately dismissive about the other. Anthropology was not so much attacked as bypassed.[24] Although the department was a pioneer in the integration of sociology and anthropology, there was no sustained engagement with the internal developments within anthropology in the West from where a powerful critique of scientism was emerging. The image of anthropology that students of that time carried was that of a colonially tarnished discipline steeped in racial stereotypes.

Students in the 1990s had a different image and were more receptive to anthropology courses. Marxist approaches became less prominent. There were new teachers in the anthropology positions.[25] Other new faculty have since joined whose research interests and theoretical frameworks are more diverse. Their impetus for critiquing objectivity and the scientific method in sociology has come largely from outside the discipline: from feminism, critical caste studies, sociology of marginalised communities, sociology of science, sociology of environment. Several of them have also engaged in ethnographic fieldwork that best brought out the specificity of their research problematique. The additions of new faculty members in the last decade holds the potential for a critical tribal studies to emerge in parallel with critical caste studies in the UDS. Their engagement with the traditional anthropological discourse of tribe as well as ethnography holds the promise of opening new doors.

Sociological and anthropological methods have mingled in much of this recent research along with an eclectic, interdisciplinary approach to method in general. Sociology of science is an exciting area, raising questions upfront about method and methodology.[26]

With all the arguments made and reservations expressed about unifying sociology and social anthropology in India, I take the position that although the two lineages are distinct, they are closer to each other than any other pair among the social science disciplines. There is no doubt in my mind that this enabled the distinct methods of each to be available to both. It created a broader base for integrating the productive practices of different methods. The sometimes chaotic coexistence and accommodation of contrasting, even opposing approaches has the saving grace of keeping practitioners open to a range of ideas. The self-reflexivity of sociology has been much discussed. Chaudhuri and Jeyachandran (2013: 89) aver that it is not incidental but central to the nature of the discipline. I would argue that of the methods prevalent, ethnography had the greatest potential to precipitate this and create a climate where the imitative application of scientific methods could be queried. But in an era of interdisciplinarity and transdisciplinarity, and for that matter, even for earlier times, fixating on a single method and source for tracing lineages of an idea runs contrary to the reality of the messy interaction of giving and taking of ideas and making them one's own.

***

216

Acknowledgements

The Institute of Social and Economic Change, Bangalore, for giving me the Prof. M.N. Srinivas Chair Professorship in 2021, during which I wrote this chapter and presented it to faculty and students; editors Gita Chadha and Renny Thomas for valuable inputs; Indra Munshi and M.T. Joseph for sharing memories of their student days; and Ashish Upadhyay for meticulous and comprehensive research assistance.

Notes

1 https://www.americananthro.org/ConnectWithAAA/Content.aspx?ItemNumber=1650

2 https://www.americananthro.org/StayInformed/NewsDetail.aspx?ItemNumber=13032

3 In the first two sections of this chapter, I use the term 'anthropology' for social anthropology in most places. Occasionally, when referring to early phases of the discipline, the usage includes physical anthropology, as per the common practice in the Anglo-American literature. In the sections on India, unless stated otherwise, I use the term sociology to include social anthropology as well, as per the common practice here.

I use 'anthropology' for the discipline and 'ethnography' for an important research method which originated within it. There is a view though that ethnography is also a methodology and a theoretical approach in itself. Ethnography as method is now widely used by various disciplines and fields in the social science and humanities.

4 As an example, 'The Group for Debates in Anthropological Theory' founded by Tim Ingold in Manchester organises yearly meets of leading social anthropologists to debate threadbare, in parliamentary style, a motion at the heart of current theoretical developments. For the very first debate in 1988, the motion was 'Social anthropology is a generalizing science or it is nothing'. Not surprisingly it kindled passionate discussion (Ingold 1996).

5 My lack of familiarity with theory and practice outside the Euro-American and Indian contexts – not an excuse but a surrender to an unplanned academic trajectory – has regretfully limited the scope of my discussion only to these regions.

6 For example, in Latour and Woolgar's pioneering ethnography in 1979, scientists are studied in their natural environment in the Salk Institute in California as though they were a remote tribe.

7 See Bell (1993), Caplan (1992), Ardener (1992), and Lamphere (2006). My own interaction with feminist anthropologists from various continents through my association with the International Union of Anthropological and Ethnological Sciences gave me an opportunity to relook at my own earlier research. See Ganesh (1993, 2016). Collaborative relations among feminist anthropologist globally was not uncommon at that time.

8 A research that I did jointly with another Indian sociologist and a Dutch anthropologist tried to implement some of these ideas. It was on the Retracting welfare state in the Netherlands and its impact on vulnerable categories. My ethnographic fieldwork (Ganesh 2005) was with the elderly in Leiden living without institutional care. In the late 1990s, it was still a rare example of 'reverse research' (Risseeuw et al. 2005).

9 For an insightful analysis of the reasons, see Chaudhury and Jeyachandran (2013: 87–90).

10 The mutual interaction between these two prominent departments in JNU and Delhi University and how it impacted on the sociology – social anthropology unity project deserves a serious exploration which regretfully I have not under-taken within the scope of this chapter. The autobiographies and memoirs of senior sociologists, taken collectively, would give a clue to these dynamics. For example, Oommen's (2018) autobiography reveals how his own preference for avoiding social anthropological approaches cost him professionally.

11 Thus, feminist anthropology is distinct in its methodology and arguments from both feminism and anthropology. So too sociology of science, sociology of envi-ronment, and so on.

12 Whether or not the unification is 'good' or 'valid' is a normative question. Whichever position one takes, there is, as Patel (2011: xvii) argues, a need to rec-ognise and enumerate the historical and political processes that divided the two disciplines in the West and how and why scholars in India conflated them. The ramifications and outcomes of this unification are not homogeneous and deserve nuanced analysis in the present interdisciplinary era.

13 Sundar et al.'s (2000) report on the national workshop held at Institute of Economic growth on this theme records a range of views expressed. See also Deshpande (2018) and Patel (2011: 81–86).

14 In the UDS, the faculty positions were named as Cultural Anthropology rather than Social Anthropology, for reasons I have not been able to trace.

15 Irawati Karve, the distinguished anthropologist, did her Master's under Ghurye's guidance.

16 A few random examples from the JASB Index (1937), in the Old Series (1886–1936): articles on aborigines, age of consent controversy, ancestor worship, ancestral property among Hindus, child marriage, *devi* worship, folklore on the peacock, funeral ceremonies of Nagar brahmins, Hindu coronation rites and ideas of government, marriage songs, methods for interpreting Vedic texts, sta-tistical analysis of suicides in Bombay city, and study of anthropology in the West.

17 Later renamed as the Asiatic Society of Mumbai, it continues to be a thriving institution, with which I was closely involved for several decades. As a Learned Society, its approach of drawing the city's elite into its broad canvas of Indology, social sciences, and humanities has continued over the decades, with some mod-ifications. Within the Asiatic Society, ASB retained a separate identity and its journal was published in its own name, distinct from the *Journal of the Asiatic Society of Bombay* (Ganesh 2002). A.M. Shah, with Lancy Lobo embarked on a venture to republish select articles from the *JASB*. Two volumes were brought out in 2019 and 2021. It was my privilege to have assisted Prof. Shah in contact-ing and communicating with the Asiatic Society of Mumbai where the journal issues are housed and in acquiring some issues.

18 *Societas Verbi Divini* or 'Society of the Divine Word'.

19 Since then St. Xavier's College has offered courses in anthropology. Technically, it is clubbed with sociology and called the Department of Sociology and Anthropology. But it has an independent existence and students can choose anthropology electives without choosing sociology.

20 Looking at his anthropological research as a whole, Pflug (2018: 21) argues that the scholar in Fuchs had always the upper hand though this hand was tied to the task of mission.

21 Three other anthropologists joined much later as faculty. Except for me, the other two – S.M. Michael, now retired, and M.T. Joseph, who is part of the cur-

rent faculty – are both ordained priests of the SVD order from the IIC. Thus, the links of the USD with IIC initiated by Ghurye continue till the present.

22 Ferreira influenced the type of anthropology students of the department were exposed to. I registered with Ferreira for a PhD in 1976. He confessed that he was not familiar with South India nor with gender issues that I wanted to study. Still, he agreed to be my guide. He intervened on overall approach and presentation but did not interfere with my analysis. For my ethnographic fieldwork, he suggested I consult 'Notes and Queries'.

23 Ganesh (2013: 301).

24 Munshi was one of the few who did intensive ethnographic fieldwork at that time. Her research was on a tribal community in Thane district. Although her theoretical approach was influenced by Desai's wider political framework, she could not get much guidance from him for the nitty-gritty of fieldwork and drew from the ethnographic writings of Srinivas, Beteille, and many others (personal communication).

25 M.T. Joseph, who was a student of the department in the mid-1990s, remembers that his teachers – S.M. Michael and myself – interwove sociology and anthropology in all the papers that they taught and that he never saw the two disciplines as separate (personal communication). I have written elsewhere on my own experiences in anthropology and in the UDS (Ganesh 2013, 2016).

26 Gita Chadha's doctoral work (2005) on the creative process in science is an example of the mingling of sociology and anthropology that got revived in the department As an MA student in the UDS in the 1980s, she was wary of anthropology. Later she critically re-examined her reservations, anchored herself in feminist perspectives, developed an interdisciplinary outlook situating her work at the intersection of sociology of science and sociology of knowledge, and actualised it through an ethnography of scientists at the TIFR, Mumbai.

References

American Anthropological Association. 2009. AAA Statement of Purpose 2009. https://www.americananthro.org/ConnectWithAAA/Content.aspx?ItemNumber =1650 (accessed on November 3, 2021).

American Anthropological Association. 2010. AAA Responds to Public Controversy Over Science in Anthropology. https://www.americananthro.org/StayInformed/ NewsDetail.aspx?ItemNumber=13032 (accessed on November 3, 2021).

Ardener, Shirley. 1992. 'Introduction.' In Shirley Ardener (ed.) *Persons and Powers of Women in Diverse Cultures*. New York: Berg Publishers Ltd, 1–10.

Bell, Diane. 1993. 'Introduction.' In Diane Bell, Pat Caplan, and Wazir Jahan Karim (eds) *Gendered Fields: Women, Men and Ethnography*. London: Routledge, 1–18.

Campa, Riccardo. 2008. 'Making Science by Serendipity. A Review of Robert K. Merton and Elinor Barber's the Travels and Adventures of Serendipity.' *Journal of Evolution and Technology*, 17 (1): 75–83.

Candea, Matei. 2016. Science. https://www.anthroencyclopedia.com/entry/science (accessed on November 2, 2021).

Caplan, Pat. 1992. 'Engendering Knowledge: The Politics of Ethnography.' In Shirley Ardener (ed.) *Persons and Powers of Women in Diverse Cultures*. New York: Berg Publishers Ltd, 65–88.

Chadha, Gita. 2005. *Gender, Genius, Scientific Intuition: A Feminist Perspective.* PhD thesis, Department of Sociology, University of Mumbai.

Chaudhuri, Maitrayee, and Jesna Jayachandran. 2013. 'Theory and Methods in Indian Sociology.' In Yogendra Singh (ed.) *Indian Sociology: Emerging Concepts, Structure, and Change.* New Delhi: OUP.

Chaudhury, Maitrayee. 2011. 'Looking Back: The Practice of Sociology in CSSS/ JNU.' In Sujata Patel (ed.) *Doing Sociology in India: Genealogies, Locations and Practices.* New Delhi: OUP.

Clifford, James, and George E. Marcus (eds). 2010. *Writing Culture: The Poetics and Politics of Ethnography.* Oakland: University of California Press.

Deshpande, Satish. 2018. 'Anthropology in India.' In Hilary Callan (ed.) *The International Encyclopedia of Anthropology.* New Jersey: John Wiley & Sons.

Evans-Pritchard, E. E. 1950. 'Social Anthropology: Past and Present.' Marett Lecture, *Man*, 50: 18–124.

Ganesh, Kamala. 1993. 'Breaching the Walls of Difference: Fieldwork and a Personal Journey to Srivaikuntam.' In Diane Bell, Pat Caplan, and Wazir Jahan Karim (eds) *Gendered Fields: Women, Men and Ethnography.* London: Routledge, 128–142.

Ganesh, Kamala. 2002. 'Maps for a Different Journey: Introduction to the Themes in the Work of Leela Dube.' In Leela Dube (ed.) *Intersecting Fields: Chapters in the Anthropology of Gender.* New Delhi: SAGE Publications, 13–32.

Ganesh, Kamala. 2005. 'Made to Measure: Dutch Elder Care at the Intersections of Policy and Culture.' Risseeuw (*et al.*) *Care, Culture and Citizenship: Revisiting the Politics of the Dutch Welfare State.* Amsterdam: Het Spinhuis, 116–158.

Ganesh, Kamala. 2013. 'New Wine in Old Bottles? Family and Kinship Studies in the Bombay School, Special Issue on The Bombay School of Sociology: The Stalwarts and Their Legacies.' *Sociological Bulletin*, 62 (2): 288–310.

Ganesh, Kamala. 2016. 'No Full Circle: Revisiting My Journey in Feminist Anthropology.' *Contributions to Indian Sociology*, 50 (3): 293–319.

Garson, Johan George, and Charles Hercules Read. 1899. '*Notes and Queries on Anthropology*.' Edited for the British Association for the Advancement of Science. London: Anthropological Institute.

Hirst, K. Kris. 2019. Is Anthropology a Science? https://www.thoughtco.com/is -anthropology-a-science-3971060 (accessed on November 2, 2021).

Ingold, Tim (ed.). 1996. *Key Debates in Anthropology.* New York and London: Routledge, 3–34.

Jabalal, Lughat. (2017). Butterfly-Collecting: The History of an Insult. http://lughat .blogspot.com/2017/10/butterfly-collecting-history-of-insult.html (accessed on November 3, 2021).

Lamphere, Louis. 2006. 'Foreword.' In Pamela L. Geller and Miranda K. Stockett (eds) *Feminist Anthropology: Past, Present, and Future.* Philadelphia: University of Pennsylvania Press, ix–xvi.

Latour, Bruno, and Steve Woolgar. 1979. *Laboratory Life: The Construction of Scientific Facts.* New Jersey: Princeton University Press.

Merton, R. K., and E. Barber. 2004. *The Travels and Adventures of Serendipity: A Study in Sociological Semantics and the Sociology of Science.* New Jersey: Princeton University Press.

Oommen, T. K. 2018. *Trials, Tribulations and Triumphs: Life and Times of a Sociologist.* Konark: Delhi.

Pandey, Ajit Kumar. 2021. 'A. K. Saran's Sociology: Towards a Critique of Modernist Mode in Social Sciences.' *Sociological Bulletin*, 70 (2): 232–251.

Patel, Sujata. 2011. 'Introduction.' In Sujata Patel (ed.) *Doing Sociology in India: Genealogies, Locations and Practices*. New Delhi: OUP, xi–xxxviii.

Pepper, George B. 1961. 'Anthropology, Science or Humanity?' *Anthropological Quarterly*, 34 (3): 150–157.

Pflug, Bernd. 2018. 'Between Ethnography and Mission in India: The Anthropology of Stephen Fuchs.' *Sociological Bulletin*, 67 (1): 20–34.

Risseeuw, Carla, Rajni Palriwala, and Kamala Ganesh. 2005. *Care, Culture and Citizenship: Revisiting the Politics of the Dutch Welfare State*. Amsterdam: Het Spinhuis.

River, W. H. R. 1913. Cited in: Urry, James. 1972. *Notes and Queries on Anthropology and the Development of Field Methods in British Anthropology, 1870–1920*. Proceedings of the Royal Anthropological Institute of Great Britain and Ireland, No. 1972 (1972): 45–57.

Shah, A. M. 2014. 'Anthropology in Bombay, 1886–1936.' *Sociological Bulletin*, 63 (3): 355–367.

Shah, A. M., and Lancy Lobo (eds). 2018. *Essays on Suicide and Self Immolation*. Delhi: Primus Books.

Shah, A. M., and Lancy Lobo (eds). 2021. *An Ethnography of the Parsees of India 1886–1936*. London: Routledge.

Srinivas, M. N. (ed.). 1958. *Method in Social Anthropology: Selected Essays by A. R. Radcliff Brown*. Chicago: University of Chicago Press.

Sundar, Nandini, Satish Deshpande, and Patricia Uberoi. 2000. 'Indian Anthropology and Sociology: Towards a History.' *Economic and Political Weekly*, 35 (24): 1998–2002.

The Anthropological Society of Bombay. 1937. *Index to the Journal 1886–1936*. Bombay: The Times of India Press.

Uberoi, J. P. Singh. 1968. 'Science and Swaraj.' *Contributions to Indian Sociology*, 2 (1): 119–123.

Uberoi, Patricia, Nandini Sundar, and Satish Deshpande (eds). 2008. *Anthropology in the East: Founders of Indian Sociology and Anthropology*. London & New York & Calcutta: Seagull. 2–3.

Upadhya, Carol. 2007. 'The Idea of Indian Society: G. S. Ghurye and Making of Indian Sociology.' In Patricia Uberoi, Nandini Sundar, and Satish Deshpande (eds) *Anthropology in the East: Founders of Indian Sociology and Anthropology*. London & New York & Calcutta: Seagull.

Urry, James. 1972. 'Notes and Queries on Anthropology and the Development of Field Methods in British Anthropology, 1870–1920.' *Proceedings of the Royal Anthropological Institute of Great Britain and Ireland*, No. 1972 (1972): 45–57.

Vogt, Evon Z. 1954. 'Notes and Queries on Anthropology.' *American Anthropologist (New Series)*, 56 (6): 1154–1156.

Wade, Nicholas. 2010. 'Anthropology a Science? Statement Deepens a Rift.' *New York Times*, December 9, 2010. https://www.nytimes.com/2010/12/10/science/10anthropology.html (accessed on November 2, 2021).

10

ECONOMICS, FEMINIST ECONOMICS, AND WOMEN'S STUDIES

Methodological Orientations and Disciplinary Boundaries

Neetha N.

Introduction

Among the various social science disciplines, economics has a problematic acceptance and is often seen as a bridge between natural science and social science. Closeness to scientific/positivist methodologies and use of quantitative data have definitely given an impression that economic analysis are more rigorous and objective, thus claiming a superior status. The desire to stay close to hard sciences led to the development of the neoclassical paradigm[1] in the 19th century which dominates economic thinking, largely referred to as mainstream economics. Over time, neoclassical economics became the dominant or universal thinking with other streams in economics eroding.[2]

Mainstream economics falls into the intellectual camp of methodological individualism and reductionist ontologies where individuals are claimed to be behaving purposefully and their experiences are taken as the only foundation of knowledge. An important characteristic of the 'economic man' is that he is rational and also free to take all economic decisions whether related to production or consumption. The unit of analysis in economics is thus always the individual, whose behaviour is added up for understanding the economy.

Given this larger understanding of the economy and individual behaviours, in economics, research is about testing predictions based on a theory which is mostly bound by assumptions. Friedman dominated the discussions on the methodology of conventional economics, and his 1953 essay is probably the best-known writing on the methodology of mainstream economics. Formulation of hypotheses based on existing theories that can be tested using data collected scientifically following statistical principles is a central principle of this methodology. Theories are abstractions based on

DOI: 10.4324/9781003298908-17

prior knowledge of the data, and hypotheses are formulated based on these theories. If the data collected when fitted into the model based on the theory does not support the hypothesis, the hypothesis is rejected. If hypothesis is accepted based on the empirical data, the hypothesis is proved to be 'confirmed' by experience (Friedman 1953: 8–9).

Ceteris paribus clauses are used in all sciences for analysing a given reality and its link to the complex larger world. Assumptions or *ceteris paribus* clauses are central to the falsification process in economics which are about economic and social realities. These assumptions do not have anything to do with the theory and should ideally be based on the analysis of past data. Past data is always taken as given and considered to be objective. Thus, economics makes claims about realities through formulating hypotheses based on existing theories with the aid of these assumptions.

The claim of falsification,[3] possibility of testing through empirical data, gives economics a scientific label as in the natural sciences, and this is what distinguishes economics from other social sciences. Though falsifiability is an important way to check the scientific nature of the theory, as per conventional understanding, it cannot verify the theory in terms of its ability to capture reality or truth.[4] Thus, factual evidence cannot prove a hypothesis but can only confirm or reject the hypothesis. Going by this logic, it is possible to make statements based on economic theory which can be falsified by empirical data which does not lead to the rejection of the theory but only the hypothesis. Thus, it is possible 'at the same time to claim the truth of all possible statements that can be expressed in the language of economic theory and the falsity of the consequent (namely the given prediction)' (Keppler 1998).

The deductive method of verification or falsification with all its assumptions drives economic research even now where progress in economic research is about developing better theoretical models for verifications and predictions. The methodological position which Friedman made explicit is what that keeps research in economics distinct from other social sciences and the reason for its claim of objectivity, as in other natural sciences. The dispensation of social variables in the neoclassical understanding of the economy led to the expansion of specific branches of economics such as mathematical economics and econometrics and the disappearing or neglect of political economy questions, including development economics.[5] Whether individualism and falsificationism are relevant methodologies even for economics needs to be analysed in terms of the possibilities of it being useful for critical explorations. There have been many developments within economics which have exposed the methodological limitations of mainstream economics which have resulted in heterodox approaches. Rather than accepting the limitation and addressing the methodological issues and challenges, many mainstream economists have chosen to remain silent as a way to avoid critical questions.

The Heterodox Approaches

There presently exists a number of societies and associations of economists and other social scientists, all of which are united by their concern about the theoretical and practical limitations of neoclassical economics. In addition, they share the conviction that the current dominance of the subject by mainstream economics threatens academic freedom and is contrary to the norm of methodological pluralism.

(ICAPE 2021)

What distinguish these heterodox schools of thought from the mainstream are not their principles or positions but largely the questions raised by them and the specificity of their focus and concerns. The quest for an objective analysis of contemporary socio-economic phenomena and thus a defined knowledge underline these schools of thought. There are a range of interpretations on the nature and directions of development within heterodox economics and many are very much within the neoclassical paradigm, such as Keynesian and institutional economics (Dow 2000).

Keynesian analysis and institutional economics, major developments within the heterodox tradition, did raise challenges to the neoclassical approach of individualism and the *ceterus paribus* assumption. What these schools does is a recombination of core classical or neoclassical ideas with pragmatist ways of thinking and analysis and structuralist conceptual elements. Thus, most of the developments within the heterodox tradition do not have a distinctive existence in terms of epistemology, ontology, ideology, and ethics as they are largely embedded within the neoclassical paradigm.

There have also been attempts to add on to the postulate of falsificationism bringing in the descriptive account of experience and economic conventions and practices. Though such efforts helped in acknowledging the importance of descriptive aspects of economic questions, beyond prices and quantities, these were incompatible with the core method of falsificationism. Most of these studies were developed not only as an alternative to neoclassical economics but also in the context of discussions on specific issues in other social sciences such as gender and race that required a new recombination of structuralism with pragmatist ways of thinking. Feminist economics, one among the heterodox schools of thought, is instrumental in highlighting the gender blindness of economics and its dominant methods. The next section will provide an overview of the various feminist engagements with conventional approaches in economics and challenges that it poses.

Another way of distinguishing developments in economics is with respect to their approaches –closed-system[6] and open-system[7] approaches, respectively. As opposed to the assumption that all variables and their relationships are knowable and predictable and thus could be represented as a mathematical formulation, the open system is based on the understanding that variables are not known in many contexts. To claim that all heterodox economists follow open-systems approach in their ontological, epistemological, and methodological stances is also questionable. Heterodox economists with pluralist methodological orientation may even follow a closed-system approach in terms of their ontological commitments (Davis 2008) and materialism underlines their ontological position.

Heterodoxy though encompasses marked internal divisions and have methodological differences, these approaches are unified in terms of their differences with mainstream economics and thus could be identified as a single discourse. Because of the lack of much conceptual or concrete methodological challenges that heterodox traditions could raise to counter the neoclassical turn of the discipline, teaching of economics worldwide is still dominated by neoclassical theories with the rest having secondary positions. As Davis (2008) summarises, though there are new ideas and approaches within economics, either these are still at the bottom or more or less identical to the mainstream neoclassical thought.[8] In research, the period since the mid-1940s is marked by theoretical approaches towards developing mathematical models. This took a turn in the 1990s with a shift in emphasis from theory to empirical (Fourcade et al. 2015) with statistical analysis and mathematical modelling still continuing to enjoy special importance in economic research. Thus, no non-neoclassical approach has been able to change the neoclassical domination even when pluralistic environments exist in many academic institutions.

Feminist Economics in Economics

An important development within the heterodox traditions, which is relatively recent, is feminist economics. Feminist economics though is sometimes taken as a separate school of thought, it is mainly a lens through which economic analysis is done like that of other orthodox and heterodox–methodological approaches. Gender inequality, in relation to other social inequalities, is the main concern of this approach.

Feminist economists[9] have in the past many years, especially since 1991 after the formation of the International Association for Feminist Economics (IAFE) and the initiation of the *Journal of Feminist Economics* in 1995, established a space for themselves. Feminist economics, of course, did not start with the association that was set up in the early 1990s or the journal. The perspective has much earlier roots, well traced by Amartya Sen (2005) to Mary Wollstonecraft.[10] What is today called feminist economics, or gender analysis of the economy, used to be referred to as work on women's economic position before the development of this stream of study in economics.

The existence of gender inequality alongside other social inequality and the need to incorporate gender into the analysis of mainstream economics has gained much momentum now. Feminist economics, going beyond sex-disaggregation, have exposed the claim of gender neutrality and shown the value-embedded nature of mainstream economic analysis. Such analysis has shown that the implicit gender differentiations in mainstream economics are not value neutral, thus merely pointing at differences between men and women is not enough. Implicit gender meanings often involve value judgements in which masculinity tends to be valued higher and femininity lower.

One of the most important interventions of feminist economists is the questioning of the concept of the 'rational economic man' (REM), who is solely driven by self-interest, ignoring the existence of a gender-stratified economy and women's roles (Folbre and Hartmann 1988; Grapard 1995). Challenging the underlying hierarchical and dichotomous distinctions in economics such as self-interest/altruism, rational/emotional, and efficiency/equity, feminist economists have rejected gender-based differentiation and dualisms (England 1993; Jennings 1993; Nelson 1996).

Though feminist engagement is now diverse and extends to many fields of economics, women's work is central. National income, the key variable to the understanding of economic growth, is one of the critical concepts in economics. By classifying all activities into two categories – those within production boundary and those outside the production boundary – and by only acknowledging what is counted as production as its subject matter, economics has tried mostly not to acknowledge the interrelationship between the two. The market definition of production is inbuilt in this understanding, distinguishing it from non-productive activities. This setting aside of the unproductive activity, a reflection of the dominance of neoclassical economics, is a central feminist criticism. Feminist's writings have argued for long that women's work, even though not performed for the market, is not only unproductive but also critical for the economy. Because of these interventions, it is now well-acknowledged that an important segment of the economy and of women's work was kept outside and rendered invisible by mainstream economists.

The existence of many forms of inequalities and their continuation such as labour market segmentation, differentiated wages for men and women, and division between leisure, unpaid work, and paid work were issues that received considerable attention. Feminist economists have argued that the concept of gender has wider analytical implications than helping to explain male–female differences in measured economic outcomes. Gender together with other structures of constraint such as class, race, age, and sexuality (Folbre 1994) is also a way in which agency, power, and social relations are articulated in the economy.

The need to understand the origin, continuation, and subversion of a wide variety of inequalities and expressions of power in the economy was highlighted through many studies undertaken by feminist scholars. Even when women work in the market economy, they are treated by the mainstream as economic man and thus their specificities are overlooked. Women are also assumed to be rational in economics and thus concerned only about wealth maximisation behaviour. The low labour force participation of women, feminist economists have argued, thus cannot be understood fully using the leisure and wage arguments given the larger norm of women as unpaid workers in the household. Human capital theory, which supported wage differences between categories of workers, legitimises differences in wages or earnings between men and women attributing it to differences in education and skill. The residual, often left even when men and women of the same human capital was compared, challenged this theory but was mostly ignored or seen as having issues/errors that ignore the historical and social underpinnings of such differences. Many analysis in the context of developing countries have specifically highlighted how cultural and social expectations have restricted not only women's economic participation but also the nature of such participation.[11] As women have a comparative advantage in terms of reproductive work, they are assumed to be preferring to stay at home looking after men and children.

The deficiencies inherent in the use of the 'unitary' household model, an analytical category in economics, has also been a subject of feminist intervention. The individuals and their relationships within and between households, which were ignored in economic analysis, by taking the household head as an altruistic though selfish player in the market, has been pointed out as a critical mistake (Bergmann 1995; Kuiper 2001; Nelson 1996). The study of intra-household relationships through the household-bargaining approach adapted from game theory was developed in this context by feminist economists (Agarwal 1997; Kabeer 2001; Katz 1997). The bargaining approach, however, has also been found to simplify household relations where, as in the real world, there are a variety of gender norms that influence households, some of which are contradictory and limit their usefulness. In many households, though women may be economically active or own property or assets, the control is with men which such models could not explain. Further, the assumption of self-interested motivation of different individuals in the bargaining process is also a debated issue as many women because of their care-giving role do not represent a self-interested economic agent. Women's decisions, it was highlighted by many Indian feminist scholars, including economists, are often determined by cultural and social norms as well as economic factors apart from state policies and laws (Agarwal 2001).

With these developments, the institutional approach to the analysis of the household and economic decisions, with an increased realisation that households share boundaries with other institutions such as markets and firms, gained acceptance. Households are now increasingly understood as sites for production, with a direct or indirect link with the market, and reproduction of labour force through unpaid work. Given this, members of the household could represent multiple sometimes overlapping and contradictory roles as an employer, worker, care-giver, or care-receiver. Thus, the desire to spend more by one household member may be constrained or undermined by another member who may want to save or invest. Further, expectations of members also vary as it is influenced by a host of factors, including their individual attributes such as age and education as well as differences in roles, exposure, and experiences.

Though considerable developments happened within the Marxist feminist literature which highlighted the importance of social reproduction and its linkage to capitalist production, mainstream economists largely bypassed these challenges. Feminist economists have argued that unpaid work and caring are critical parts of the economy, as it is directly or indirectly linked to production, consumption, saving, investment, labour supply, productivity, and so on, which are the concerns of the mainstream. The implicit assumption that unpaid work and care work and its categories are separate from the monetised economy and do not affect the monetised economy, however it is still central in the teaching and researching within economics.

Challenges of Feminist Economists

As mentioned above, core theories and ideologies that frame the subject of economics has been resistant to all criticisms, including that of feminist's economists. Thus, there are no significant changes in the core concepts and ideas of the subject even with many critical analysts both within and outside the discipline exposing the problems of scientific approach and the myths of objectivity which are considered as defining its superiority among other social sciences.

Feminist's analysis in economics have exposed the limitations of the framework which only deals with material reproduction. Further, feminist economists were successful in pointing out 'the inability of Economics to fulfill its desire of totality' (Khan 1993 cited in Keppler 1998[12]), thereby forcing it to accept non-economic arguments. However, there has not been any change in the framework and also in the related concepts to incorporate any of these concerns.[13] The assumption of an *'economic man'* whose choices and decisions are all governed purely by rationality is central to the study of economics even now, though the exclusion of women is often taken care by the assumption that economic man represents a genderless economic unit. Cult of domesticity of women which underlie the neoclassical notion is still the foundation of many theories and frameworks in general economics, but particularly in gender economics.

National income calculations are still governed by a separation of production and consumption with production defined in terms of market, excluding what many women are engaged in which are central to the economic survival of households, especially in many agricultural economies. How have methodological debates contributed to changes in the national income statistics and the sphere of production? The requirement for data that are required to compare countries led to the establishment of the United Nations System of National Accounts (UNSNA) in 1953 which guides national income calculations and thus the basis of the definition of production. UNSNA is used by countries to construct their measures of economic growth, the gross domestic product (GDP). All that activities which are performed for household consumption are not counted in UNSNA. The 1980 United Nations Women's Conference in Copenhagen in 1980 initiated the discussion on the need to account for the economic contribution of women in the national income data.

The resistance to changes in the accounting of activities was largely from the point of comparison over time and across countries. In 1993, UNSNA was revisited making clear distinction between production and consumption. The changes though acknowledges many domestic and personal services produced and consumed by households, these are included in the 'consumption boundary' and a distinction was drawn between SNA, extended SNA, and non-SNA with production strictly restricted to activities under SNA. Extended SNA includes all activities that members of household undertake outside the wage economy mostly for self-consumption and social reproduction. Though UNSNA is not binding on countries and a few countries have their own system of accounting, GDP figures are important not only for national comparisons and policies but they are also the basis by which international aid agencies such as the World Bank and International Monetary Fund (IMF) determine whether countries should receive aid or loans, how much, and on what terms. Thus, these calculations are binding to all countries, though seemingly it appears as a flexible option. As has been argued by May,

> The assumption that work, a fundamental ingredient in the management of the household, is not truly work unless done in a market place, is as much testimony to the pervasive influence of gender bias as it is testimony to the influence of the market mentality.
>
> (May 2002)

An alternative to incorporate the criticism of the UNSNA that the feminist's economists have negotiated is the proposal of having a satellite account which captures all those activities that accounts for unpaid labour (Waring 2003). The dual analyses where unpaid and care work are analysed as distinct from economic work meant for the monetised market economy are still normalised and accepted. This is despite the works of many feminist economists who have argued for the interconnectedness of the two spheres of work showing how the unpaid economy is linked to production such that markets cannot function without unpaid work.[14] A system of satellite accounts though have been seen as a solution, the fact is that this has not changed the invisibility issues of unpaid labour and thus the contribution of women in any serious manner with macroeconomic policies completely bypassing these concerns.

Family is the only institution that few economists like Becker (1981) have acknowledged. Even, in his work, to quote May, 'Becker's monumental contribution to our understanding of choice within the family reflects little more than the imposition of patriarchal assumptions on a distinctly patriarchal institution in society' (May 2002). Thus, economists have constructed an economic woman who is irrational and unpredictable which are opposite to the qualities of the economic man. However, the operation of these unpredictable women are in private and not in public domains where production and other economic transactions happen. On the contrary, women's liberation through economic independence is one of the key concerns of feminism, historically as well as in the contemporary context. Feminist demand for economic independence of women brought in related debates around social reproduction, its gendering, and the need for an integrated analysis of production and reproduction, as discussed earlier, which has also been ignored completely by the discipline.

Thus, feminist concerns are largely unheard in the larger discipline or even when it is acknowledged these concerns have been taken up in a sketchy way and sometimes as add-on variable. Economics continues to be one of the last social sciences to accept and integrate a feminist reading, as Diana Strassmann points out, 'core economic theories as yet fail to incorporate the insights of this [feminist] scholarship' (Strassmann 1993). Many economists do not even want to acknowledge criticisms which are probably considered as coming from lesser beings who are not interested in such serious economic issues and its analysis. Further, the radical notion of equality which defined many feminist approaches was clearly against the spirit of competition that defined existence of neoclassical analysis and thus the foundation of contemporary economics. It could be because of this reason that the contributions of feminist economists are rarely acknowledged or appreciated.

Though feminist economists have raised fundamental issues with regard to the mainstream teaching of economics, they are largely unsuccessful here too. No student of economics, even now, surely till a decade or more, hardly would have come through any gendered reading of economics or discussions on women or social groups as specific categories requiring separate analysis. In many institutions of economics, even now there is no acknowledgement of these categories as part of their regular course structure and even when they are referred to, they are dealt as variations to the normal.

Thus, feminist interventions have mostly been parallel and the mainstream has conveniently neglected the questions raised by feminist economists. Though feminist economists have made entries into many conventional areas of economics, it is striking to see that almost all publications with feminist concerns related to economic issues are mostly in the field of labour market studies. Feminist analysis is often taken as micro-level analysis as it brings aspects of women suggesting interdependence, thereby non-autonomous agency and identity. However, there are diversities among feminist economists with many working within the neoclassical tradition. Even those who do not identify themselves as a neoclassical economist and have exposed the problems with the standard notion of 'rational economic man' and thus need for alternatives may combine the idea of individual choice with interdependence and social determination.

To be fair to the contributions of feminist economists, though their interventions could not redefine some of the fundamental foundations of the discipline, these helped in exposing many myths that existed about the discipline. Since it was coming from those who are internal to the paradigms, this helped in posing fundamental challenge to the categories of discourse that mainstream economist based on neoclassical framework takes for granted. Household and women's contribution to the economy and the interrelationship between women's work and capitalist society, the fundamental feminist concerns, got a new life with feminist scholars extending the framework to include such dimensions. Conceptualisation of work to only those connected to the market was one of the main criticisms which exposed the neoclassical notion of productive activity opening up the debate around unpaid economic work, national income, and its accounting.[15]

In macroeconomics, branches such as national income, public finance, fiscal policies, and international trade are full of metaphors through which economists have maintained superiority. Feminist economists though have exposed and challenged the problematic conceptualisation of many of these specific branches, they have not been able to influence these branches in any major ways unlike their engagement with women's work. Overemphasis still exists in terms of traditional economic themes such as growth, finance, and trade in economics. This neglect of feminist concerns is evident in the composition of many institutions and organisations for the teaching and researching of economics. Even now many economics-centric organisations have very few women. But this is not, however, the reason for the unchallenged continuation of the subject. Even when women economists are present, many in their efforts to protect their dominance as acceptance of its weakness and subsequent revisions are thought to dilute the subject.[16] There is also an ongoing struggle among feminist economists resulting from the contradiction between the need to ensure scientificity/objectivity in feminist economic research and the underlying gendered social reality. Feminist economists have not been able to develop a coherent theoretical framework yet. This is very clear from the discussion on GDP, the components of GDP calculations such as unpaid economic or reproductive work.

Economists with a gendered understanding of issues are often blamed for their simplistic understanding of the subject and thus lacking rigorous economic analysis. This is evident in the number of women economists taking up women/gender studies as a field of specialisation. Many PhD students in economics-oriented institutions and universities are discouraged from taking up women studies as these issues are not 'considered economics enough'. Students from the economics stream rarely take up women studies as a subject of preference for higher studies as the subject is never taught in economics and thus often considered as a soft subject unlike/distant from their discipline, economics.[17] As Olson (2002) has commented, 'a talented woman economist would do better to be an expert in some other field in economics than in feminist economics' (Olson and Emami 2002).

The resistance of the discipline for a gendered reading of its theories and frameworks have distanced women studies scholars at large with very few feminist scholars engaging with the discipline. Given this larger context of a secondary status among the community of economists, the next section captures my experience of teaching and researching women's work from the perspective of feminist economics in a women studies centre. The account highlights the contestations, negotiations, and compromises arising out of disciplinary rigidities and orientations of women studies which is an interdisciplinary field of studies.

Neither Here nor There: Feminist Economists in Women Studies

There is no doubt that feminist economists who have critically looked at their discipline have contributed to the discipline of economics and also to women studies. Their contributions though are acknowledged in women studies, in the actual teaching of women studies, such contributions do not find sufficient space. Women studies has now emerged as a separate discipline not only at the international level but is also taught across the country with the emergence of women studies centres in many central and state universities. The initial position that women studies is a perspective and not a separate discipline is now no longer a matter of discourse and every year a number of students are entering university centres to study and research in the discipline of women studies. It is in the teaching of women studies where a serious methodological discussion is required as this determines/influences further research. In these emerging centres of women studies, scholars have mostly ignored economics and quantitative data and analysis because of two reasons. One is the absence of faculty from an economics background; the other and the most important is the dominant understanding that quantitative data is subsidiary or inadequate to study women's issues. This is particularly true as the discipline attracts mostly non-economics students who consider economics as an abstract social science with a heavy load of quantitative data. This puts economists like me in such institutions, who believe in the importance of an engagement with economic data, in a strange situation. Quantitative methods could not only enrich feminist analysis but critical feminist insights could also strengthen quantitative methods. Reflections on my journey as a researcher and related exposure to other methodologies may help in the understanding of following an open approach to methodological possibilities in women studies.

My training in economics got challenged once I joined the National Labour Institute where I was asked to specialise in women's work. My first few readings on women's work exposed the methodological limitation of my training in economics to study the subject. The theoretical and empirical writings on female labour within the conventional economics paradigm, though few, did generate a distinct discomfort arising out of the realisation of its narrow and partial engagement. The casual treatment of structural factors in the human capital theory which was an accepted model in the study of labour market segmentation was striking as I started relating it to real-life experiences. The more I read on writings from other disciplinary perspectives, the more the limitations of my training in economics became evident to study women's work. The work of Indian feminist economists (Jain and Chand 1982; Jain 1983; Sen and Sen 1985; Hirway 1991) questioning and exposing various data sources and arguing for alternative systems of data collection was encouraging.[18]

The empirical research work I carried out at the National Labour Institute on women workers in the garment industry, domestic workers, and women migrants further exposed me to the complexity of women's work decisions and in the understanding of gendered employment relations. These studies also helped to rethink my baggage with regard to scientific sampling and sampling procedures, survey methods, and tools. The near impossibility of following scientific sampling was evident, study after study, with the size and characteristics of the population remaining unknown or messy. The number of workers to be studied was decided not on the basis of any proportionate sampling but more by practical considerations, giving due attention to social factors. The questionnaire-based survey was designed, not to make it easily amenable to quantitative and statistical analysis but to bring in as much details, with many questions being open-ended. The difficulty in accessing women and the need to engage with them outside the questionnaire to capture many underlying personal and social considerations are lessons that one learned. More than the collection of data, the analysis was a challenge as it was clear that women's work cannot be analysed using any borrowed theory from economics with its simplified approaches. Contextual and structural variables were noted to be layered and interrelated, sometimes making it difficult to come up with only an economic argument.

Yet another challenge was in terms of analysing the macro-data on women's employment. Measurement is key to quantitative data and thus the assumption that governs the collection of data is that anything and everything is definable and thus measurable. Feminist economists have contributed much to the understanding of women's work and its measurement over the years. Invisibility of women's economic contribution spearheaded an interest among women studies scholars and economists in India, resulting in many methodological debates around measuring women's work. Unpaid work, both within and outside the production boundary, was the subject that attracted and continues to attract much attention in India. Over time, time use survey[19] has gained acceptance not only for visibilising women's economic contribution but also in engaging with house work social reproduction.

Data has become an important tool for feminist campaigns both for drawing attention to the specific issues of women's employment and also for campaigning for better sources of data. Data on women's employment, with all its limitations, have been important in highlighting the ongoing crisis in female employment, evident in the falling work participation rates of women. Further, the data has also helped in pointing to larger social dimensions of this decline in female employment. The gendered nature of the labour market, the class, and the social group dynamics of women's employment are also critical issues that the macro-data has helped to underline in the context of an overall economic change. The data is based undeniably on a market-oriented definition of work underlined by the deductive understanding of employment outcomes. However, an engagement with the data, exposing the gendered nature of employment, is important for feminist politics at large. My grounding in quantitative data analysis did really help in undertaking research on the analysis of macro-data, be it employment or migration, which I think is an important contribution of economists.

With this experience at a labour institute, my shifting to the Centre for Women's Development Studies (CWDS), which is the first and foremost women studies research institute, was an opportunity to work with women studies scholars from diverse disciplinary orientations. Though the small size of the faculty and the pressures of mobilising funding for research were concerns, the opportunity to work with a small team at CWDS and the possibilities of interdisciplinary dialogues helped in furthering my analytical skills. However, silos of specialisation exist even in such places, which provided little opportunities for methodological conversations beyond peripheral aspects.[20] When discussions on a collaborative MPhil/PhD programme were initiated after a few years of my joining CWDS, one was excited about the possibility of teaching an interdisciplinary subject with my orientation in economics. When the teaching programme was finally initiated in 2013, as a natural choice, I became part of the Research Methodology Course. This compulsory course was offered as part of the course work by a team of faculty from CWDS and Ambedkar University, Delhi.[21] The course was designed as a theme-based, reading-oriented one and given my specialisation and also being the only economist in the whole team, quantitative methodology and women's work was my share in the course.

With almost all students coming from either women studies, sociology, or literature backgrounds, teaching quantitative research methods was a challenge. The fact that not even a single student, after teaching the course for about five years, came from economics stream made the discussion on quantitative data and empirical research on women's work often a monologue. There is a general understanding among these scholars that quantitative research uses anti-feminist methodologies and is thus not relevant for women studies. The fear of engaging with numerical data was also another factor that led to an aversion. The challenge was to teach data and its reading explaining the specific issues of various concepts used, linking it to methods of data collection and analysis to completely uninterested group of women studies scholars. As the course was designed as a reading-based one, relevant and select literature on women's work were circulated much in advance to the interactive sessions that followed. Since discussion was primarily to address the methodological aspects of researching women's work and the importance of quantitative data analysis, one or two sessions were devoted to provide an overview of quantitative research and the various macro-data sources on women's work in India. The epistemological orientation of the concept of work as used in the data sources cannot be explained without discussing economics and its positivist orientation. These meant elaborate discussions on the concepts and definitions used to capture women's work and the distinction that the data makes between economic (productive) and non-economic (outside production) activities.

The major challenge was to engage with the stereotyped understanding that anything to do with quantitative data which has methodological orientation to positivists as non-feminist. By reading the works of feminist economists, a case for using quantitative methods to capture measurable aspects of women's life which help in describing and analysing spatial and temporal inequalities was also taken up for discussion. Writings of various feminist economists such as Nancy Folbre (Folbre 1994, 1991) and in the Indian context Devaki Jain, Saradamoni, Gita Sen and Indira Hirway (Jain and Chand 1982; Jain 1983; Sen and Sen 1985; Saradamoni 1987; Hirway 1991, 2008) and other scholarly work were used to expose the market-oriented understanding of work in data systems and its gender and other social biases.

The literature on visibilising women's work which is mostly entrenched in the measurement trap is drawn into these discussions to highlight the nature and relational dimensions of the issues of invisibility. Unpaid economic activities, which account for a good proportion of what women do in rural areas or in many informal sector activities, are often the entry point into these discussions exposing the possible errors in capturing women's work in its entirety. The relational dimensions and interconnectedness of women's work at large (paid/unpaid) are explored further highlighting the fact that these are mostly ignored by the data collection agencies in the actual framing of surveys. Through these discussions, a case is built for an active engagement with the data, its conceptualisation, collection, and analysis. The reading of and discussion on existing literature on women's work helped in suggesting the need to focus more on the research questions and the field contexts before deciding on the methodology of the study. The possibility of a mixed-method approach to women studies research which helps in linking the power of the general and their larger contexts to the specific diversities of women is discussed based on field-level studies. Though there are reluctant signs of some agreement on the usefulness of quantitative data[22] or for an integrative approach by the end of these sessions, in their individual research, engagement with any quantitative data is rarely noticed.[23]

Unlike my experience of teaching women's studies, in the larger group of women studies researchers, feminist economists are well-acknowledged and their work duly recognised. However, the larger acceptance and acknowledgement of economists' training to engage with quantitative data often leads to an othering of the few feminist economists. Many feminist economists are not located in women studies centres and hence the othering may not be a regular feature, though there could be other forms of academic tensions and negotiations. It is clear that not only are feminist economists isolated from the larger group of economists, but they are also not accepted as one among women studies scholars.

Conclusion

Though feminist economists are not homogeneous in their responses to the discipline, they are united in exposing the gendered approaches in economics and have made significant contributions to the subject. Feminist economics continues to challenge the status quo of mainstream economics by questioning the methodological foundations of mainstream economics and offering alternative explanations to economic outcomes by bringing women's experiences. This helps in exposing the methodological weakness of economics as a social science discipline. Cross-fertilisation of ideas among feminist economists and their engagement with other disciplines have resulted in novel combinations of ideas and methodologies. Through their interventions, feminist economists are able to increase the descriptive realism of economics highlighting the ambivalent relationship between description (historical or contemporary) of real-life and scientific objectivity in economic research. Feminist interventions have also made economics closer to social sciences in general, including women studies. Though women studies is an interdisciplinary area of investigation, the qualitative nature of most research in women studies is in contrast to the quantitative methods that characterise the work of many feminist economists. The feminist standpoint on quantitative data for advancing women's interest though is slowly getting accepted in women studies; however, it is yet to be integrated fully into the methodological discussions within women studies. Women studies, given its interdisciplinary profile as opposed to other traditional disciplines, has the potential to be the platform for methodological pluralities paving way for reconciliations amongst social science disciplines, including economics.

Notes

1 William Jevons and Alfred Marshal are the forefathers of this tradition.
2 There were diversities among countries and regions in teaching and also in economic practices before the domination of neoclassical economics. For example, in India, economics had much historical and sociological thrusts which got eroded (Rammohan and Ramakrishnan 2020).
3 Milton Friedman's in his 1953 article, 'The Methodology of Positive Economics', talks about falsificationism as the dominant methodology of economic research. This was and is still the only methodology that is taught in many departments and centres of economics teaching and researching, though not as ready-made prescription but as an ideal procedure.
4 The influence of positivist methodology and falsification, especially that of Popper (1959) is clear in the writings of Freidman.
5 Very few institutions or universities, both at the international level and in India, offer/teach courses on political economy or development economics, though now development and policy studies departments and courses have come up in many new private universities in India as an interdisciplinary subject.
6 Neoclassical approaches are referred to as closed systems, while most heterodox approaches in different degrees are open systems.
7 Marx, Keynes, and Veblen are some of the earlier heterodox thinkers with whom open-school approach was initially associated. For a discussion on open and closed systems in economics, see Chick and Dow (2005).
8 Austrian School of Economics, though a part of the heterodox school, is based on methodological individualism, which makes it closer to neoclassical economics.
9 Feminist concerns could be traced to the early stages of the development of economics, the origin of economics from the Greek word Oikonomia, meaning 'managing the household'.
10 Mary Wollstonecraft (1759–1797) was a moral and political philosopher whose analysis of women's economic position challenging existing social, economic, and political order is considered radical. Her writing, *Vindication of the Rights of Woman* (1792), is an important contribution to feminism.
11 The contributions of Indian feminist economists like Indira Hirway, Jayati Ghosh, and others on this debate is particularly striking.
12 Khan (1993). On the irony in/of economic theory, *Modern Economic Notes*, Volume 108 (759–803) as quoted in Keppler (1998).
13 There are no significant changes in the core concepts and ideas of the subject even with many critical analysts both within and outside the discipline exposing the problems of scientific approach and the myths of objectivity which are considered as defining its superiority among other social sciences.
14 Studies on the gender impacts of economic and financial crises show considerable increase in unpaid work, especially of women in order to tide over reduced household income during such periods (Beneria and Feldman 1992; Elson 1995; Floro and Dymski 2000; van Staveren 2002).

15 The work of Indian feminist economists such as Devaki Jain and Indira Hirway are important interventions.

16 However, feminist economists do not fall into one category, though they all have differences with the mainstream. Dolfsma and Hoppe (1996) classify feminist economists into three streams: (1) feminist constructionists who are concerned with bringing gender dimensions into economics theories, methods, and value judgements; (2) affirmative action–based feminists who do not embrace the position of constructionists and are largely concerned with women's representation and thereby improving women's overall economic position by incorporating gender within existing frameworks; (3) the third stream, feminist empiricists, is also uncritical of the discipline and holds the position that economic theories are gender neutral and objective and women's lower economic status are because of wrong application in economics.

17 It is assumed that economists who specialise in women studies have shifted to the subject because of their inability to cope with the specialised demands of economics, which are difficult for any average student. This also result in the elimination of such economists from all 'economic forums' as they are assumed to be now incapable of comprehending discussions on core issues of the subject such as growth, macroeconomic policies, business, and trade.

18 The contribution of Indian feminist economists cannot be underscored even at the international level.

19 Time use survey, which records activities of women for different time slots during fixed number of hours, usually 24 hours, emerged as an alternative to Census and NSS data.

20 I was the only economist at CWDS and was mostly consulted by colleagues, to give inputs on data sources, quantitative data collection, and analysis.

21 The programme was in collaboration with the Gender Studies Centre of Ambedkar University, Delhi.

22 The realisation that many women studies scholars are ignorant and uninterested in the declining work participation of women, which was one of my topics of research, was a major eye-opener and also a source of disappointment in the course of my teaching. The only topic that interested research scholars was sampling methods, as many are anxious about the number of women/respondents that they should be studying in their own research work.

23 Many students, as I understood, found the course not having much relevance in a research methodology course for women studies scholars, where scholars are from non-economics background.

References

Agarwal, B. 1997. "Bargaining' and Gender Relations: Within and Beyond the Household.' *Feminist Economics*, 3 (1): 1–51.

Agarwal, B. 2001. 'Economics and Other Social Sciences: An Inevitable Divide?' *Contributions to Indian Sociology*, 35 (3): 389–399.

Becker, Gary. 1981. *A Treatise on the Family*. Cambridge: Harvard University Press.

Beneria, Lourdes, and Shelley Feldman. 1992. *Unequal Burden: Economic Crises, Persistent Poverty, and Women's Work*. Boulder: Westview.

Bergmann, Barbara. 1995. 'Becker's Theory of the Family: Preposterous Conclusions.' *Feminist Economics*, 1 (1): 141–150.

Chick, V., and S. Dow. 2005. 'The Meaning of Open Systems.' *Journal of Economic Methodology*, 12 (3): 361–381.

Davis, Kathy. 2008. 'Intersectionality as Buzzword: A Sociology of Science Perspective on What Makes a Feminist Theory Successful.' *Feminist Theory*, 9 (1): 67–85.

Dolfsma, W., and H. Hoppe. 1996. 'The Challenges of Feminist Economics.' *Freiburger FrauenStudien*, 2: 59–72.

Dow, S. C. 2000. 'Prospects for Progress in Heterodox Economics.' *Journal of the History of Economic Thought*, 22 (2): 157–170.

Elson, Diane. 1995. *Male Bias in the Development Process*. Manchester: Manchester University Press.

England, P. 1993. 'The Separative Self: Androcentric Bias in Neoclassical Assumptions.' In M. Ferber and J. Nelson (eds) *Beyond Economic Man: Feminist Theory and Economics*, pp. 37–53. Chicago: University of Chicago Press.

Floro, Maria, and Gary A. Dymski. 2000. 'Financial Crisis, Gender, and Power: An Analytical Framework.' *World Development*, 28 (7): 1269–1283.

Folbre, Nancy. 1991. 'The Unproductive Housewife: Her Evolution in Nineteenth-Century Economic Thought.' *Signs*, 16 (3): 463–484.

Folbre, N. 1994. *Who Pays for the Kids? Gender and the Structures of Constraint*. London: Routledge.

Folbre, Nancy, and Heidi Hartmann. 1988. 'The Rhetoric of Self-Interest: Ideology and Gender in Economic Theory.' In Arjo Klamer (ed.) *The Consequences of Economic Rhetoric*, pp. 184–204. Cambridge: Cambridge University Press.

Fourcade, Marion, Etienne Ollion, and Yann Algan. 2015. 'The Superiority of Economists.' *Journal of Economic Perspectives*, 29 (1): 89–114.

Friedman, Milton. 1953. *Essays in Positive Economics*. Chicago, Chicago University Press.

Grapard, Ulla. 1995. 'Robinson Crusoe: The Quintessential Economic Man?' *Feminist Economics*, 1 (1): 33–52.

Hirway, Indira. 1991. *Women's Work in Gujarat: An Analysis of the 1991 Census Data*. Ahmedabad: Gandhi Labour Institute.

Hirway, Indira. 2008. 'Expanding Statistical Paradigm for Macro Policy Formulation.' In *Towards Mainstreaming Time Use Surveys in National Statistical System in India, Ministry of Women and Child Development*. New Delhi: UNDP and World Bank.

International Confederation of Associations for Pluralism in Economics (ICAPE). 2021. 'ICAPE's History-ICARE's Original Statement of Purpose.' https://icape .org/?page=icape_history (accessed on 28th October 2021).

Jain, Devaki. 1983. 'Co-Opting Women's Work into the Statistical System: Some Indian Milestones.' *Samya Shakti: A Journal of Women's Studies*, 1 (1): 85–99.

Jain, Devaki, and Malini Chand. 1982. *Report on a Time Allocation Study – Its Methodological Implications*. Technical Seminar on 'Women's Work and Employment, 9–11 April, Institute of Social Studies Trust, New Delhi.

Jennings, Ann L. 1993. 'Public or Private? Institutional Economics and Feminism.' In M. Ferber and J. Nelson (eds) *Beyond Economic Man. Feminist Theory and Economics*, pp. 111–130. Chicago: University of Chicago Press.

Kabeer, N. 2001. 'Family Bargaining.' In Neil Smelser and Paul Bates (eds) *International Encyclopedia of the Social and Behavioural Sciences*, Vol. VIII, pp. 5314–5319. London: Elsevier Press.

Katz, E. 1997. 'The Intra-Household Economics of Voice and Exit.' *Feminist Economics*, 3 (3): 25–46.

Keppler, Jan Horst. 1998. 'The Genesis of 'Positive Economics' and the Rejection of Monopolistic Competition Theory: A Methodological Debate.' *Cambridge Journal of Economics*, 22 (3): 261–276.

Khan, M. A. 1993. 'Economics and language. '*Journal of Economic Studies*, 20 (3): 51–69.

Kuiper, E. 2001. *The Most Valuable of All Capital. A Gender Reading of Economic Texts*, Tinbergen Institute Research Series, No. 244, University of Amsterdam, Amsterdam.

May, Ann Mari. 2002. 'The Feminist Challenge to Economics.' *Challenge*, 45 (6): 45–69.

Nelson, J. 1996. *Feminism, Objectivity and Economics*. London: Routledge.

Olson, P., and Z. Emami. 2002. *Engendering Economics. Conversations With Women Economists in the United States*. London: Routledge.

Popper, K. R. 1959. *The Logic of Scientific Discovery*. London, Hutchinson, the English Translation of Logik der Forschung. 1934. Vienna: Julius Springer.

Rammohan, K. T., and A. Ramakrishnan. 2020. 'The Making of Economics as a Discipline in India: Universal Theory and Local Tensions.' *Orissa Economic Journal*, 52 (2): 4–26.

Saradamoni, K. 1987. 'Labour, Land and Rice Production – Women Involvement in Three States.' *Economic and Political Weekly, Review of Women Studies*, 22 (17): WS-2–WS-6.

Sen, Amartya. 2005. 'Human Rights and Capabilities.' *Journal of Human Development*, 6 (2): 151–166.

Sen, Gita, and Chiranjib Sen. 1985. 'Women's Domestic Work and Economic Activity, Results From the National Sample Survey.' *Economic & Political Weekly*, 20 (17): WS49–WS56.

Strassmann, Diana. 1993. 'Not a Free Market: The Rhetoric of Disciplinary Authority in Economics.' In M. Ferber and J. Nelson (eds) *Beyond Economic Man. Feminist Theory and Economics*. Chicago: University of Chicago Press.

Van Staveren, Irene. 2002. '*Social Capital: What is in it for Feminist Economics?*' Working Paper 368, ORPAS - Institute of Social Studies, The Hague, The Netherlands.

Waring, Marilyn. 2003. 'Counting for Something! Recognising Women's Contribution to the Global Economy Through Alternative Accounting Systems.' *Gender and Development*, 11 (1): 35–43.

11

METHOD, OBJECT, AND PRAXIS
Marx and the Historians of Science

Rahul Govind

Introduction

Marxism has engaged with the question of scientific method in its historical investigations, while 'historians' of science, from Bachelard to Foucault and Popper to Feyerabend, have used the historical method to contest philosophical paradigms on the very nature of science, scientific method, and its advancement. This chapter argues for bringing together Marxism and these historians by proposing that both 'methods' critique a certain understanding and image of science: an understanding and image that is ascribed a logical structure which simultaneously exhibits scientific advancement and history as an essentially progressive force perfecting itself in a linear and cumulative chronology. This subordination of time and its contingencies to logical structure is often sought to be replicated in the social sciences and history which take science as a paradigmatic method. In such an image, the object, whether nature or history, remains a self-subsistent whole characterised by perfectly consistent laws which merely have to be discovered by the human mind.

While Marx critiques the science of classical political economy on these grounds of attempting to subordinate history to the laws of the economy, historians of science critique conceptualisations of science where the laws of nature – logic as world – are revealed in a cumulative history of advancement. Both these approaches critique an axiomatic separation of method (or theory) and nature (or history). Together they see scientific 'progress' as a process where the subjective method and objective material, theory and practice, form in effect an indivisible whole with a concrete criterion of validation that cannot be derived from either any 'immediate' observation (or historical document) or *a priori* theory (a logic independent of its application or formulation in specific material). They recognise that as indivisible wholes, theories (or methods) will have anomalies or countervailing instances (or anomalous facts); such instances do not in themselves suffice to disprove or undermine method or theory. They would themselves have

DOI: 10.4324/9781003298908-18

to be integrated in a coherent theory that provides a better explanation, in practice, of the problem at hand, to undermine any existing theory. In this sense, both approaches require and recognise a method in the human sciences, in the minimal sense that research can be improved upon, defended, and acted upon; unlike more contemporary forms of relativism that see in anomalies nothing but the impossibility of method and theory, thereby, contradictorily, deriving a negating value from an anomaly, for a destruction (of 'theory'), that has already been accomplished *a priori* ('theory' denial at the outset in presupposing that one can begin without a theory).

The following section discusses these problems through a reading of Marx and two traditions of the history of science that evolved independently: an 'Anglo-American' tradition embodied in the work of K. Popper, T.S. Kuhn, and P. Feyerabend along with a 'French' tradition embodied in the work of G. Bachelard, G. Canguilhem, and M. Foucault.[1] However, it will be argued that these two traditions articulated a shared set of theoretical problems on the nature of knowledge and knowledge of nature as much as the relationship between method, logic, and time: problems that were also of direct concern for Marx and Marxists. The focus will be on delineating and weaving an argumentative thread of these specific problems and responses to them, so as to better shed light on the stakes and promise of the historical method.

Karl Marx: Praxis as Progress in the Metabolic of History and Nature

This section explores Marx's critique of an incorporation of a certain model science conceived of as the study of set laws, whether of nature, history, or historical periods. It takes the following steps. Firstly, Marx employs the science of history as a critique of the science of classical political economy and its scientific laws, i.e. the labour–capital relationship described as a natural phenomenon (or law) expresses a particular history. Secondly, history serves the purpose of diagnosing scientific laws as a concealment of time/change as well as a concealment of the interests of those who gain by treating the laws (of the present) as immutable. Thirdly, the present condition of 'free labour' is diagnosed as an alterable present which tells the story of the human subject who is characterised as an active social subject. That is to say, history and its periodisation are intelligible in terms of such an active (social) subject rendered legible through a double lens: labour as purposive activity and the social valuation of labour. Finally, this allows for a particular interpretation of history that is at odds with a standard reading of Marxist history as divided into a deterministic periodisation, with each period being subject to a larger structure of historical laws.

An entry point into Marx's reconfiguration of science and history could well begin with Lukacs's critique of Engels[2] and what he argued was

Engels's conflation of scientific practice and praxis (Lukacs 1971: 131–133; Lukacs 2000: 94–137). Engels writes of 'inner general laws' in and of history and nature, where the subjective and objective appear as mental cognition and physical nature (or history), respectively. The task was to discover these 'general laws' and 'dialectics [is] the conscious reflex of the dialectical motion of the real world', 'the general laws of motion, both of the external world and of human thought'.[3] The direct comparison between human action and natural motion and the acknowledged success of the natural sciences suggested that the same method ought to be approximated by the human sciences. In such a context, the task is the cognitive one of 'discovering' and 'applying' laws. For Lukacs, on the other hand, far from praxis, the scientific practice was caught in the antinomy of creating artificial conditions so as to correlate the 'material substratum' of phenomena to its own mathematical reason. 'Matter' was rendered legible only in the mathematico-epistemological language of the scientist, while the duality of matter and mathematical language remained; what was derived was not 'nature' but (artificially produced) scientific laws. For Lukacs, praxis did not operate like science with the irredeemable antinomy of scientific practice and nature (and its laws), where the latter is systematically and progressively unravelled. Praxis was a process and procedure, a dialectic, in which what appeared as subjective and objective are moments in an unfolding historical process. The changing correlations between method (scientific language) and its object are historically produced and cannot be derived from either nature or the human mind, treated as separable and separate entities.[4]

Marx had already written, 'the weakness of the abstract materialism of natural science, a materialism which excluded the historical process are immediately evident from the abstract and ideological conceptions expressed by its spokesmen whenever they venture beyond the bounds of their chosen speciality' (Marx 1990: 494). This would, as we will see, require a thorough reconceptualisation of the human cognitive subject as human social activity. In the context of Marx's writings, what appear as objective scientific laws might well be traced to a particular subjectivist-epistemological position, the law of value, revealing but a capitalist perspective and determination. But conceptualising the law as subjectivist is not so much a negation of the objective laws but the normative articulation of the possibility – and need – to change what appear as objective conditions.

A crucial dimension of Marx's critique of the classical political economy – incarnated in its most developed form in Ricardo – was its inability to conceptualise labour as 'actively' value producing and not merely labour-time (Marx 1990: 173–177; Marx 2010: 43–50, 114).[5] In merely discerning the law of value, the 'bourgeois' political economist treated the present state of political economy as though it consisted of 'natural laws', 'eternal', and therefore independent of time and action (Marx 2010: 97, 112–113; Marx 1993: 85–88). The 'bourgeois' law of value was in one sense true as an accurate description of the workings of the contemporary political economy, but equally untrue in masking their contingency by taking them to be eternal. This was linked to an understanding of labour which was based on conceptualising the human individual as a self-sufficient trans-historical category unit: in principle independent of society (Marx 1993: 83–85). By contrast, for Marx, individuation takes place only in society, and so characterising the former requires an understanding of the latter.

How is one to understand the 'deception' of such a law of value as to be found in Ricardo? It was true that under the conditions of the contemporaneous economy, humans had possession of their own person; they were neither slaves nor serfs nor members of a 'communal mode of production'. So they were 'free' and independent in one sense as recognised by classical political economy. A law of value was in operation and was reflected in the system of wages and profit. Yet this was partial since it did not recognise the contradictions *both* in the 'objective' conditions of capital as well as the 'subjective' condition of the human being as free worker and also organic interrelations in which the two emerge. Marx forcefully argues that, in fact, the individual was free in another sense – apart from the (self)ownership of his person. He was also free of objectivity; i.e. he lacked in the ownership of the objective conditions that would have enabled him to live (Marx 1990: 270–280). Under these conditions, he would thus be required to sell the only possession he had, labour power, so as to survive. This argument goes on to speak to the possibility of altering and overcoming this present condition and therein voiding the present law of value which is treated by the political economists as eternal as well as necessary. Marx therefore finds it epistemologically essential to historicise. Documenting a historical emergence establishes the possibility of transcendence. Yet considering the 'naturalised' and ideologically entrenched tendency to view the human individual in abstract – 'dot like' – isolation, Marx has to first fundamentally clarify his conception of the human individual.

In *Capital*, Marx characterises the labour process as an activity, a 'metabolic exchange' with nature, characterised by purposive unity in the shape of the will (Marx 1990: 283–284). The human body as much as the earth qualifies as his organs, and neither he nor Nature remain unchanged in the labour process; nature is as much material as instrument (Marx 1990: 284–286). This feature is 'common to all forms of society', across history (Marx 1990: 290). This is one standpoint, but there is another equally essential standpoint from which labour has to be examined, i.e. the valorisation process. Under conditions of capital, labour power and the means of production (past labour) combine as 'things' producing for the capitalist. The product is the capitalist's.[6] *Capital* stages the valorisation process in terms of the free/unfree exchange between labour (without objective conditions such as the means of production) and capital (possessing the means of production as past labour well as the labour power which it buys). Political economists observe valorisation without conceptualising the (active) labour process and therefore devise laws that they consider 'natural' and 'eternal'.

It is in such a context that one can speak of historical method. Unlike the naturalisation of the present, Marx's method involves historical investigations which demonstrate not only that capitalist forms are historically constituted but that this allows for its transcendence by organised action and sufficient will.

> Just as on one side the pre bourgeoisie phases appear as merely historical i.e. suspended presuppositions so do the contemporary conditions of production likewise appear as engaged in suspending themselves and hence in positing the historical presuppositions for a new state of society.
>
> (Marx 1993: 461)

The past is revealed only in the recognition of the contradictions within bourgeois society, in the 'self-criticism' of the present (Marx 1993: 106) and any naturalisation of the present, in the form of the eternal laws of political economy, blocks any understanding of the past by treating it as a necessary line cumulatively culminating in the now, obeying 'scientific' laws.

Defining the human in terms of his metabolic exchange with nature, Marx demonstrates that the conditions of capital emerge when humans lose access to their objective conditions of production, including land or the earth as their workshop. The chapter on so-called 'primitive accumulation' documents the violence with which this severance was politically organised and accomplished; composed of politics, history does not unfold as the necessary (Marx 1990: 873–940). The conditions in which primitive accumulation and the centralisation of power were to have taken place involved 'communal property', which 'lived on under the cover of feudalism' (Marx 1990: 885). 'Communal property' formed both a historical and an epistemological standpoint from which to critique capital (Anderson 2014), and must not be seen some romantic exemplar to be imitated.

Capital and the *Grundrisse*, as well as the late *Critique of the Gotha Programme*, had underlined the fact that the active unity between man and nature as characteristic of human activity also required his self-conception – as well as the objective condition – of membership in a community where life is reproduced and lived. This self-perceived membership is 'subjective', while the relationship with one's bodied nature is treated as 'objective'; human life historically 'metabolises' in their active unity (Marx 1993: 490). The analogy with language illuminates the 'subjective' and 'objective'. Humans relate to their own distinctive voice only in a language or that while it is shared. In the same way, property, including 'ownership' of one's body as much as any other instrument used for life which creates the objective conditions for its reproduction, is one's own only in the membership of a community and subject to its laws (Marx 1993: 471–472).

By focusing on activity and work as a metabolic between himself and nature, Marx delineates a normative as much as a communal dimension that is treated as essential. Purposive work is as much a material as a norm establishing activity that is necessarily 'communal' or 'social', and therefore the reference is neither to the empirical individual nor to society as such, which is always arrived at through abstractions and generalisations. Notwithstanding the great deal of attention in Marxist circles to the serial staging of history, in the *Grundrisse*, in a densely abbreviated form, Marx's diagnosis of the contemporary condition of capital takes the form of a history of the *separation* between the 'inorganic conditions of human existence' and his 'active existence'(Marx 1993: 489) or in other words between capital and labour, where the later has been deprived of his inorganic conditions, his means to live life independently, the necessary objective conditions. Slavery and serfdom are episodic modifications of the conditions of 'communal property' where one section or class of society is treated as the 'inorganic natural conditions' of the other, the former having lost even possession of their own labour. In *Capital* Marx had argued that 'communal property' lived under the 'cover of feudalism' even after the period of serfdom. Primitive accumulation, as the violent class and state-led attack on, and destruction of, such 'communal property' conditions the emergence of capitalist production (Marx 1990: 876–904). From here, the late Marx's investigations into Russia as much as his anthropological studies on 'communal forms' as a possible point of departure for contemporary critique and socialist imaginings are not necessarily anomalous.[7]

Periodisation and time is the terrain of struggle, and each phase of the past (and present) moves in an uneven and contradictory way; contradictions concealed in the enunciation of abstract, 'natural', or 'eternal' laws. In fact, as Marx explains, the establishment of the 'normal working day' and the *law* of work hours is a 'product of a protracted and more or less concealed civil war between the capitalist class and the working class' (Marx 1990: 412–413). Categories such as the peasantry or slavery are meaningful only in terms of the relations within which they 'work' and the struggles that map history (Marx 2010: 104, 1998: 126–129).[8] There are fundamental differences on the kind of obligations of abstract categories such as 'peasants' ('rent' or 'debt', or a combination) and slaves (colonial slavery is not a residue or relic of the past but the 'pivot' of bourgeois industry). What the young Marx had diagnosed as the juridical promise of freedom is in itself revealed as the capitalist conditions of 'free labour' (Govind 2015: 514–525).

The method of presentation must differ from that of inquiry. The latter has to *appropriate* the material in detail, analyse its different forms of development, and to track down their *inner connection.* Only after this work has been done can the *real movement* be appropriately presented.

The '*life* of the subject matter is now reflected back in the ideas ... and we have before us an *a priori* construction' (Marx 1990: 102) (emphases mine).[9] The historical critique of the present acquires force only when what presents itself as 'natural' is shown to be contingent. Perceiving the worker–capitalist exchange as embodying the 'scientific' laws of exchange, population and value found in 'bourgeois' economics is revealed to be a habituated ideologically coloured vision that occludes the histories of expropriation as much as actively demands the accomplishment of equality and freedom as norms. Historical 'method' therein proves to be a moment of praxis.

Karl Popper and Gaston Bachelard: The Endurance of Concepts as Social Practice

In this part, we re-explore the dense entanglements of method and object in a different idiom, i.e. the debates around the history of science. The year 1934 saw the publication of both Karl Popper's *Logic of Scientific Discovery* and Gaston Bachelard's *New Scientific Spirit.* Both these texts and figures helped initiate a constellation of problems, concepts, texts, and figures in the shape of two distinct traditions, the 'Anglo-American' and the 'French', of writing the history of science. It will be established here that there is a remarkable set of affinities that both texts share as well as – not surprisingly – the constellations spawned.[10] Hence, Bachelard and Popper are here treated together, before treating the distinct traditions separately in the form of two pairs: Kuhn–Feyerabend and Canguilhem–Foucault.

For both Bachelard and Popper, scientific practice and scientific progress could not be articulated or explained either in terms of a logic (abstracted from specific practices) or a cumulative history of results (that were meant to continually approximate to a defined truth). Since 'immediate' sense perception or observations were themselves intelligible only from a conceptual system, they could not be independently taken as definitive proofs or negations of 'theories'. On the other hand, theories were themselves intelligible and demonstrable only in terms of their applications, real or imaginary (Bachelard 1984: 5; Popper 2012: 464–467). Drawing on the science of the early 20th century as well as the history of science, both Bachelard and Popper showed that specific rationalities which mixed theory-formation and experiment and did not obey any larger cumulative historical law or logical structure. In this sense, they would both agree with the Marxist argument that scientific practice is inherently social and as a form of practice, there is no *a priori* logic or nature which is to be described and explained; scientific growth required 'inter-subjective' experiments (Popper 2012: 22–26)[11] for

Popper while Bachelard wrote about the 'social aspects of proof' (Bachelard 1984: 12). No grander independent laws of nature could be delineated outside of scientific practice. The history of this practice – rather than any putative logic – demonstrated its character.

Popper's critique of logic takes the form of a critique of what he perceives to be a nexus between inductive logic and the idea that sense-perceptual experience is independent of – and in turn verifies – scientific theories (Popper 2012: 74–88). The problem of induction as the problem of moving from singular cases to universal theories assumes that such cases can be legible without any theory. In contrast to this, Popper argues that all such 'cases' and even sensory experience is legible only in the light of a theory; there is no 'immediacy' (Popper 2012: 22, 312–318). The experience of 'this glass of water' requires a 'universal' understanding of glass and water, i.e. universal in the sense that one has conceptions of glass and water in terms of their law-like behaviour that always exceeds the 'immediate' perception of a particular glass of water (Popper 2012: 76). The critique of induction is a critique of the collapsing of scientific and everyday common knowledge, for what characterises the former is the possibility of the 'growth' of knowledge. If knowledge was merely the bringing together of specific instances, one could not differentiate forms of knowledge from each other. In such a case, knowledge of any sort is merely verified by a corresponding instance or observation, everyday knowledge being no different from scientific theory. On the other hand, scientific knowledge grows, has clear criteria for such growth, and cannot be either confirmed or denied merely by an instance of observation.

In certain ways, Bachelard's critique of sensory perception is even more radical, a critique that is thrown into relief by what he calls his critique of Cartesian epistemology. In the second meditation, Descartes had argued that the changing appearance of wax in different circumstances – whether close to or away from the fire – meant that the senses were deceptive and one needed an 'intellectual seeing' to grasp the 'idea' clearly and distinctly. Bachelard counters by arguing that science's 'progressive objectification' means that one would have to forgo the Descartian ideal of a clear, distinctive, unified, and simple idea like 'wax' (Bachelard 1984: 166–171). Scientific experimentation will break up this 'unity' into a range of fields and subfields, each 'reality' discovered will be a function of a rational-experimental method. So what the Cartesian takes as 'wax' conceals the different molecular structures and spectrograms, the differentiation between the surface structure and the deeper structures, each of these possibly leading to more and more complex directions that have shown the 'fleeting wax' to have but been the non-scientific mode of sense-perceiving. In his study of mathematical spectra, Bachelard again argues that one cannot employ a Cartesian method that moves from the simple to the complex. Scientific research revealed that the hydrogenated image far from being simple was

in fact a 'simplified' instance of the Alkaliniod spectrum. So while historically one moved from hydrogen to the alkaliniod, the former's complexity
was better understood only through studying the latter (Bachelard 1984:
149–154). Microphysics showed that the object cannot be divorced from
motion, and so a concrete description from ordinary experience has to be
replaced by an abstract mathematical description.

While both Popper and Bachelard thereby critique any notion of an immediate experience or observation, they speak of the progress of science as 'radical' and 'revolutionary', distinguishing it from ordinary forms of knowledge.
In such a context, Popper argues for falsifiability (rather than verification) as
a criterion for scientific knowledge which is itself characterised by growth
– or the possibility of growth. Merely verifying a theory by a corresponding instance is insufficient, because the latter instance is itself theory-laden.
It is falsifiability that characterises a theory because falsifiability sketches
the scope of the theory in terms of claims and deductions. Statements or
theories – such as 'there are black ravens' – cannot be called scientific since
it is impossible to refute and negate such a statement. However, far from
dismissing 'empiricism', Popper brings the empirical and experience as the
experiment into the heart of scientific practice. In his terms, scientific laws lie
at the conjunction of universal statements ('whenever a thread is loaded with
a weight exceeding that which characterizes the tensile strength of the thread
it will break') and the initial conditions consisting of a particular statement/
case (this thread of 1 lb has something weighing 2 lb put on it and the thread
breaks) (Popper 2012: 38). The strictly universal statement or scientific law
is intelligible through particular empirical conditions, but it is impossible
to fully verify it by collating together numerically the indefinite number of
instances where an object with a greater weight than the thread is put on it.
Thus, though in conjunction with particular instances, no single instance
of the latter establishes its truth conclusively and scientific truth cannot be
established as such; this was the mistake of the inductivists. The success of
the theory lies both in the repeatability of the phenomena and the consequences that may be deduced through various forms of experimentation;
experimentation that itself might be triggered by attempts at falsification.

A countervailing instance is insufficient to negate the scientific theory,
since it requires to be framed in turn a falsifying hypothesis, itself falsifiable
and having its own corroborating instances (Popper 2012: 66–67).[12] This
reiterates Popper's critique of a 'raw' independent sense data that can serve
either as a verifying or negating instance. Critique therefore has to take the
form of an alternative hypothesis that in its turn describes and explains phenomena to a greater extent and understanding than the previous hypothesis;
it is insufficient to merely point to an 'anomaly' to overturn a scientific theory. The truth of a scientific theory is thus validated and expanded only if it
is falsifiable and therein continuously tested giving rise to newer and newer
deductions and knowledge. Its truth lies in these series of 'experiments' and

not in approximating to a transcendent trans-historical logical structure. The nature of such experimentation as constitutive of the truth of scientific practice exhibits the fact that it is not a question of a scientific practitioner discerning a pre-existing logical structure of the world or its eternally opera-tive laws. Popper argues that the scientific truth of the theory is established by continuous testing – enabled by its characteristic falsifiability – and its value is not determined by psychology and individual motivations or pre-dilections (Popper 2012: 23–24, 80–81). Intersubjective experiments in the establishment of the truth of the scientific theory have no place for such phenomena (Popper 2012: 23–26).

Bachelard too insists that mere 'anomalies' do not disprove a scientific the-ory even as he recognises that scientific knowledge does indeed grow. Certain experiments and their results were never enough since they had to *justify* their opposition (Bachelard 1984: 9). And like Popper, he too is dissatisfied with the 'conventionalist' argument that all anomalies can always be explained by making suitable readjustments to the theory or adding auxiliary hypotheses. A complete refutation would thus have to be theoretical. Science was revolu-tionary because there were many moments in science where one had to 'start afresh' (Bachelard 1984: 173). In this discussion, Bachelard anticipates the dis-cussion around incommensurability in Popper, Kuhn, and Feyerabend. In his study of Newtonian mechanics and relativity, he argues that while the experi-mental results could approximate to one another, the basic concepts such as mass and velocity in Newton and Einstein were incommensurable. Of course, this did not prevent a reconstruction of the Newtonian system within the theory of relativity (Bachelard 1984: 48–49, 54). Yet this was the 'revolution-ary' if paradoxical nature of scientific change. When several inconsistencies and anomalies forced a relook, a new synthetic theory emerges in conjunc-tion with the old and reverses many fundamental notions. The scientific spirit could only be captured in these leaps where none of the words/terms – such as force – meant the same thing anymore. This was a fundamental change and not incremental. Bachelard too resorts to language to explain this shift where it was not a question of the change of the meaning of the word while preserving the syntax, or changing the syntax and preserving the terminologi-cal definition (Bachelard 1984: 53). The change was more fundamental. The tenacity of the older model is also beautifully captured by a poetic image; in a reverie, we might spot a unicorn in the clouds as the clouds change, and when we are shaken awake, we can no longer see the unicorn; perhaps the tenacity of the unicorn is what Kuhn will later call 'normal' science.

Thomas Kuhn and Paul Feyerabend: The Nub of Progress as Conceptualising Contingency

Kuhn follows Popper in the critique of 'immediacy' (Kuhn 1996: 125), the necessary interrelations between theory and practice (Kuhn 1996: 52), and the non-cumulative nature of scientific advancement in general. However, explicitly using a history of science, Kuhn makes a distinction between 'normal science', which proceeds 'cumulatively' (Kuhn 1996: 52), and revolutions or paradigm shifts, which are not cumulative (Kuhn 1996: 84). This is contrasted to Popper's conceptualising of scientific change as proceeding through falsifying hypotheses. 'Normal science' is a 'puzzle-solving practice' which proceeds through resembling or modelling what are considered paradigms, which they either further articulate or extend but never question. Even while normal science involves the discernment of anomalies or discrepancies, these are worked upon rather than seen as evidence of any deficiency in the current paradigm.[13] Anomalies may be attributed to deficiencies in the particular scientist's practice (his experimental methods, circumstances, derivations and inferences, etc.) rather than the inadequacy of the current paradigmatic theory (Kuhn 1996: 80). The paradigm functions like a common law decision (Kuhn 1996: 23) which is further refined and articulated in the practice of normal science. A practice whose unity does not lie in the conscious application of rules but is better described in terms of Wittgenstein's 'family resemblance' (Kuhn 1996: 45). It is only where anomalies remain, and greater interest is taken in relation to them, leading to further research, that in time another alternative to the current paradigm may present itself. This is seen as a 'crisis' where there is a re-evaluation of fundamentals, and in certain quarters a paradigm shift occurs. Kuhn terms this 'revolutionary' because such changes are non-cumulative and cannot be accounted for with reference to either a neutral observation language or a pure logic. Like Bachelard, Kuhn argues that for specific purposes and by certain measures, the succeeding paradigm does progress and the succeeding paradigm can reconstruct its predecessor in its own terms even if the concepts of each paradigm would all the same remain incommensurable: mass in classical mechanics and relativity (Kuhn 1996: 99–101). Kuhn denies that the successive paradigm approximates more closely to an objective truth than its predecessor.

But how is one to explain this alteration of paradigms? It is here that the difficult problem of the 'growth' of scientific knowledge reappears since Kuhn argues that paradigm shift has to be attributed to the particular scientific community's choice to accept the new paradigm or remain with the old (Kuhn 1996: 150–159). While much of Kuhn's critique is in terms of characterising scientific growth as something that happens without an explicit (Popperian) falsification or 'testing', the distinction both between normal science and revolutionary breaks (and the nature of the latter) is attributed to 'ideology' or the scientific community's choice, which does not contain any trans-historical logically verifiable criteria. That ordinary scientific practice works with anomalous observational instances is what makes it difficult to come up with a global rational formula in terms of which a paradigm shift can be measured as progress. In response to Popper's argument that 'incommensurability' was disproved merely by the fact that one could learn a foreign language, Kuhn states that the problem of incommensurability lay not so much in learning a language as in translation. He argues, in almost Benjaminin–Pannwitzian fashion, that in translation, neither the original nor the receiving language remains untouched and no meta-language – neutral observation language – is resorted to. That it requires effort and produces a new idiom points to the challenges involved in incommensurability, which is not meant to signal a mental paralysis.

Feyerabend is simultaneously critical of both Popper and Kuhn, but at the same time 'dialectically' transfigures their arguments. He does believe that science 'advances', but not through the Poperrian principle of falsifiability which he doesn't find evidence of in the history of scientific practice (Feyerabend 2002: 172–179). At the same time, he is critical of what he sees as Kuhn's sequencing of normal science and revolution, treating them as though they were distinct temporal moments (Lakatos and Musgrave 1970: 207–214). From the history of science, he argues that at *any* given moment, there existed multiple competing theories (or paradigms) with their own corresponding observational facts-fields. It is not merely that 'facts' are theory-laden, but rather that the refuting facts of one theory may become available only because of another theory; thus plural theories existing at the same time also allow refutation of each other through the unearthing of new facts through experiments. All of this is again proven through historical examples such as the refutation of Brownian motion through Einstein's use of kinetic theory (Feyerabend 2010: 17–24). For Feyerabend, Kuhn's characterisation of 'normal science' as the success of a single paradigm over a period of time was neither true in history nor desirable as a norm. His own pluralist model at a given moment veers towards Lakatos and Popper (Lakatos and Musgrave 1970: 119), but he moves over to Kuhn's side by emphasising the crucial importance of 'incommensurability' and the implied lack of any global notion of global progress. For Feyerabend too, the history of science presented situations where the greater predictability of consequences did not entail a commensurability of concepts such as the concept of mass in Newtonian mechanics and relativity as argued by Kuhn and Bachelard (Lakatos and Musgrave 1970: 219–222).

Moreover, any cumulative history of science was negated by the fact that classical mechanics did not definitely negate Aristotelian arguments regarding space and matter, and the former could be established as closer to more current quantum notions (Feyerabend 2002: 239). Arguing that any past theory could prove helpful in the advancement of science, Feyerabend writes that it was only through a detailed study of the past – or any existing marginalised tradition – that one could rid oneself of the ideology of progress which blindsided one into believing that there existed a single rational model that propelled all scientific progress. This shielded science as much as history from democratic scrutiny and would be no more than a version of the victor's justice. Both Kuhn and Feyerabend join in their critique a cumulative reading of science by pointing to the fact that each 'advance' brought up problems unknown to previous theories, while the history of science was also littered with abandoned problems, which might only be taken up and resolved much later in different contexts. This essential complexity in science seems to resemble Bachelard's argument regarding the duality of science in each moment of its practice (Bachelard 1984: 15–16), but Bachelard also veers towards a notion of science where the current paradigm reconstructs its past and therein provides new insight therein too (Bachelard 1984: 54–55). There is no study showing the vitality of a past science which can inform and reorient its successor.

It is precisely such a historical set of studies that Feyerabend undertakes, studies which do the work of both voiding retrospective illusions of progress and proving the genuine and irreducibility multiplicity of science in terms of method and value. A historical example he often returns to is that of Galileo's refutation of Aristotelian notions where it is argued, for instance, that the conceptual vitality of Aristotle's conceptualisation of matter and motion that stands up to Galileo's critique, and still provides fundamental insights (Feyerabend 2002: 219–247). In the face of the dangers of single theories becoming dominant, and such dominance edging out competitors, Feyerabend argues for a hedonism and multiplicity that he sees completely missing in Popper's 'critical rationalism' and Kuhn's description sequencing of 'normal science' and 'paradigm shifts'.

However, Feyerabend goes much further in drawing the larger implications of his critique than anything that may be found in Popper, Bachelard, or Kuhn. He argues that science itself should not be given any special sanctity, pointing not merely to the imperialistic tendencies of dominant scientific models, but also to their theoretical and practical impoverishment and their alienating effects (Feyerabend 2010: 27–33, 223–265, 2002: 90–162). Rather than a general scepticism or relativism (a label he rejects),[14] it is in the advocacy and practice of detailed work which questions the dominant paradigms with regard to their rationality as well as the general values of progress that they lay claim to that Feyerabend seems to distinguish himself.

Georges Canguilhem and Michel Foucault: Normative Multiplication and the Rebellion of Norms

Notwithstanding important differences, Canguilhem's work provides several intersections with that of Feyerabend. For him too, history operates as a 'laboratory' for epistemology and several theories operate simultaneously and in contestation. While the latter is a feature that Canguilhem takes explicitly from Bachelard, he also attributes a lack of a genuine heterogeneity to Bachelard because the latter worked primarily on mathematical physics; the 'epistemological recursion' at work in mathematical physics was not as evident in the biological sciences (Delaporte 2000: 33). There is no plea for hedonism or anarchy, but all scientific research in biology is meant to take biological normativity, the fact of the human being as a being that establishes norms, as its point of orientation: an argument that we will elaborate below. This 'normative structure' as a methodological principle – even if reduced to the treatment of constitutions of units like enzymes and genes (Delaporte 2000: 214) – of life and death, success and error finds little equivalent in physics and chemistry (Canguilhem 1991: 127–129). After a discussion of Canguilhem, we will examine the creative treatment of his legacy in the much more familiar work of Michel Foucault.

Rather than any putative object or nature, the history of science, as distinguished from the practice of science, is seen by Canguilhem, as a history of an 'axiological activity' (Canguilhem 1991: 30). The nature of crystals is the science of crystals (crystallography, crystal optics, and inorganic chemistry), but the object of a history of science is essentially 'incomplete' (Canguilhem 1991: 28) and should not be the mere documentation of the history of the results of science – the terminology again is close to Popper, Kuhn, and Feyerabend. To treat the history as a chronicle of results would be to determine the past in terms of present standards, taken as true and eternal; and so history would be nothing but a continuous self-determination of a motorised or immortalised present (Delaporte 2002: 41–42). It is in this light that Canguilhem carries out his detailed historical studies where there is no self-evident or given relationship between the present and the past, and one moment of the past and its predecessor. Mendel's break would be obscured if we focused merely on the general problem of heredity rather than the precise fact that he was concerned with the problem without looking for a specific agent of transmission as was previously done, thereby redefining the concept of 'character' (Canguilhem 1981: 22–23). This closely anticipates the discussion of the difficulty of pinpointing the 'discoverer' of oxygen, oxygen (Kuhn 1996: 53–56) being more akin to a 'project' rather than a naturally defined 'object' (Canguilhem 1991: 41).

History serves as 'epistemology's laboratory' and shows how ideas even as they were discredited did play a role in the development of the science (Canguilhem 1991: 45). Astrology was at some point not considered scientific, but it all the same provided a whole set of observations that were crucial to the development of astronomy (Canguilhem 1981: 21). And the development of any science always modifies an existing theory and cannot be immediately derived from facts (Canguilhem 1991: 161–164), as though the latter were themselves outside of a theoretical framework. While all of these problems are shared with those studying the history of physics and chemistry, Canguilhem argues that the biological sciences have their own specific objects (or 'projects'), and one cannot transplant mechanical laws from the other sciences into the sphere of biology. In fact, *The Normal and the Pathological* is a critique of just such an attempt which tries to unify the normal and the pathological basing their distinction merely on quantitative variations. In such an approach of examining quantitative variations, the norm is scarcely distinguished from the average (Canguilhem 1991: 122).

By way of critique, Canguilhem argues that this unification of the normal and pathological in terms of laws does not give us any precise criteria with which to distinguish the normal and the pathological, or even define the 'normal' (so as to in turn define the pathological), since both refer to each other. More strongly put, no amount of quantitative data and 'objective' research would yield anything without a notion of the human being as himself a norm-establishing being (Canguilhem 1991: 126–127). He attempts to establish that the normal emerges only through the consciousness of infractions, and the norm is ultimately derived from the individual (or the patient) and cannot be derived abstractly in terms of uniform laws (Canguilhem 1991: 118, 126). Whether an arm is now 'working' or has fully recovered is determined by the individual and not with mere reference to an objective research on averages since the norm includes a (variable) sense of what ought to be rather than merely an average determined by research. At what point the average transits into the pathological varies from situation to situation. Even physiological conditions whether glycaemia, calcaemia, or blood PH are affected by social and environmental factors from hygiene to altitude to nutrition (Canguilhem 1991: 162). Human practices such as meditation abolish physiological distinctions between striated and smooth muscle systems (Canguilhem 1991: 165–168). Thus, an 'objective' normality of the physiological is itself intelligible only in terms of a more fundamental biological normativity. This parallels Marxist arguments in Lukacs and Trotsky that deny that class consciousness or conditions that are meant to be opportune for revolutionary action can be measured in any statistical or objective manner, and would have to account for concrete actions, organisation, and events.[15]

The significance of anomalies introduces a particular dimension of time and therefore cannot be determined from their distance to averages in any particular bracket of time. Biological normativity exists at the intersection between environmental conditions and the living being with neither term in itself being either pathological or normal. The haemophiliac can lead a perfectly normal life if an environment without injuries can be designed and the benefits of vestigial wings in the Drosophila depends on whether it lives in a closed or open environment (Canguilhem 1991: 141–142). Laboratory experimentation and the resulting data are seen as so crucial to objective research themselves effect and alter the bodies/organs being examined and therefore undermine 'objectivity' (Canguilhem 1991: 147). Canguilhem wryly comments that one does not have to wait for wave mechanics to understand the indeterminate relationship between the procedures and results of knowledge (Canguilhem 1991: 147).

For Canguilhem, the psychic anomaly (insanity) is not fundamentally distinguished from the somatic illness, because in both cases, health (biological normativity) is always constituted with reference *through* diseases and anomalies. If the norm is an 'ordering structure', error is not a correctable feature of judgement, but germane to the morbid life form itself, which is constantly struggling in the establishment of norms (Canguilhem 1991: 278). Foucault's *History of Madness* pushes the problem of biological normativity to the extreme, by studying the heterogeneous historical alterations through which discourses both constitute – and are marked by – the object 'madness' (as madmen). Like any of the historians of science that we have discussed, he refuses to take the psychiatrist's history of his own discipline as a transparent rendition of 'madness'; as though there existed a unified object, the disease 'madness', whose obscurity was slowly unpeeled by a history of the progressive genius of psychiatry and its heroes. Rather, Foucault shows that the object called the madmen and the disease madness, did not have a unified history and converged only at a late moment of time.

The object presented as the mad in their madness to Pinel, Tuke, and Wagnitz in the Hospital General and houses of correction in the early 19th century had been herded and confined to that place already for over a century and a half. In the Classical Age, the 'mad' and 'madness' were not a distinctive object but perceived and treated within the horizon of the disorderly poor: the poor who needed to be taught through gruelling punishment the harsh values of labour and be proscribed from 'infecting' the peace of the cities. Foucault meticulously documents the mixing of various kinds of peoples subject to confinement – from the subjects of venereal diseases to those out of the bonds of morality, to others considered 'mad'. This 'Great Confinement' in the middle of the 17th century Europe that appeared to have affected much of Europe, itself indexed a shift in the conception and figuration of madmen. Descartes sets the trend by excluding the mad so as to prove the value of reason through an elaborately orchestrated doubting process. The earlier resemblance and exchange between reason and madness, found in figures like Montaigne, led to a 'critical consciousness' where the mad were no longer treated as a privileged sign of the cosmos in a religiously suffused consciousness, a sign febrile with meaning, as was the case previously (Foucault 2006: 21–43). We thus see Foucault's minute attention to a range of sources revealing a heterogeneous set of fragmented figures and functions – through time and space that is only roughly captured by the inherently fragmented category of the 'mad'. Only their physical confinement in the 17th century began a process that ended in the constitution of madness and the mad deposited by the violent forces of history for the psychiatrist to train his gaze: a gaze which obliterated any sign of its object's dispersed (pre)history.

Foucault therein refutes the standard history of the origins of rational psychiatry. While the confinement of the mad with others took place under the horizon of a religious framework of punishment for the poor and the idle, even within this aggregated population the mad did have a faintly distinctive profile (Foucault 2006: 55–108). In a rough and necessarily schematising way, one can say that from the 17th century onwards, with and through confinement, the magistrates and the juridical establishment decided whether the 'mad' would be confined, influenced by social pressures, by families, and by neighbours. Thus. social perception played an important role in the confinement of the mad in hospitals and correction houses.

While in the 18th century there was a detailed taxonomic medical analysis of madness, such an analysis did not converge with concrete space where the 'mad' (Foucault 2006: 393–394) were confined. The separation of the mad from other populations in the late 18th century took place because the newer colonial, plantation, and industrial economy required labour. A new form of thinking arose which criticised the earlier policy of confining and enforcing work and strenuously argued for the release of the poor who, it was argued, should be unfettered in their search for work. This was to help the new economy in reducing prices and at the same time save the financial burdens that were involved in the earlier policies regarding the confinement and regulation of work. It was this insertion of the poor in the economy that left the mad as a residue for the psychologist's gaze. That a separation took place had to thus be attributed to the new economic conditions and correlative discourses, and this was what left the 'mad' as an object (Foucault 2006: 398). It was only at this point that a convergence took place between medical analysis and the actual examination of the mad in the confined institutional space in which they were interred (Foucault 2006: 381–419). Neither humanity nor science as trans-historical phenomena could explain the constitution of such an object.

The detailed tracking on specific norms – like the nosology of medical analysis or the determining double framing of madness through the febrile and the imbecilic – and their intermeshing with domains today conventionally considered separate (medicine and jurisprudence) remain Foucault's signature throughout his work. Madness is the painfully fragmented site through which Foucault irradiates the complex movements of discourse, actions, and events until it becomes hollow in which psychiatry deposits its analyses. What does such a study do to Canguilhem's 'biological normativity'? It becomes increasingly difficult to disentangle the traces of the subject from that of method, the discourse from its own mutations, norms from sub- and super-norms. Yet Foucault insistently writes, not to reveal the mad in its 'rawness', but so as to trace and account for the silencing in the psychiatry that claims to study it. In later work, such as *Discipline and Punish*, the detailed tracing of discourses and events may well form images of the individual that are targeted through specific apparatus, whether they be disciplinary or sovereign; such a form of the individual cannot be confused with individualism as a trans-historical value of freedom. On the other hand, discourses and techniques may target what are called populations, a form of power that acquires particular significance in modern times. Norm formation continues to inform the late Foucault's work on antiquity, which have to be described in scrupulous detail within their contexts, contexts that constantly spill through the leaking crack of time. Historical study once again challenges that which presents itself as natural, whether it be 'madness', 'crime', or 'pleasure', disturbing the present through a uniquely scrupulous rummaging of the 'repressed' past.

Conclusion

The conviction of sense certainty often provides the aura of presently experienced, ideologically determined categories and their laws, science and natural laws, the laws of the market, the laws of progress. That the sensorially perceived involves concepts therein detonates any easy route to confirming or refuting theories through observations or sources. Marx's conceptualisation of praxis – and vice versa – negates both the world as an independent static structure and the mind as self-subsistent, independent of body and society, allowing one to *see* the 'free' worker, as in turn, the exploited and revolutionary worker through a historical method that is at once normative and political. Such active forms of conceptualisation are indeed 'play-acts' (Feyerabend) and 'normative' (Kuhn) where 'objects' are 'projects' (Canguilhem).

The critical discussion in this chapter may well shed light on – and contribute to – certain current discussions on historical method. If abstract laws pose one threat, the claimed exhibition of nugatory detail or presumptions of cultural and political identities poses another. Historians at times argue for the refutation of historical conceptualisation by providing what they take to be empirical counter-instances or the irreducibility of 'culture' to 'theory', as though no theoretical act was required to define a culture. Such arguments themselves depend on dubious theory-evidence, theory-experience or theory-identity divides and assume theories to be necessarily complete and whole ready to break down in the face of anomalous 'evidence' or what is construed to be an innate immaculate cultural or political identity. These problems assail all too quick critiques of 'European theories' as inherently inadequate to 'Indian experiences' therein leaving un-clarified the nature of predication (that leaves no 'fact' untouched) as well as casting 'theory' and 'experience', of whatever sort, with the pseudo-rigid, impermeable fact-like aura of the sensorially perceived. Secondly, on the question of details and sources, the critique of 'verification' is a salutary reminder that 'verification' as such does nothing to help us understand phenomena. Put into question are all those newer approaches to history writing that believe historical understanding is advanced merely by the uncritical exhibition of newer sources, whether literature or the age-old anthropological interviewing that has recently taken the guise of 'oral history'. To emphasise, 'newness' becomes a way of speaking of the source as uncontaminated (concept free), thereby rendering redundant conceptual work. To speak of something as 'concept'-free is to render impossible teaching and learning, research and its progress, communication and reflection, enquiry and solidarity, transformation and self-transformation, in fact, the world as we know it.

In the context of scientific practice and society at large, Feyerabend goes furthest, amongst his peers, in pitching his scholarship as an argument for the greater 'democratic control' over scientific activity. Despite Kuhn's emphasis on paradigm shifts and community interests, the social function or purpose of science is left unthematised. This reticence on the larger social function or purpose of science allows for the spell of science creates an aura of invincibility around the laws of the economy, the nature of progress and scientific research for the world at large charmed by a success that it takes to be self-evident. That much of this science is enabled by an economy of funding and capital completely outside the direct participation or questioning from those it is meant to serve spurs thought to act. In places where millions lie on the threshold of bare life, the issue is urgent. Here Feyerabend touches the Marxist argument that scientific research in any form has to be conceived as a social category and no 'nature' or law awaits a description of itself as a logical structure; no amount of 'objective research' can nullify, or exclude, the demanding norms of freedom and justice that give history both dimension as well as life. Revealed, with Foucault in agreement, is the fact that the juridically free worker, the mad, and the criminal are not simply functions of the disciplinary matrices of economics, psychiatry, and law, but historical precipitates and historical actors; any given time (order) or 'a priori' logic could well be hammered into as much as out of joint.

Notes

1 Notwithstanding the affinities that we will explore between these distinct traditions, there is little cross-referencing. Kuhn interestingly refers to E. Meyerson but not to Bachelard when listing figures he found influential, in the preface of the *Structure of Scientific Revolutions*. Foucault states that though he doesn't refer to Kuhn in his *Order of Things*, he refers to Canguilhem who he argues influenced Kuhn. See "Foucault Responds to Steiner". Kuhn doesn't ever seem to acknowledge such an influence, though Canguilhem appears to agree with the criticism of Kuhn that the latter ultimately reduces conceptual issues to matters of sociological, *Vital Rationalist*, 46. In relation to Marxism, while Popper is explicitly critical of what he sees as Marxist determination (or historicism), Feyerabend has favourable, if scattered remarks, on Marx, Lenin, and Trotsky. See, for instance, *Against Method*, 105-7.

2 See also Marcuse's critique of Engels, *Soviet Marxism*, pp. 142-6, and the subsequent influence on Stalin and, in turn, a range of Marxist writers. More recently, Banaji attributes the same 'vulgar' conception of a theory – that merely required application to independent 'facts' – to Stalin and other Marxists. While exempting Engels from this conception, he critiques influential Marxist historians such

as Perry Anderson, and their standardised law-like, linear periodisations. See Jairus Banaji, *Theory as History*, 9, and 45-49.

3 See Engels, *Ludwig Feurbach and the End of Classical German Philosophy*.

4 A theoretical amplification of this dialectic with 'nature' can be found in Sartre's *Critique of Dialectical Reason* I, emerging from the arguments of 'need', 'scarcity', and historical intelligibility (notwithstanding the famous confusion between Engels and Marx, p. 27), Deleuze's analysis of Eisenstein and montage in *Cinema* I as well as the whole field known as 'ecological Marxism'.

5 See, for example, Laurent Baronian (2013) on 'active' and 'living' labour in his *Marx and Living Labour*.

6 Not so much to the individual capitalist but the capitalist as the 'bearer' of 'class-relations and interests'; Marx, Capital I, 92

7 For an important and scholarly argument on the importance of non-Western societies for Marx's mature conceptualisation, notwithstanding having earlier retained forms of Eurocentrism, see Kevin B. Anderson, *Marx at the Margins*.

8 For an elaboration, see Katz (1992), "Class in Itself or Class Struggle".

9 See the remarks on method, in the *Grundrisse*, 100-108.

10 See note 1 above.

11 Notwithstanding Popper's critique of Marx in *Poverty of Historicism* and *The Open Societies and Its Enemies*, the chapter will argue that he shares with Marx more than might meet the eye.

12 This is important in the context of Kuhn's criticism of falsification even though Popper is not named, *Structure of Scientific Revolutions*, 77.

13 These arguments are substantiated by historical examples, such as when Kuhn tells us of Newton's laws were not questioned despite 'long recognized discrepancies between the predictions from that theory and both the speed of sound and the motion of mercury'. See *Structure of Scientific Revolutions*, 81.

14 Feyerabend repeatedly resists being characterised as a relativist, a position that he finds all too 'abstract' and not allowing for systematic research and argumentation. See *Against Method*, 283-9.

15 See Lukacs, *History and Class Consciousness* 46-83; Trotsky, *History of the Russian Revolution*, 29-43a.

References

Anderson, K. B. 2014. *Marx at the Margins*. New Delhi: Pinnacle Learning.

Bachelard, G. 1984. *The New Scientific Spirit*. Boston: The Beacon Press.

Banaji, J. 2013. *Theory as History*. Delhi: Aakar Books.

Baronian, L. 2013. *Marx and Living Labour*. New York: Routledge.

Canguilhem, G. 1981. 'What is Scientific Ideology.' *Radical Philosophy* 129: 20–25.

Canguilhem, G. 1991. *The Normal and the Pathological*. New York: Zone Books.

Delaporte, F. (ed.). 2000. *A Vital Rationalist: The Selected Writings of Georges Canguilhem*. New York: Zone Books.

Feyerabend, P. 2002. *Farewell to Reason*. Verso: London.

Feyerabend, P. 2010. *Against Method*. Verso: London.

Foucault, M. 2006. *History of Madness*. New York: Routledge.

Govind, R. 2015. *Infinite Double*. Shimla: Institute of Advanced Study.

Katz, C. 1992. 'Marx on the Peasant Struggle: Class in Itself or Class in Struggle.' *The Review of Politics*, 54 (1): 50–71.

Kuhn, T. S. 1996. *The Structure of Scientific Revolutions*. Chicago: The University of Chicago Press.

Lakatos, I., and Musgrave, A. (eds). 1970. *Criticism and the Growth of Knowledge*. Cambridge: Cambridge University Press.

Lukacs, G. 1971. *History and Class Consciousness*. Cambridge, MA: The Merlin Press.

Lukacs, G. 2000. *A Defence of History and Cass Consciousness*. London: Verso.

Marx, K. 1990. *Capital* (Volume I). New York: Penguin.

Marx, K. 1993. *Grundrisse*. New York: Penguin.

Marx, K. 1998. *The Eighteenth Brumaire of Louis Bonaparte*. New York: International Publishers.

Marx, K. 2010. *The Poverty of Philosophy*. New Delhi: Peoples Publishing House.

Popper, K. 2012. *The Logic of Scientific Discovery*. London: Routledge Classics.

Trotsky, Leon. 1980. *History of the Russian Revolution*. New York: Pathfinder.

12

PSYCHOLOGY IN INDIA
Knowledge, Method, Nation

Sabah Siddiqui

What is intelligence?

Intelligence is what intelligence tests measure.

What sounds like a joke was offered very seriously by Edwin Boring in a paper titled *Intelligence as the Test Tests It* (1923). Boring, an experimental psychologist, was writing about the difficulty of defining intelligence, a scientific concept developed by the discipline of psychology. Almost a hundred years later, there are several definitions of intelligence characterised by multivarious dimensions of understanding the phenomenon. Nonetheless, Boring's early attempt at an objective psychological definition of intelligence encapsulates the methodological dilemma of the discipline.

Psychology is located between the objectivity and subjectivity debate, a space of limbo that defines most methodological debates. In particular for this subject, 'the theoretical discussions are as old as the discipline itself and are in good health' (Narciandi 2019: 31). The self-definition of the discipline has been rewritten at every turn as a new set of psychologists have taken upon themselves to construct a synchronous area of study and method of study. Hence, the question of what to study and how it has changed so frequently; it appears to me injudicious to hold psychology to one definition now. At present, psychologists explore and interrogate the psychological basis of being human, from behaviour to neurocognition, from consciousness to the unconscious, from normal to pathological (and even the paranormal), and from individual to social. This forced binary listing is not exhaustive, but to summarise the subject matter of psychology is a tall task, and I will not attempt that. The spread of the psychological dimensions of being human is too broad, but also this is a discipline with an aptitude to psychologise everyday life and thus ingest further dimensions of what we understand to be human. Today we have subdisciplines as diverse as the psychology of war and the science of happiness. Applications of psychology can be seen in areas from business and HR to mental health, from

DOI: 10.4324/9781003298908-19

advertising to AI, from sports to criminality, and from spirituality to army morale. Over the last 150 years of its existence as a specialised discipline, what characterises psychology is its push towards a positivist methodology that provided the initial impetus to a fledgling area of research and the pull of critical theories that became its most prominent critique in the latter half of the 20th century.

Constructing an *Introduction to psychology*

With a history of less than 150 years, the discipline of psychology is so new that the record of its progress is well-preserved. However, as we can see from the historical vignette in defining intelligence that I presented above, definitions of broad ideas are both reductive and unsatisfactory. Fortunately, rather than attempting a definition that can satisfy adherents of all schools of psychology, *Introduction to Psychology* is also an undergraduate course taught in all Indian universities that does the same by way of pedagogy. The curriculum of this course constructs a definition that is spread over many months. In fact, *Introduction to Psychology* is a core course in psychology taught in universities around the world. I have taught this course over the last decade in different universities in India, and the syllabus of the course has retained the 'core' modules. Thus, the syllabus of this undergraduate course can serve as an elaborate definition of psychology, especially since this is how psychology students are inducted into the discipline.

According to the University Grants Commission's[1] proposed syllabus in 2015 for *Introduction to Psychology* for BA (Honours), the objective of this course is 'to introduce students to the basic concepts of the field of psychology with an emphasis on applications of psychology in everyday life'. In the same year, the UGC's introductory course for the BA (Programme) is titled Foundations of Psychology with a similar objective and syllabus; the objective of this course is 'to understand the basic psychological processes and their applications in everyday life'. The UGC syllabi for both streams have four units. Table 12.1 lists the syllabus from both UGC streams that serves as the introductory course to psychology.

The first module in both syllabi is an introduction to the field with a focus on method. Within the first couple of hours of this course, psychology gets introduced as the science of behaviour and cognitive processes (Baron and Misra 2015). In fact, this was my own introduction to the field as an undergraduate student two decades ago, with the same textbook by Robert Baron, albeit an ancient version. This emphasis on science has been noted as psychology's insistence on producing measurable descriptions of observable individual differences. Thus, right from the start psychology gets placed within the fold of a positivist scientific method with an insistence on experimentation and statistical analysis. The subsequent units in the syllabi

Table 12.1 The Introductory course on psychology in the UGC syllabi

UGC BA (Honours) syllabus	UGC BA (Programme) syllabus
Introduction to psychology	Foundations of psychology
1. Introduction: What is psychology? Perspectives on behaviour; Methods of psychology; subfields of psychology; Psychology in modern India	Unit 1: Introduction: Psychology as a science, perspectives, origin, and development of psychology, psychology in India, methods; experimental and case study
2. Perception: Perceptual processing, Role of attention in perception, Perceptual organisation, Perceptual sets, Perceptual constancies, depth perception, distance and movement; Illusions	Unit 2: Cognitive processes: Perception – nature of perception, laws of perceptual organisation; learning-conditioning, observational learning; memory – processes, information processing model, techniques for improving memory
3. Learning and motivation: Principles and applications of classical conditioning, operant conditioning, and observational learning; Learning strategies; Learning in a digital world; Self-regulated learning; Perspectives on motivation, types of motivation, motivational conflicts	Unit 3: Motivation and emotion: Motives: biogenic and sociogenic Emotions: nature of emotions, key emotions
4. Memory: Models of memory – Levels of processing, Parallel Distributed Processing model, Information processing, Reconstructive nature of memory; Forgetting, Improving memory	Unit 4: Personality and intelligence: Personality: nature and theories Intelligence: nature and theories

touch upon topics such as perception, learning, memory, motivation, and emotion. In the case of the BA (Programme) in psychology, personality and intelligence are also included in the syllabus. It is obvious both from the textbook definition and by scanning the syllabi that these topics are positioned as behavioural and/or cognitive processes.

This course teaches new entrants to the university the origins of the discipline from structuralism and functionalism in the late 19th century to the developments of artificial intelligence (AI) and neuroplasticity a hundred years later. In the march of this progress, the course lays out the journey of psychology as a scientific endeavour from the early methods of introspection of Edward Titchener and psychoanalysis of Sigmund Freud to experimental methods of John Watson and computational modelling of cognitive neuroscience. The history of the discipline is narrated as the movement towards an objective, reliable, and universal science of human behaviour and cognition. Through the pedagogic exercise of the *Introduction to Psychology* course, quantitative and statistical methods of generating psychological knowledge are privileged over qualitative and constructivist methods. Psychology is thus taught to every new entrant to the field as falling short of scientific rigour only due to a lack of the will to commit to the scientific method. The insistence of scientific rigour, involving some exposure to both biology and statistics, comes as a surprise to many students, who believed that this would be the place in the university where they could discuss what is moral or evil behaviour, while developing the abilities to read faces and futures. The introduction of psychology as a scientific enterprise comes as a surprise, and some young scholars turn away, either towards a discipline with a more established claim on the scientific method or towards one with a 'softer' emphasis on biology or mathematical reasoning.

A Very Short History of the Discipline

The Scientific Method

As mentioned before, psychology is a discipline caught between objectivity and subjectivity. The particular conundrum here is specific to the historical problem of laying out its methods, the route to constructing scientific theories, and applying theories to real-world contexts. On the one hand, psychologists are aware that they deal with human phenomena that are spread over a wide range of variables and responses. On the other hand, psychologists are required to capture this wide range of responses according to the scientific principles that emerge from the strict rationality of 'the scientific method', which is an idea already in the want of a better definition. José Carlos Loredo Narciandi suggests it is not the scientificity of psychology that needs to be interrogated, but rather the idea of the scientificity of the scientific method when he writes that 'it is not evident that there is such a thing as the scientific method' (2019: 31). However, psychology has worked with the scientific method in a discipline-specific way that cannot be generalised to the idea of science in other disciplines. This chapter will take a look at the trajectory taken by psychology in developing its notion of method, validity, and rigour.

The history of the discipline is usually traced to the first experimental lab established by Wilhelm Wundt in Germany dating back to 1879, and thus what distinguished psychology from philosophy was the experimental method. 'The ideas of which psychology seeks to investigate the attributes, are identical with those upon which natural science is based' (Wundt 1896: 2). Wundt is making a case for relating psychology to the natural sciences, rather than philosophy, and his scientific method conducted experimental introspection on sensory data humans acquire through the subjective activities of feelings, emotions, and volition, which he deems to be 'immediate reality'. Edward Titchener brought this method to the United States, where psychology started developing at an accelerated pace, especially after the experimental introspection method was criticised as not scientific enough, primarily by the behaviourist John Watson.

'Psychology as the behaviorist views it is a purely objective experimental branch of natural science. Its theoretical goal is the prediction and control of behavior. Introspection forms no essential part of its methods' (Watson 1913: 158). Watson brought a lasting change to the field of psychology when he discarded all unobservable, subjective phenomena for what could be empirically located and directly studied: human and animal behaviour.[2] He brought to the fore the insistence on uniformity in experimental procedures, and it is impressive that his ideas carry greater weight today as *the* method of psychology than it did more than a century ago.

If behaviourism set the subject matter of psychology, another behaviourist expanded its scope by bringing language into the picture as linguistic behaviour or 'verbal behaviour' (Skinner 1957). B.F. Skinner used the experimental data he had collected on operant learning processes in humans and animals to interpret language learning processes. However, this was an interpretive method rather than an experimental one:

> I could not find experiments for the greater part of the analysis. I was still the empiricist at heart, but I did not think it would betray that position if my book were not a review of established facts. I was interpreting a complex field, using principles that had been verified under simpler, controlled conditions.
>
> (Skinner, cited in Michael 1984: 364)

It was a move away from the strict experimental method of Watson's behaviourism, and it led to a critique that has been credited for the rising prominence of the cognitive turn in psychology.

Within two years of its publication, Skinner's *Verbal Behaviour* drew a response from Noam Chomsky. In a review paper, Chomsky decries the limitation that behaviourists impose on themselves by solely studying observable behaviour when

> [o]ne would naturally expect that prediction of the behavior of a complex organism (or machine) would require, in addition to information about external stimulation, knowledge of the internal structure of the organism, the ways in which it processes input information and organizes its own behavior.
>
> (Chomsky 1959)

Chomsky is a linguist, not a psychologist, and there were cognitive-minded psychologists before his critique of behaviourism, such as John Piaget However, with Chomsky's response to Skinner's behaviourism, cognitivism became a force in the 1960s and connected psychology to explorations in

(AI) and cognitive neuroscience, introducing methods such as computational modelling and neuroimaging. Although cognitivism makes psychology take heed of internal structures, the method of psychology remains anchored in new forms of empiricism that continue to be dominant today.

The Medical Method

Alongside the discussion on how perception, memory, or language are developed and understood, there is another strand of psychology that defines it in both scope and method. That is its alignment with psychiatry and the medical model. This area of psychology is concerned with psychopathology and its treatment. While psychiatry has located human pathology or illness within physiology, psychology has looked at pathology as abnormality, as a (statistical) deviation from both reality and normality. 'Abnormality' requires the conception of normal individuals that make up the core of society. Psychopathology constructs abnormal as deviance and insanity, afflicting those members of society unable to 'fit in' or 'bear up' to the pressures of ordinary life. Nikolas Rose (1996) traces the development of the idea of the normal individual in the 19th century. It instigated the construction of therapies of normality that must bring individuals to a greater sense of freedom and autonomy, which can only mean inclusion in or adaptation to the average world of the normal individual. Thus, psychologists took upon themselves the task to develop technologies of the self (Foucault 2007) that police the borders of normality.

Historically, psychology's association with medicine is strongly referenced in the founding of psychoanalysis by Sigmund Freud, a medical doctor interested in psychological processes underlying hysteria while working with Josef Breuer (1895). Freud constructs the idea of the unconscious (1915) to work with clinical symptoms that are both stubborn and repetitive, and thus comes up with the method of free association that is allowing the individual to speak without constraints or censorship. This is the beginning of the talking cure, where speech becomes the central unit of analysis (rather than the behaviour of behaviourism). While psychoanalysis is moving away from physiology to psychology, the medical method persists by locating the problem within the individual.

> The individual in medical approaches is assumed to be responsible for aiding the process of cure and then to accept diagnosis and medication in suffering silence. The psychoanalytic version is that it becomes the responsibility of the individual to speak as they seek 'cure'. The common core of both medical and psychoanalytic variants of psychiatric practice, though, is that the abnormal is experienced as something which is internal to the person.
>
> (Parker et al. 1995: 13)

While psychoanalysis produces a radically different methodology in the talking cure enacted between analyst and analysand, it contributed to the excavation of the human by producing an interiority of the psyche that is in need of scientific treatment. Although Freud's purpose is therapeutic, he leaves us with the sense that the psyche is trapped by the contradictions of its own internal workings (Hutton 1988: 125). The interiorisation of the individual would be replicated and intensified as clinical psychology developed an empirical nomenclature for diagnostic categories in psychopathology that came into public prominence with the Second World War. Clinical psychology is the medical branch of psychology that has historically been linked to psychiatry,[3] but started emerging as an important subdiscipline in the 1950s.

> While clinical psychology owes its early development to its links with medical practice, its modern format and success since the Second World War owes a great deal the manner in which it has incorporated and internalised a coherent narrative about itself, namely that of the scientist practitioner.
>
> (Woolfe 2016: 13–14)

The method of the scientific-practitioner in clinical psychology is the RCT: randomised controlled trials, which have a greater claim of producing empirically supported treatment protocols.

Two critiques arise from the medical model of psychoanalysis and clinical psychology. The first is humanistic psychology of the 1960s that wished to discover the actualising tendency of the individual that could be harnessed by psychotherapy, such as in the client-centred approach pioneered by Carl Rogers. After the mainstreaming of clinical psychology, soon the subdiscipline of counselling psychology also emerged. It distinguishes itself as being invested in reflective practice, rather than the much-too-harsh evidence-based practice of clinical psychology (Woolfe 2016). The reflective practice model of counselling psychology is beholden to the humanistic tradition of Rogers. Nevertheless, it continues to be embroiled with the classificatory and diagnostic compulsions of the medical method.

The second critique of the medical model has been developed by positive psychology that took off in the first decade of the 21st century and continues to gain popularity. Positive psychology moves away from the disease model of psychology to work on what it deems makes life worth living; its founder Martin Seligman positions it as 'a science of positive subjective experience, positive individual traits, and positive institutions [that] promises to improve quality of life and prevent the pathologies that arise when life is barren and meaningless' (Seligman and Chikszentmihalyi 2000: 5) and proposes areas of research such as happiness, hope, optimism, strength, creativity, and mindfulness. In both branches of psychology, the individual self (Rose 1996) reigns supreme: it is the responsibility of the individual to pull oneself out of pathology, unhappiness, stagnation, or discontentment. It would seem that the method remains consistent, even when the subject matter or areas of research in psychology change.

The Other Side of Psychology

This brief sketch of the debates within psychology demonstrates heterogeneous concerns synchronised with coeval methods. Each wave of psychologists has reconfigured the subject matter of psychology and concurrently produced a methodology to study it, even if obeisance to rational, empirical, and objective science must be preserved at the core of the new branch. Nonetheless, within psychology, there have been multiple instances of resistance, both epistemological and practical. One such instance that contends with the individualising tendency inherent in the field is social psychology. Social psychology studies the subject as a social being (Squire 1990) and corresponds to a democratising turn in the discipline:

> Social psychology as a complex of knowledges, professionals, techniques, and forms of judgement has been constitutively linked to democracy, as a way of organizing, exercising, and legitimating political power. For to rule subjects democratically it has become necessary to know them intimately.
>
> (Rose 1996: 117)

What followed from the 1940s onwards were researches on the behaviour and responses of individuals in group situations, including psychology classics of Solomon Asch's studies on conformity to group opinion (1951, 1956), Stanley Milgram's research on obedience to authority (1963), and Philip Zimbardo's work on the effects of social expectations on behaviour (Haney, Banks & Zimbardo 1973). Several of these experimental studies were later criticised for using deception and manipulation of research subjects, and thus not conforming to ideas of ethical practice in research. Thus, we should read Rose's formulation of the democratising principle in social psychology as an acknowledgement that psychology itself has been linked to the exercise of power and the government of people in undemocratic ways.

Despite its early impulse to reframe the individual in terms of its social environment and thus produce methods more suited to this interdisciplinary field of study, social psychology had fallen foul of a well-known methodological trap; 'Whilst early European work in social psychology explicitly drew upon work in other humanities such as anthropology, by the 1950s psychological social psychology was dominated by a natural science orientation towards experimentation and quantification' (Brown and Locke 2008: 374), and this is apparent even in the quick sampling of social psychological research cited above. This kind of experimental research in social psychology was getting remagnetised to the individualising tendency inherent in the methods of mainstream psychology, since 'any text which implies a focus on simply individualistic explanations, without consideration of the influence of collectivity, does not suffice to convey the actuality of a social psychological concept such as conformity' (Marquez 2019: 222). It is not surprising that what followed has been dubbed as 'The Crisis'; 'Psychology's crisis is a crisis of its relations with the social world. This crisis of the "social" emerges with particular force and persistence in social psychology' (Squire 1990).

By the 1970s when the crisis peaked, the behavioural-cognitivist approach of psychology was vying with the interpretivist and constructionist approach of sociology. The next couple of decades were periods of methodological radicalism when the hypothetico-deductive methods of psychology were critiqued and deconstructed by juxtaposing them what could be loosely understood as qualitative methods of doing and interpreting research.

> There was nonetheless always a parallel tradition of research in psychology that was devoted to the 'meanings' that people produced, and to an investigation of experience through gathering accounts from people. This parallel tradition of what we now call 'qualitative research' voiced its concerns about the mechanistic dehumanizing laboratory-experimental paradigm.
>
> (Parker 2015: 3)

The study of meaning-making was understood to be at once an individual and a social way of constructing subjectivity and produced analyses of the experience of the individual within the sociopolitical reality. Qualitative researchers were interested in exploring and describing the quality and texture of this experience, rather than the identification of cause–effect relationships enshrined in the hypothetico-deductive method (Willig 2013). Various offshoots of the qualitative turn in psychology have flowered into areas of feminist psychology, discursive psychology, narrative psychology, and psychosocial studies.

It is best to be warned that qualitative methods do not necessarily lead to less reductive and more ethical methods of describing the complex social realities of people who occupy diverse subject positions in the world.

> There is a clear need for a more nuanced discussion of power dynamics, representation, and how best to capture the sociopolitical grounding of experience. Unfortunately, however, the quantitative/qualitative divide tends to pit one side against the other in a debate solely or predominantly about methods.
>
> (Cosgrove et al. 2015: 19)

The quantitative and qualitative methods of doing research and psychology can mask an agenda (conscious or unconscious) to decouple knowledge and power. Drawing upon KumKum Bhavnani (1990), a starting point to look at the nexus between knowledge and power is by analysing the differential access to power between the researcher and the researched in the context of conducting and writing up research, and the inequality of treatment and attention given to phenomena with socially ascribed characteristics. The unequal distribution of power inherent in mainstream psychology (and in fact other sciences) replicate older imperial relation to the subjugated other, and now beyond the colonially anticipated rights on body and labour to their mind and knowledge.

> This neocolonialization takes place in psychology through the under-representation of, for example, African experiences in psychological literature, the normalization and universalization of white, middle-class male characteristics, the reproduction of particular understandings of what psychology is about (or what Montero and Christlieb call 'symbolic colonialism'), the provision of tools of measurement that are complicit in modes of (neo)colonialist forms of regulation and governance, and the skewing of the production of knowledge through dependence on 'Western' sponsoring agents and publishing houses, unequal research resources, conference locations, and publishing practices (Macleod and Bhatia 2008).
>
> (Macleod and Bhatia 2008: 587)

Qualitative research produced through neocolonial mechanisms of acquiring knowledge and producing theory are no better guarantee of empowerment and democracy in psychology, and it is only through sustained ethical practice that psychology, in all its branches, can acquire an emancipatory potential for the subjects that it studies. Rather than fall prey to an overestimation of the dichotomy between quantitative and qualitative methods, there is a call within psychology to disengage as a maturing field of study from its 'physics envy' (Fish 2000) and critically interrogate and reflect on its own position in the interdisciplinary and intersectional fields of knowledge.

Culture and Psychology

One route to the politics of location in psychology takes the cultural turn. While cultural analyses have been a salient feature of early contributors of psychological theory, such as Wundt's *Völkerpsychologie* (1900), Freud's *Totem and Taboo* (1913) and *Civilization and Its Discontents* (1930), Vygotsky's sociocultural theory of learning (developed in the 1920s), or Frederick Bartlett's research on memory and culture (conducted in the 1930s), mainstream psychology was focused on universal tendencies in human behaviour and thought (Kashima and Gelfand 2012). Sustained attention to the significance of culture in psychological studies began only in the 1950s, as psychologists began to attribute cultural meanings to account for the wide range of individual differences in psychological phenomena around the globe. Almost every topic of investigation in psychology was re-examined through a cultural lens, and some classic studies were conducted on perception, personality, intelligence, cognition, language, and social identity. Neighbouring specialisations such as cross-cultural psychology and cultural psychology gained traction in the 1970s, the difference between them being methodological rather than epistemological.

> Cross-cultural psychologists are more likely to use questionnaires and collect data in many cultures, whereas cultural psychologists are more likely to use intense ethnographic work or experiments and collect data in a few cultures. As a result, cross-cultural psychologists consider differences in meaning barriers to overcome with elaborate psychometric methods, whereas cultural psychologists focus their research on differences of meaning.
>
> (Triandis 2007: 68)

Thus, the natural scientific approach of cross-cultural psychology was contrasted with the anthropological approach of cultural psychology, yet once more (re)creating a methodological binary within the discipline, this time within the discussions on culture and psychology.

The critique brought by analyses of culture did not break the hegemony of the Global North in delineating the research agenda and the beneficiaries of this knowledge production. While the turn to culture did reference the regions and people of the developing world or the Global South, and it did not try to develop psychology that it is culturally appropriate and receptive to the sociocultural realities in the Global South. Durganand Sinha (1997) opines that indigenising psychology is inevitable as scepticism of the Western system of knowledge production grows, and different parts of the world look inward to address the lacuna of the discipline of psychology. Heeding K.C. Bhattacharya's call for a 'swaraj in ideas' (1954) or liberation by throwing off the yoke of subjection in the field of ideas (Siddiqui 2016), Sinha exhorts psychologists to indigenise psychology. His summary of research in cultural psychology collating work from Africa, Latin America, China, India, Japan, Pakistan, the Philippines, and Turkey show the wealth of indigenous theories and methods across the world. Sinha points out that behaviour is to be understood and interpreted not in terms of imported categories and foreign theories, but in terms of indigenous and local frames of reference and categories (Sinha 1997: 132). Thus, the goal of indigenous psychology is more forthright in its political aspirations, in that 'it emphasizes the research conducted by people with their native cultural background, using their indigenous cultural concepts, and for the people of that culture facing their political, economic, and social circumstances' (Kashima and Gelfand 2012: 512).

Indigenising psychology has the consequence of directly taking from and contributing back to the ground-level realities of its subject(s). However, Sinha points out that the objective of indigenous psychology is to root theory in local contexts while securing the establishment of universal psychology (1997: 160). In short, the cultural particular must eventually add to the tome of the universal science of 'global psychology', and even if there is the use of indigenous methods, these must be attuned to the scientific method. It would seem that this process of indigenisation looks to be another dismal debate composed of fixed binaries and determinisms (Burman 2007). This time the binary is between cultural specificity and universalism, and indigenous psychology has yet to construct an epistemic bridge between the two. What we witness instead is *methodological nationalism* that lies on the assumption that the nation or state is the natural political form of the modern world (Wimmer and Schiller 2003: 302). Erica Burman has critiqued the methodological nationalism of indigenous psychology, which may defend some notion of cultural authenticity without interrogating if everything cultural must be defended, and it may be that 'the privileging of "culture" occludes other dimensions of political and psychological significance, such as gender, class, and sexuality' (2007: 181). Furthermore, indigenous psychology has another problematic to resolve: in foregrounding local culture, it may well reify the components that are already hegemonic within that cultural context, and perhaps even undemocratic.

> The problem with this viewpoint comes not from its critique of Euro-American ideas, this critique indeed offers a good diagnostics; but the trouble is inherent in the nature and kinds of the alternatives it suggests as remedial measures (such as their call to 'go back to Ancient scriptures and texts' for spiritual and intellectual enlightenment). In the work of indigenous psychologists, this 'going back to origins' is marked with a concurrent blurring of history, political developments and history of development of ideas.
>
> (Kumar 2005: 239)

It is within the history of ideas that one can find the theoretical links between the origins thesis and the contemporary moment, and the move from theory to politics is immediate here. This is because discussions on indigeneity are inherently political in nature: indigeneity gestures at ethnicity and group identity on the one hand, and citizenship and legal identity on the other. However, discussions on the political associations of the 'indigenous' or the 'original' implicit in indigenous psychology are missing in the literature.

The past is rife with epistemic and political violence, for which reform and reparatory work is crucial. Nonetheless, the present is a lived reality where new political developments must not be ignored. Furthermore, an analysis of the political in psychology is also a reconfiguration of its disciplinary knowledge by bringing to light the psychological dimension within politics. This is the critical approach to psychology since 'it is not enough to conceptualize how politics impacts psychology; we need an awareness also of the psychological working of power, of how subjectivity repeats, reiterates, reinforces the political' (Hook 2005: 484). Thus, when indigenous psychology comes home, to India, in the form of Indian Psychology, one must proceed with theoretical rigour and political precision.

Psychology, Method, Nation: Homecoming

The first psychology department in India was inaugurated at the University of Calcutta in 1915. In less than a decade, the institutionalisation of the discipline was evident as psychology was included as a separate section in the Indian Science Congress in 1923, the Indian Psychological Association founded in 1924, and the *Indian Journal of Psychology* was started by 1925 (Dalal 2014). In 1924, the second department of psychology was started at the University of Mysore. Pre-Independent India saw the inauguration of only one more psychology department in 1946 at the University of Patna. Nonetheless, the pace of the institutionalisation of psychology as an academic discipline picked up thereafter, and by the end of 1975, there were 51 universities that offered psychology courses (Prasadarao and Sudhir 2001). By the turn of the century, psychology was fully institutionalised in the Indian higher education sector.[4] In the last decade, in my perception, psychology has become one of the most popular options at an undergraduate level of study, whether it is placed in the division of sciences, or in the social sciences and humanities.

Despite more than a century of it being introduced as an academic discipline, research and publications from India have remained meagre in output and insubstantial in influence within global psychology. Keeping aside the material causes for a deficient performance,[5] are there psychological factors that have impeded the progress of the discipline in India? Do Indian psychologists suffer from an inferiority complex? Is the lack of originality in psychology in India a sign of the continuing 'colonization of our minds' (Nandy 1983)? Have we failed to attain a *swaraj* in ideas? If psychology in general suffers from physics envy, does psychology in India suffer from Western envy? According to K. Ramakrishna Rao and Anand Paranjape (2016), this is indeed the case. Psychology in India has either replicated the research originating in Euro-American psychology or has tried to study Indian subjects with the support of Western theories and methods. It led to the proposal of developing Indian Psychology (Safaya 1975) that will free itself from Western influence and build a psychology that is attuned to the national ethos of India.

In 2002, during the National Conference on Yoga and Indian Approaches to Psychology, conference delegates produced a manifesto titled the *Pondicherry Manifesto of Indian Psychology* that presented the urgent need for an Indian Psychology that would be rooted in classical Indian thought. The manifesto, published thereafter in *Psychological Studies*, is clear that 'by Indian psychology we mean a distinct psychological tradition that is rooted in Indian ethos and thought, including the variety of psychological practices that exist in the country' (Cornelissen 2002: 168). One of its aims was the introduction of Indian psychology into university education, and a committee was constituted to look into the logistics of introducing Indian psychology in Indian universities and colleges. Building on the momentum produced by this conference, two handbooks were published: *Handbook of Indian Psychology* (2008) and *Foundations and Applications of Indian Psychology* (2014). These introduced to the world stage new topics such as Yoga psychology, Buddhist psychology, Jaina psychology, and Spiritual psychology, as well as an Indian psychological approach to topics such as consciousness, self, healing, and transformation. Thus, the epistemological position of Indian psychology is that it is not a psychology of Indian people but a psychology from Indian thought. It also holds the position that Indian psychology is universal and its principles are for the transformation and perfection of humankind (Dalal 2014).

This impulse to decolonise has sometimes conceptualised the problem as the psychological effects of colonialism and foreign rule that has divorced 'our people' from 'our truth', necessitating a revival of old philosophical and intellectual traditions and a return to an authentic self. As an adjacent argument, I will make a case for a postcolonial analysis: postcoloniality as a position of no return; a history of colonialism has left indelible marks on our minds and societies that cannot be reversed – without suggesting that the processes of colonialism or the impact of colonisation is complete and in the past. Thus, the 'post' in postcolonialism does not signify that the historical period of colonialism is finally over,[6] but an acknowledgement that the social and psychological effects of colonialism from the past are evident in the present, and that both colonisers and colonised carry the weight of its violence and brutality into this 'emerging time' (Mbembe 2001). Postcolonialism topicalises subjectivity in analysis, for 'to postulate the existence of a "before" and an "after" of colonization could not exhaust the problem of the relationship between temporality and subjectivity' (Mbembe 2001: 15). In the time between the before and the after, there was the formation of new subjectivities that have persisted into the present, and these are the subjects of contemporary psychological investigation.

In this postcolonial analysis, science and the scientific method get implicated as a Western hegemonic practice of knowledge production that strangles traditional/cultural systems of knowledge (Nandy 1988). Psychology in postcolonial India is caught between these dual impulses. In making a place for India in a psychology stricken by dual impulses, a *dual critique* is necessary, i.e. one that is a critique of the Global North's hegemony in constructing both subject and method, as well as a critique of ideas that hegemonise the Global South that may emanate from the north or the south. The research agenda of psychology in India must include not only a critique of Western hegemony, but also a critique of the hegemonic construction of Indian society built on the intersections of class, caste, gender, indigeneity, ethnicity, and subjectivity in South Asia.

The Way Forward

Scientific psychology privileged the method over the phenomenon, so what could not be captured by the method was relegated to the unsavoury and unscientific part of the discipline. A hundred years ago, behaviourism emerged victorious in the battle of fixing the method of scientific psychology, and its dominion over the discipline continues till today, not the least in India. However, psychology in India must contend with both historical and political forces on the production of knowledge in the discipline. While India attained its political sovereignty in 1947, it is today home to political subjects in a modern nation-state with a mandate of developing scientific temper to break away from regressive social and cultural practices of the past. Concurrently, the nation is caught in a postcolonial prism; over several decades, a political vision has been produced to break the shackles of continuing colonialism and neocolonialism. While this chapter has critiqued the hegemony of the Global North in setting the agenda for psychology, there is work yet to be done at the other pole. As the Global South develops both critique and theory, a double take on psychology in India is important, premised on a dual critique of knowledge, method, and nation.

Notes

1 The University Grants Commission (UGC) came into existence in 1953 and became a statutory organisation of the Government of India in 1956, for the coordination, determination, and maintenance of standards of teaching, examination, and research in university education.

2 With observable behaviour as the central focus, the functionalist perspective in psychology developed by William James of looking at consciousness as a human attribute was discarded. In fact, Watson makes no distinction between human and animal behaviour, 'The behaviorist, in his efforts to get a unitary scheme of animal response, recognizes no dividing line between man and brute' (Watson 1913: 158).

3 The association between psychoanalysis and clinical psychology is no longer remembered very well. In 1942, the first Diagnostic and Statistical Manual of Mental Disorders (DSM) was written by the American Psychiatric Association,

which conceptualised psychopathology by heavily relying on psychoanalytic theories. Even the second version of the DSM (DSM II) in 1968 continued to borrow from psychoanalytic conceptualisation of psychic conflict. It is not until its third iteration in 1980 that the DSM unshackles itself – almost completely – from psychoanalysis.

4 For a historical overview of the development of psychology in India, see Dalal (2014).

5 Lack of adequate funding for research and development, for building necessary infrastructure such as labs and institutions, and lack of support for constructing channels for dissemination of available knowledge.

6 The critiques of postcolonialism are well-received. For instance, Rajagopalan Radhakrishnan (1993) historicises the term with reference to its site of production in the First World and warns us that the 'theoretical metaphorics of the "post" conflates politics with epistemology, history with theory, and operates as the master code of transcendence as such' (p. 751), and thus performs as a floating signifier. Nonetheless, within the same paper, he demonstrates its relevance for shifting from a politics of return to a politics of solidarity.

References

Amedeo, Marquez S. 2019. 'Towards a more social psychology.' In R. Beshara (ed.) *A critical introduction to psychology*, pp. 209–232. Hauppauge: Nova Science Publishers.

Asch, S. E. 1951. 'Effects of group pressure upon the modification and distortion of judgment.' In H. Guetzkow (ed.) *Groups, leadership and men*. Pittsburgh, PA: Carnegie Press.

Asch, S. E. 1956. 'Studies of independence and conformity: I. A minority of one against a unanimous majority.' *Psychological Monographs: General and Applied*, 70(9): 1–70.

Baron, R., and Misra, G. 2015. *Psychology (Indian Subcontinent Edition)*. New Delhi: Pearson Education Ltd.

Bhattacharya, K. C. 1954. 'Swaraj in ideas.' *Visvabharati Quarterly*, 20: 103–114.

Bhavnani, K. 1990. 'What's power got to do with it? Empowerment and social research.' In I. Parker and J. Shotter (eds), *Deconstructing social psychology*, pp. 141–152. London: Psychology Press.

Boring, E. G. 1923. 'Intelligence as the tests test it.' *New Republic*, 36: 35–37.

Breuer, J., and Freud, S. 1895. 'Studies on hysteria.' In J. Strachey (trans.) *The standard edition of the complete psychological works of Sigmund Freud, volume II*. London: Hogarth Press.

Brown, S. D., and Locke, A. 2008. 'Social psychology.' In C. Willig and W. Stainton-Rogers (eds) *The Sage handbook of qualitative research in psychology*, pp. 373–389. Los Angeles: Sage Publications.

Burman, E. 2007. 'Between orientalism and normalization: Cross-cultural lessons from Japan for a critical history of psychology.' *History of Psychology*, 10(2): 179–198.

Chomsky, N. 1959. 'A review of B. F. Skinner's verbal behavior.' *Language*, 35: 26–57.

Cornelissen, M. 2002. 'Pondicherry manifesto of Indian psychology.' *Psychological Studies*, 47(1–3): 168–169.

Cornelissen, M., Misra, G., and Varma, S. (eds). 2014. *Foundations and applications of Indian psychology* (2nd ed.). Delhi: Dorling Kindersley.

Cosgrove, L., Wheeler, E. E., and Kosterina, E. 2015. 'Quantitative methods: Science means and ends.' In I. Parker (ed.) *Handbook of critical psychology*, pp. 15–23. London: Routledge/Taylor & Francis Group.

Dalal, A. 2014. 'A journey back to the roots: Psychology in India.' In M. Cornelissen, G. Misra, and S. Varma (eds) *Foundations and applications of Indian psychology* (2nd ed.), pp. 18–39. Delhi: Dorling Kindersley.

Fish, J. M. 2000. 'What anthropology can do for psychology: Facing physics envy, ethnocentrism, and a belief in "race".' *American Anthropologist*, 102: 552–563.

Foucault, M. 2007. *Ethics: Subjectivity and truth*. New York: New Press.

Freud, S. 1913. 'Totem and taboo.' In J. Strachey (Trans.) *The standard edition of the complete psychological works of Sigmund Freud, volume XIII*, pp. 1–161. London: Hogarth Press.

Freud, S. 1915. 'The unconscious.' In J. Strachey (Trans.) *The standard edition of the complete psychological works of Sigmund Freud, volume XIV*, pp. 159–204. London: Hogarth Press.

Freud, S. 1930 [1963]. 'Civilization and its discontents.' In J. Strachey (Trans.) *The standard edition of the complete psychological works of Sigmund Freud, volume XXI*, pp. 58–146. London: Hogarth Press.

Haney, C., Banks, W. C., and Zimbardo, P. G. 1973. 'A study of prisoners and guards in a simulated prison.' *Naval Research Review*, 30: 4–17.

Hook, D. 2005. 'A critical psychology of the postcolonial.' *Theory & Psychology*, 15(4): 475–503.

Hutton, P. 1988. 'Foucault, Freud, and the technologies of the self.' In Michel Foucault, Luther H. Martin, Huck Gutman, and Patrick H. Hutton (eds) *Technologies of the self: A seminar with Michel Foucault*, pp. 121–144. London: Tavistock Publications.

Kashima, Y., and Gelfand, M. J. 2012. 'A history of culture in psychology.' In A. W. Kruglanski and W. Stroebe (eds) *Handbook of the history of social psychology*, pp. 499–520. New York: Psychology Press.

Kumar, M. 2005. 'Rethinking psychology in India: Debating pasts and futures.' *Annual Review of Critical Psychology*, 5: 236–256.

Macleod, C., and Bhatia, S. 2008. 'Postcolonialism and psychology.' In C. Willig and W. Stainton-Rogers (eds) *The Sage handbook of qualitative research in psychology*, pp. 576–589. Los Angeles: Sage Publications.

Mbembe, A. 2001. *On the postcolony*. Berkeley: University of California Press.

MichaelJ. 1984. 'Verbal behavior.' *Journal of the Experimental Analysis of Behavior*, 42(3): 363–376.

Milgram, S. 1963. 'Behavioral study of obedience.' *Journal of Abnormal and Social Psychology*, 67: 371–378.

Nandy, A. 1983. *The intimate enemy: Loss and recovery of self under colonialism*. Delhi: Oxford.

Nandy, A. (ed.). 1988. *Science, hegemony and violence: A requiem for modernity*. Tokyo: United Nations University.

Narciandi, J. C. N. 2019. 'Psychology as a technology of subjectification.' *Psychologist Papers*, 40(1): 31–38.

Parker, I. 2015. *Psychology after the crisis: Scientific paradigms and political debate.* London: Routledge/Taylor & Francis Group.

Parker, I., Georgaca, E., Harper, D., McLaughlin, T., and Stowell-Smith, M. 1995. *Deconstructing psychopathology.* London: Sage Publications.

Prasadarao, P. S., and Sudhir, P. M. 2001. 'Clinical psychology in India.' *Journal of Clinical Psychology in Medical Settings*, 8(1): 31–38.

Radhakrishnan, R. 1993. 'Postcoloniality and the boundaries of identity.' *Callaloo*, 16(4): 750–771.

Rao, K. R., and Paranjpe, A. C. 2016. *Psychology in the Indian tradition.* New Delhi: Springer.

Rao, K. R., Paranjpe, A. C., and Dalal, A. K. (eds). 2008. *Handbook of Indian psychology.* New Delhi: Foundation Books.

Rose, N. S. 1996. *Inventing our selves: Psychology, power, and personhood.* Cambridge, England: Cambridge University Press.

Safaya, R. 1975. *Indian psychology: A critical and historical analysis of the psychological speculations in Indian philosophical literature.* New Delhi: Munshiram Manoharlal Publishers.

Seligman, M. E. P., and Csikszentmihalyi, M. 2000. 'Positive psychology: An introduction.' *American Psychologist*, 55(1): 5–14.

Siddiqui, S. 2016. *Religion and psychoanalysis in India: Critical clinical practice.* London: Routledge.

Sinha, D. 1997. 'Indigenizing psychology.' In J. W. Berry, Y. H. Poortinga, and J. Pandey (eds) *Handbook of cross-cultural psychology: Theory and method*, pp. 129–169. Boston: Allyn & Bacon.

Skinner, B. F. 1957. *Verbal behavior.* Acton, MA: Copley Publishing Group.

Squire, C. 1990. 'Crisis, what crisis?: Discourses and narratives of the 'social' in social psychology.' In I. Parker and J. Shotter (eds) *Deconstructing social psychology*, pp 33–46. London: Psychology Press.

Triandis, H. C. 2007. 'Culture and psychology: A history of the study of their relationship.' In S. Kitayama and D. Cohen (eds) *Handbook of cultural psychology*, pp. 59–76. New York: Guilford Press.

Watson, J. B. 1913. 'Psychology as the behaviorist views it.' *Psychological Review*, 20(2): 158–177.

Willig, C. 2013. *Introducing qualitative research in psychology* (3rd Edition). New York: Open University Press.

Wimmer, A., and Schiller, N. G. 2003. 'Methodological nationalism, the social sciences, and the study of migration: An essay in historical epistemology.' *International Migration Review*, 37(3): 576–610.

Woolfe, R. 2016. 'Mapping the world of helping: The place of counselling psychology.' In B. Douglas, R. Woolfe, S. Strawbridge, E. Kasket, and V. Galbraith (eds) *The handbook of counselling psychology (*4th edition), pp. 5–19. New Delhi: Sage Publications.

Wundt, W. 1896. *Outlines of psychology* (2nd English edition). Translated by Charles Hubbard Judd. Leipzig: Wilhelm Engelmann.

Wundt, W. 1900. *Völkerpsychologie: Eine Untersuchung der Entwicklungsgesetze von Sprache, Mythus und Sitte*, 10 Volumes. Leipzig: Kröner.

13

GEOGRAPHY IN INDIA

Gendered Concerns and Methodological Issues[1]

Saraswati Raju

Geography is, very simply put, about the interrelationship between humans and their environment, both physical (natural) and built (including social and cultural), expressed through spatial practices in terms of adaptation, changes, and modifications. The discipline of geography is thus unusual in that it spans over the natural and human-created domains. Cartesian dualism – the separation of mind and body, of immaterial mind and the material body are treated as ontologically seemed to be at work in geography assuming that the natural and the human are distinctly different and separate. Consequently, various branches of geography – physical and human geography – have largely developed as separate branches, independent of their dialectic interlinkages. Also, the initial tag of 'science' to the discipline of geography had meant that the discipline developed a certain tenor, in terms of both content and methodologies, adopted by geographers.

One of the major problems of method in geography is the intellectual legacy of the discipline. Till date, it is unresolved whether geography is part of 'science' or 'social sciences'. In many universities and colleges, geography is part of the science faculty and included in the departments of earth sciences (Progress in Indian Geography, a Country Report 2004–2008: 2008). One of the expectations, of being a part of science, is having rational and objective viewpoints – the 'enlightened' way of seeing the world as orderly, neatly organised, and causally connected and generalisable. In this rational worldview, the realm of social is seen as following a definitive evolutionary track much like the physiological properties of matter as well as the forms and conditions of existential life. In other words, scholars look for spatial regularity of function, hierarchy, and norm in society similar to that found in the body. That was what made Brian Berry (1970) assert (an infamous claim, negated by the author in a later publication) that in terms of the residential arrangement vis-à-vis the social status, in their 'equifinal' avatars, Chicago and Calcutta would look alike! In a quest for essentially rational orders, production of 'truth' meant use of methods which would lead to

DOI: 10.4324/9781003298908-20

overarching theorisation without any recognition of biases and/or the role of the observer's persuasion in the perception seeing the reality. It was not surprising, therefore, that objectively produced knowledge, which could withstand the demands of scientific rigour, scrutiny, and validation, took precedence over other forms of knowledge. Geography was no exception to this rule. The discipline came to be seen as nomothetic (a science of general or universal law) rather than ideographic (place-specific/one of a kind work), as used by geographers during the heyday of the quantitative revolution; 'statistical fix' was the order of the day.

It is well-known that Indian geography imported this model without much introspection and reflection. Moreover, the quest for 'evidence' (read data)-based objectivity has been further helped by the colonial legacy of data production in the form of extensive reports, gazetteers, and the record-keeping system. In practice, it meant that what was not quantifiable and could not be put to rigorous statistical tests could not be geography! While the academic world, particularly the Anglophone, has moved away from the quantitative revolution and resultant positivistic tradition to a much evolved and complex mix of methodologies in questioning many of the tenets of 'scientific' knowledge, this debate remains in a nascent stage in geography in India.[2]

Given this backdrop, which still persist today, it is not very difficult to see why despite vibrant feminist struggles and scholarship in India, the place that geographers, who study gender, have been able to carve out in studying women/gender issues is far from impressive. Recently, geography as a discipline seems to have been losing its popularity not only in India, but also elsewhere. In order to check such a trend, the discipline has to respond to newer subfields if it has to regain its position. As such, reorienting some of the androgynous learnings of the discipline to reclaim what can be termed as 'missed opportunities' may not be a bad idea.

At the outset, it cannot be denied that academia is a part of a larger world and thus cannot be completely free from deeply entrenched, hegemonic, and institutionalised gendered biases. However, in the contemporary sociopolitical environment wherein metanarratives are being questioned and subalterns including women are creating their own spaces, it will not be an overstatement to say that it is high time to engage in equity issues and social transformation. Moreover, as discussed later, the 'spatial turn' is reclaiming the power of space/place in social science research in terms of bringing back geographical concerns in decoding existential realities. As such the gender concerns in combination with a plea for inclusive development can easily be woven in broader issues of social change and transformation – a role that geography must play if it has to survive.

This discussion is thus an attempt in the manner of stocktaking, tracing the challenging journey that the emerging subfield of the geography of gender has taken to reach where it has as much to argue for positioning gender firmly within the orbit of geographical knowledge in terms of research and teaching in India. Although small, in doing so, it has definitely challenged the domination of 'scientific' assumptions of methods in geography.

Despite or perhaps because of awareness of how intermittent the growing scholarship on the geography of gender has become, I neither claim an exhaustive review of the repertoire of the literature nor do I attempt to trace every possible strand of argument there is across various locations – it is an impossible task, at least within the purview of this chapter. This chapter is simply an attempt to understand and position of gender firmly within the orbit of geographical research and teaching in India, as much as indicative of the arduous ongoing journey.

The chapter consists of two sections. The first section is about (i) defining gender and differentiating between gender and sex followed by (ii) differentiated responses of men and women to their environment, both natural and built, which places the subsumed scientific and androcentric assumptions regarding human interactions being held by the discipline in jeopardy. The briefs clearly indicate the need to engender the discipline providing a backdrop to sustain the establishment of the subfield of 'geography of gender' and/or 'gendered geographies' as someone may like to put it. These most definitely inform the debate, external and internal, on method.

Defining and Differentiating between Sex and Gender

This 'becoming one' is all about gendered outcomes. Let us first see, what the term 'gender', which seems to be omnipresent these days, means.

The term gender has a long history and the root of the word can be traced to Latin, 'la: genus' meaning 'type', 'kind', or 'sort'. It is also connected to the Greek root 'gen', meaning 'to produce'. In some languages, gender signifies grammatical usage – a type of noun-class system, which may be classified as masculine or feminine. Yet, another set of languages may apply the term 'gender' in a neuter-grammatical sense without attaching any masculine or feminine connection to the meaning of the word. The word 'linga', part of the vernacular Hindi, which originated in the classic Sanskrit language, is one such term which requires a qualifying prefix 'male' or 'female' – *pung-linga* and *stree-linga*, respectively – if it is to be used to mean biological sex. Interestingly, these two languages also do not have an equivalent term to denote 'gender'. In the absence of such a nomenclature, *'linga'* is used in an expanded way, that is, *prākritik linga* (natural/biological sex) and *sāmajik linga*. Sex is essentially the biological differences between males and females; although less explored, the notion that physical sex is also socially constructed is acquiring recent attention (Johnson and Repta 2012). Also, sex and gendered identities overlap. For example, girls are admonished for not covering their bodies appropriately at a certain age while boys go scot-free for much longer.

Since gendered roles are socially created, they are subject to change not only over time but have wider variations both within and between cultures at the same time. For example, while countries in the European Union and the United States, to name but a few, have been drafting their women in the army for quite some time now, in India the induction of women in the army has been relatively a recent phenomenon. Likewise, not only gender roles can change over life stages, there may be differences within the country in terms of how they are perceived and expressed. For example, it is well-documented that the parts of northern India have a relatively more restricted gendered domain for women as compared to the South. In fact, there are widespread variations across the different regions of the country. For example, as per the 2009–2010 NSS data, about one-third of the women in Rajasthan do not participate in the labour market because of social and religious strictures; this percentage is less than three in northeastern states.

Conventionally, gender has referred to the social interpretations and values assigned to being a woman, a man, a boy, or a girl. Gender determines the roles, power, and resources for females and males in any culture. For example, we often hear boys being admonished for crying or girls being chided for climbing a tree or flying a kite because 'boys do not cry' or 'girls do not climb the trees or indulge in kite-flying'. The fact that is lost is that these boys and girls had already done what they are not supposed to do! When the queen of Jhansi fought the battle with the British, she becomes *Mardani* – masculine (*khub ladi Mardani woh to Jhansi wali rani thi*). In the same way, men who may be performing household duties may be labelled as '*baila*' or 'hen-pecked'. Why these titles? It is because these men and women did not fit in with the stereotypical images of what men and women are socially meant to be doing!

Who decides what boys and men and girls and women can do or cannot do if they behave differently? It is the collective wisdom in a given society which encrypts the feminine and masculine roles that are associated with girls/women and boys/men. Such ascriptions are what we call gender norms – socially, superficially imposed constructs as to what boys and girls, men and women should or should not do. This essentialised behavioural construct transposed upon sexual differences is artificial and in constant flux. It is, therefore, possible for gender to be theorised as 'the cultural or social elaboration' of absolute (biological) sexual differences between women and men. This is gender – a social construct. For example, the ability to bear a child is, fundamentally, a function of biology, while expectations about the imperative to bear children, the nature of parenting, or the status associated with being a mother are more closely linked to gender roles and expectations and the societal constructs defining them.

Gendered natures/cultures

Whether men and women are inherently different is an issue that is still not fully resolved. There are those such as John Gray (1992) who begins his book *Men are from Mars, Women Are from Venus* with the assumption that men and women's constitutions are *inherently* different which is why they behave differently – women have more cognitive capacity than men who are better in spatial knowledge. These differences, it has been argued, are related to brain sizes. However, this idea is now passé as new research is showing that brain size may have something to do with use. The behaviour of men and women socially constructed lead to the development of the cognitive/spatial side of the brain, for example. There is nothing biological about associating the colour pink with women and blue with men or why women are mainly nurses and men are doctors or why is it that wives mostly look after the household and husbands run it? This is where socially ascribed roles come into the picture and get internalised both by boys/men and girls/women.

It is important to note, however, that earlier 'sex' and 'gender' were seen as mutually independent of each other. Now it has been argued that as biologically sexed bodies, male and female internalise certain values and behavioural codes as social beings. For example, growing girls may be asked to behave in certain manners and may become aware of their bodies much earlier than the boys. Thus, the gendered experiences interlock with sexual identities. Similarly, the sex-segregated data on education or labour force participation may in fact encapsulate and reveal the gendered roles that the society assigns to men and women.

There can, however, be situations where the gendered identities are not clearly put in a binary. In a more contemporary context, such binaries have been contested. People who feel that their gender identity is incongruent with their biological sex would identify themselves as 'inter-gender/ transgender'. Androgynous (also androgyne) has also been frequently used as a descriptive term for people in this category, though such people may express a combination of masculinity and femininity or neither, in their gender expression and not all identify as androgynous. Some references use the term transgender broadly in such a way that it includes non-binary people. The third gender, identified for the first time in the Indian Census in 2011, is one such example. India's Supreme Court has recognised transgender people as a third gender in a landmark ruling which talked about 'the right of every human being to choose their gender', in granting rights to those who identify themselves as neither male nor female. According to one estimate, India has about two million transgender people.

Gender is a relational concept – it is about social construction which sets behavioural norms that are equally applicable to men, a point which is generally neglected in popular understanding. Very often gender is used co-terminously with women, which is misleading. The essentialised role of men as breadwinners and homo economicus construct may constrain their participation in household chores even if there is willingness on their part, for example. And yet, in a relational domain, very often it is females, both in cross-comparison and within the categories, who occupy the subordinated position vis-à-vis their male counterparts. For example, the occupational data on women's placement in educational institutes show their disproportionate share as one moves in the hierarchy from primary, middle, higher secondary, to college and university appointments. Similarly, on average, tribal man may have lower literacy as compared to their non-tribal counterparts, but rarely, if ever, would one record statistics for tribal women which are having higher literacy than that for tribal men. These asymmetries are across the board with rare exceptions and reflect the uneven position that women and girls occupy as compared to their male counterparts. Thus, very often a gendered analysis may have to focus more on the former from stereotyping as compared to male counterparts (Raju 2013a, 2013b).

It is to be remembered that women are not a homogeneous category. Caste, class, religion, and ethnicity intercept gendered locations. For example, it is well-documented that women from higher castes face stricter spatial codes as opposed to those belonging to lower castes (Grace 2003; Raju and Bagchi 1993). Women belonging to different class locations often have interests which clash when it comes to environmental consumption and management. While women of affluent classes may be interested in converting open fields into polo grounds or any such extravaganza, poor women may like to protect them for their daily needs such as fuel-wood collection, grazing, and collection of free goods. While being fully aware of women's multiple locations and associated multilayered gendered existence in which, following Archer's (2004) argument, all axes of differentiation may not be equal, our prime concern here is with spatiality and how gender is geographically anchored.

The Gendered Outcomes of Differentiated Responses

The gendered concerns emerge because it was observed that men and women respond differentially to spatial stimulus. However, until about the 1970s, under the pressure of objectivity, geography remained indifferent to incorporating women's perspectives in the discipline. One was taught that the terms 'human' or even 'mankind' subsumed both men and women. It was increasingly seen that despite the claim, the lens through which the discipline was conceived and taught remained predominantly male-centric. However, very soon it became clear that the very category of humanity had to be critically evaluated to analyse for gender differences. Thus, the early 1980s saw concerted efforts amongst scholars towards discussing prospects of advancing studies on women and gender. The initial efforts were aimed at not to exclude 'half of the human from human geography' (Monk and Hanson 1982).

It was around the 1970s and 1980s that geographies of women's experiences began to develop in the Anglophone human geography. It was increasingly felt that they were not effectively articulated in the geographic literature. Such a realisation eventually culminated in an informal session of concerned scholars arranged by Janice Monk from the United States and her colleagues to discuss studies on women and gender through the prestigious International Geographical Union (IGU) at the Paris Congress in 1984. This was followed by the presentation of a half-day programme in Barcelona in 1986 to take note of emerging gender research internationally. At the Sydney Congress in 1988, a Study Group on Gender and Geography was formed and approved as the Commission on Gender and Geography in 1992. It has been highly active since then.

Along with IGU Commission on Geography of Gender, there were other US-based events such as the Department of Geography at the University of Southern California sponsoring a conference in March 1990 entitled 'Geography and Gender: A Feminist Geography Symposium' – the first of its kind in North America that specifically addressed the issue of feminist geography. This event was inspired by another conference on feminism and historical geography at the University of London in the previous year. More importantly, there was a move to go beyond description and start seeing women as situated in the asymmetrical power relations between men and women. Patriarchy was posited both as a theoretical construct and as a lived, existential and material reality to understand women's position in geographies of space (Raju and Lahiri-Dutt 2018).

As gender started figuring in international discussions and political agendas, countries became signatories to gender-pro declarations. With networking, funding, and exchanges around the world, 'gender' started becoming part of the politically correct vocabulary. These national and international discourses thus worked, even if indirectly, towards making gender a component of academia in several countries of South Asia. Geography was no exception (Raju and Lahiri-Dutt 2018).

One can attribute the inclusion of gender as a subject for study in Indian geography to larger issues of distributive justice at stake within the rubric of social geography. Isolated studies by individual scholars can be traced as far back as to the 1960s. However, much of the research remained descriptive and was not being placed in a larger regional or national framework. Suffice to say that gendered concerns were yet to become part of a collective effort to produce a feminist perspective in geography.

This state of affairs changed slowly and from not having a 'critical mass' in rooting for the establishment of the geography of gender as a legitimate subfield, we see today that teaching and research have been expanding with new areas of exploration being sought. And yet there are questions being raised 'what has geography got to do with gender?' 'Plenty' would be my immediate response, but let us first examine some of the differentiated gendered responses to environmental stimuli, which would support my assertion unabashedly, taking various examples largely from South Asia.

Gendered Responses to Environmental Settings

The presence of differential gendered spaces/places across India has been talked about. For example, recent development reports in India have brought out how the spatial location in which women live makes a difference to even as basic and 'objective' 'well-being' indicator as 'longevity'. As per 2010–2014, women in Kerala in southern India expect to live longer by a margin of about six years as compared to women in Haryana in the north despite, its much higher per capita income. Such spatial overlapping with specific gendered attributes cuts across cultures. Let us take some more examples.

The proposition that the process of globalisation erases geographical specificities can be questioned by taking the example of the fishing community in Kerala where gendered responses to globalisation differ as a result of the ways men and women are situated not only in fish commodity chains but also within the same local economy. This is because of different configurations of gender, identity, and work. In the Christian communities, households continue to rely on both men and women's work in the fish economy, whereas in the Muslim and Hindu communities, men outmigrate to Gulf countries in search of better emolument. This has resulted in profound shifts in women's autonomy and gender relations within the household, but in different ways than the Christian fisherwomen. Moreover, in the process of globalisation, production, processing, and distribution of fish undergo transformation with associated stresses and opportunities for men and women fish workers. For women, their added multiple memberships in terms of physical, financial, social, and cultural capital, livelihood options, and the strategies play out differently because of various configurations of gender, work, culture, identity, and economy and the different ways households and communities are connected to fish economies at diverse scales. The authors identify these differences in order to both broaden and deepen an understanding of the gendered nature of globalisation in resource-based economies (Hapke and Ayyankeril 2018).

A significant body of literature on gender and climate change shows that women and men perceive and experience climate change differently. As women are not a homogeneous group, their intersectionalities across social, economic, and cultural axes such as caste, class, and regions interact differentially in influencing their responses to vagaries of climate and weather. A comparative analysis of two case studies located in the Kumaon region of Uttarakhand and plains of Bihar in the mid-Indian Gangetic region recognises diverse and multiple adaptation responses and the link with gender in the context of climate change. Women in the more remote mountain region of Uttarakhand demonstrate their collective agency in reducing common vulnerabilities due to climatic variability and shocks, whereas in the Indian Gangetic region, relatively more rigid norms of women's seclusion and prevailing caste structure had barred the adoption of agro-biodiversity-based technological strategies in response to climate change (Ravera et al. 2016). The authors concur that adaptation to climate change is not a homogeneous process and there are risks of reproducing gender biases and inequalities in development policies and interventions if it is not carefully addressed.

Climate change debates related to gender are now shifting from a simple view of women as a homogeneous group towards a more complex view of identities like caste, class, and region within gender categories (Tschakert and Machado 2012). It can be seen that the intersectionalities of gender with social, economic, and cultural characteristics explain how and why people face and manage climate change in different ways (Carr 2008).

Another example is from Bangladesh. Although water security is being disturbed globally as one of the consequences of climate change, its impact is felt unevenly across social strata, particularly across gender and class. As pointed out by Sultana (2018a, 2018b), this relates to water-related productive and reproductive tasks as climate change is expected to exacerbate both: water shortages and water floods and cyclones having footprints on water geographies. According to her, gendered implications of climate change is particularly important as patriarchal norms, inequities, and inequalities often place women and men in differentiated positions in terms of their capacities to respond to and cope with changing waterscapes in the wake of socioecological relations.

Women's dependence on common property resources is disproportionately higher than that of men because they are the prime users of such resources. However, although women are primarily responsible for water, their needs can significantly differ from that of men. For example, although Sri Lankan and Nepali women worked on their husbands' plots with similar water needs, they had different opinions regarding water deliveries in terms of adequate water supply for growing crops. The Nepali male farmers were concerned about enough water at the start of the rice season to soften soil for land preparation was their prime responsibility. The women wanted water during the entire season in order to suppress weed growth because its clearing was their job. Yet another example can be cited about women's differential needs and responses: in the remote Himalayan village of Dungari-Paitoli, women stalled men from selling the community forest which was to be converted in potato seed farm because women would have been deprived of a free collection of fuel and fodder from the forest, whereas men would have gotten employment opportunities if the community forests were to convert into potato seed farm (Agarwal 1986).

The official statistics in India records a very low workforce participation rate for females, although they work for more hours. However, it is to be noted that the restrictive role of household responsibilities does not impact women uniformly across space in terms of constraining/enhancing their entries in the formal labour market. In fact, some region-specific sociocultural factors create the uneven terrain. The explanations are complex, but there seems to exist differential social spaces that treat women differently in different places. For example, while in the north Indian plains, more women are restricted to participate in the labour market on account of sociocultural reasons; in the southern and the north-eastern parts of India, such barriers are fewer (Raju 2013). Despite the regionally stark pattern that emerges in the case of female workforce participation rates, overall labour market outcomes may be seen as interlinked with women's spatial access to opportunities. Kerala in south India is a case in point. The state has the highest percentage of educated unemployment amongst women. In the absence of job opportunities, men outmigrate not only within India but also outside the country, whereas even educated women do not have such an unrestricted entry to labour markets as them – their lives are spatially embedded.

In 'doing gender', thus the place does not remain just a 'thing in itself' – a passive locus/container as the 'unchanging backdrop against which life is played out' (Lefebvre 1991) – but turns into a historically and socially grounded existential space. This space constitutes and is constituted by socially produced signifying aspects. As a consequence, theorisations of space lead us to constitute it both the imagined and symbolic. That is to say, space is no longer seen or defined as abstract subjectively–discursively, but has an existential reality. In other words, the boundaries between space (non-real) and place (real) become increasingly blurred in mutually interactive framework. Such an overlap opens up potentially latent sites for critical engagement with gender and patriarchal structures as to how the gendered realms get enacted, re-enacted, constituted, and reconstituted in mediation with specific spatial context, making a case for studying gendered issues with geographical lenses.

An Overview of Geography of Gender in India

Although I do not claim an exhaustive coverage of themes/seminars, research, and courses on gender that are being undertaken by Indian geographers in terms of either content or geographical coverage, given the international coverage especially in the United Kingdom and the United States, the field of geography of gender in India appears to be growing somewhat slower than wished for. Accordingly, the feminist repertoires of pedagogy, for instance, can broadly be classified into four broad categories: (i) projects and seminars, (ii) research, (iii) parts of other courses, and (iv) dedicated courses on gender.

As far as gender-related projects and seminars are concerned (as the dedicated seminars or parts of the seminars), it is rather impossible to cover all the places, a few are, however, being cited as examples. They are as widespread as in Kumaun University, Nainital, and HNB Garhwal University, Srinagar (Garhwal) in the north to the University of Madras in the south. Kurukshetra University in Kurukshetra, Punjab University in Chandigarh; Khwaja Moinuddin Chishti Urdu, Urabi-Farsi University in Lucknow, Presidency University in Kolkata, Aligarh University in Aligarh, and Delhi University in Delhi are some of the other universities in the league. The projects and seminars include a wide-ranging variety of topics such as gender issues in population dynamics; women empowerment towards gender equality and gender budgeting to contextualising gender studies in Asia; reorienting gender: geography of resistance, agency, violence, and desire in Asia and negotiating intersectionality between nature, culture, and the future, and so on.

The body of geography scholars doing research on gender-related topics and themes is much larger than the universities and colleges where independent courses on gender geography are being run. One of the earlier publications was the much-talked-about *Atlas on Women and Men in India* (Raju et al. 1999) which displayed district-level maps on various indicators along with textual interpretations to showcase how the marginalities and vulnerabilities of women and men are underlain by locational spatialities. With the involvement of established as well as younger scholars, primary concerns related to broader issues within the realm of development such as population dynamics (Das and Chattopadhyay 2012), education, health (Datta 2015; Rajeshwari 1996, 2010, 2011), work and labour force (Bhairannavar 2018; Bordia-Das 2004; Paul 2012, 2013; Raju 1993, 2013), and women's empowerments (Raju 2005, 2006, 2013) have expanded to include more contemporary themes of microfinance (Samanta 2009, 2012), crime and violence (Datta 2013–2014; Preeti and Rajeshwari 2018), agency and identity, resistance (Datta 2015, 2016), body politics, and queerness (Borisa 2018), to name but a few.

At the risk of digression, it is important to draw attention to the recent paradigm shift to 'spatial turn' in the social sciences. Spatial turn refocuses on 'questions of locality, sense of place and of identity in place [which] matter now more than ever' (Withers 2009: 638) instead of universalising tendencies and metanarratives which see the world as a seamless whole, as proposed by those who advocate globalisation. This emphasis on the existential reality of human lives, in my view, is in response to assuming the 'end of geography' associated with a 'collapse' of geographical space in the wake of the information age and the network society. This is reflected in writings of Indian geographers as well, although not many studies have directly dealt

290

with how space implicates the asymmetrical power relations between men and women and alternatively how geo-ecologically embedded institutional responses create gendered spaces (Raju 2003) or how spatiality of gendered existence overarch seemingly universalising processes (Raju 2006c; also see Datta and De 2008; Raju and Paul 2013, 2019; Paul 2013, 2018). Incidentally, it is of interest to know that most of the spatially anchored work on gender comes from non-geographers (see Deshmukh-Ranadive 2002; Deshpande 2003; Phadke 2005; Niranjana 2001).

Whether or not the geography of gender is part of the established curricula or added as dedicated courses, the PhD dissertations and MPhil theses covering various dimensions of gendered issues and vulnerabilities are definitely on the rise.

Indian geographers have been largely engaged in studying gender through their established disciplinary lens. That is to say, much of Indian geographical research is still following the proven areas of research via the orthodox methodological paths – the use of quantitative data and statistical methods – in order to carve out hitherto unwelcome niches without destabilising the established order. This perhaps is done subconsciously and also as a response to the need to obtain academic acceptance and acknowledgement. In our opinion, one must not dismiss such doings and strands as they are not only the products of their times but also to get the subversive value of such work in further propagating gender studies in geography – what we would like to term as 'incremental pragmatism' (Raju and Lahiri-Dutt 2011). However, with the younger scholars becoming interested in studying gender in geography, one witnesses more innovative ways including mixed methods, i.e. combinations of quantitative and qualitative analyses, visual sources, and in-depth case studies being tried out and accepted in geography despite its masculine legacy.

As a largely androcentric discipline seeped in positivist traditions, geography in India had difficulty in moving away from empirical/quantitative data-based analyses and adopt purely qualitative and ethnographic research. Yet dissatisfaction with empiricism is quite palpable and recent research by the younger scholars has begun to see the emergence of more innovative ways of mix-methods whereby both quantitative and qualitative approach together with visual sources and in-depth case studies have been adopted. For example, if poverty in Calcutta has to be studied, one would like to place West Bengal in an overall situation in India as a whole using quantitative data, but while for Calcutta, qualitative method such as FGDs, individual interviews, and discussions with local scholars, etc. would be adopted.

Scholars have used this mostly because the nod to some form of quantification is almost mandatory even now in most departments and examiners still routinely raise questions about representativeness and sample size.

Some have used ethnographic methods using face-to-face interaction and detailed unstructured conversations without any structured schedules (Sil

2018). Some have moved to autoethnography and collaborative autoethnographies as well as purely qualitative approaches (Datta 2021). Incidentally, auto-ethnography is one of the approaches that acknowledge and accommodate subjectivity, emotionality, and the researcher's influence on research, rather than hiding them or assuming they do not exist. It questions conventional ways of doing research; it enables researchers to view the world with a wider lens and helps them to understand, experience, influence ways of seeing and interpreting the world. One form of auto-ethnography is personal narratives, where authors view themselves as the phenomenon and write narratives specifically focused on their academic and personal lives.

Datta (2016) points out that as the junior- and mid-level faculty who have been pursuing gendered concerns gain seniority, it becomes easier for them to spread the interests in the geographies of gender. There is thus now a wider acceptance of such scholarship amongst the younger generation of scholars. Part of this is also because gender concerns are now increasingly drawing greater attention and becoming mandatory.

Relatively speaking, there are fewer universities, about 57 according to one estimate (www.indiaeduinfo.co.in/careers/geog.htm), where geography is taught at the postgraduate level. According to an earlier report by the University Grants Commission (2001), geography was taught in 83 universities (including graduate and postgraduate courses). Although it is quite difficult to comment on the precise numbers of universities having a geography department, the fact remains that very few universities with geography departments have courses in social and cultural geographies which could have legitimately included gendered concerns as part of their existing courses or introduced dedicated gender and geography courses per se. However, along with the spread of seminars, projects, and publications, geography departments in government and private universities now have independent courses on different dimensions of gendered discourses at postgraduate levels. Arunachal, Delhi, Goa, Jamia Millia Islamia, Kurukshetra, and Mumbai universities are cases in point. Jawaharlal Nehru University is amongst the first few to introduce dedicated courses in geography on gendered concerns. Others include Ch. Ranbir Singh University (Haryana), Central University of Punjab, Diamond Harbour Women's University (West Bengal), and Khwaja Moinuddin Chishti Urdu, Arabi-Farsi University (Uttar Pradesh).[4]

Another dimension added to this gamut of activities, according to Datta (2016: 143–144), is the international presence of Indian scholars-in-residence positions, participation in numerous international conferences as plenary and keynote speakers, as well as presentation of papers on gender with reference to India. Indian scholars are also members of international editorial boards, review committees, and steering groups within the International Geographical Union and other international bodies. A number of universities also have students' exchange programmes with institutions/universities abroad. The broadening of national as well as international activities is seen by Datta as 'an outcome as well as driver of the widening of the subfield and its now greater acceptability'.

To sum up, I have picked up some of the place-specific nuances and complexities of women's gendered experiences. This is to argue, through snapshots, for reorienting some of the androgynous learnings of the discipline to reclaim what can be termed as 'missed opportunities', i.e. the inclusion of gendered responses and outcomes for making geography inclusive and receptive to changing environs.

Notes

1 This chapter draws and borrows from some of my previous writings.
2 In a conversational piece on Indian geography, Datta and De (2008) engage with some of these issues and the politics of knowledge production in terms of how certain knowledge gets privileged over the other. Making a strong case for 'other' geographies that may draw upon methodologies which do not always stand to the arduous 'scientific' treatment of the so-called 'dominant' geographies, the authors argue that the dominant is also masculine.

References

Agarwal, Bina. 1986. *Cold Hearths and Barren Slopes: The Wood-Fuel Crisis in the Third World*. London: Zed Books; Delhi: Allied Publishers; Maryland, USA.

Archer, L. 2004. 'Re/Theorizing "Differences" in Feminist Research.' *Women Studies International Forum*, 27: 459–473.

Bhairannavar, K. 2018. 'Women, Environment and Migration: A Case Study of Female Construction Workers in Delhi.' *The Horizon: A Journal of Social Sciences*, 1 (9): 16–27.

Bordia, Das. 2004. 'Muslim Women's Low Labour Force Participation in India: Some Structural Explanations.' In Z. Hasan and R. Menon (eds) *In a Minority: Essays on Muslim Women in India*, pp. 189–221. Delhi: Oxford University Press.

Borisa, D. 2018. 'Imagined Spaces of Freedom: Negotiating Queer Cartographies of Desires.' Unpublished Ph. D. Thesis, Centre for the Study of Regional Development. New Delhi: Jawaharlal Nehru University.

Carr, E. R. 2008b. 'Men's Crops and Women's Crops: The Importance of Gender to the Understanding of Agricultural and Development Outcomes in Ghana's Central Region.' *World Development*, 36 (5): 900–915.

Das, A., and A. Chattopadhyay. 2012. 'Women's Status in West Bengal and Bangladesh: A Cross Country Analysis.' *Demography India*, 41 (1&2): 83–102.

Datta, A. 2015. 'Gender Discrimination in Access to Health Care Facilities in West Bengal –A Case Study of Selected Towns of Kolkata Metropolitan Area.' Unpublished Doctoral Thesis, Jawaharlal Nehru University.

Datta, A. 2016. *Voices From the Disciplinary Edge: Mapping the Development of Gender and Feminist Geographies in India, Funded by the University of Delhi Under the R and D Grants*.

Datta, A. 2021. Personal Communication, 31 July 2021.

Datta, Anindita, and Aparajita De. 2008. 'Reimagining Impossible Worlds: Beyond Circumcised Geographical Imaginations, a Play in Many Acts.' *Progress in Human Geography*, 32 (5): 603–612.

Deshmukh-Ranadive, Joy. 2002. *Space for Power: Women's Work and Family Strategies in South and South East Asia*. Rainbow Publishers in Collaboration With Centre for Women's Development Studies.

Deshpande, S. 2003. *Contemporary India: A Sociological View*. New Delhi: Penguin Books.

Grace, Daphne. 2003. 'Women's Space "Inside the Haveli": Incarceration or Insurrection?' *Journal of International Women's Studies*, 4 (2): 95–126.

Gray, John. 1992. *Men Are From Mars, Women are from Venus*. New York: Harper Collins.

Hapke, H. M., and D. Ayyankeril. 2018. Published Online, Springer-Verlag GmbH, Germany, Part of Springer Nature.

Johnson, J. L., and R. Repta. 2012. 'Sex and Gender: Beyond the Binaries.' In J. L. Oliffe and L. Greaves (eds) *Designing and Conducting Gender, Sex and Health Research*, pp. 17–37. Thousand Oaks, CA: Sage.

Lefebvre, H. 1991. *The Production of Space*. Translated by Donald Nicholson-Smith. United Kingdom: Basil Blackwell Ltd.

Monk, J., and S. Hanson. 1982. 'On Not Excluding Half of the Human in Human Geography.' *The Professional Geographer*, 34 (1): 11–23.

Paul, T. 2012. 'Impact of Gendered Spaces on Female Employment in the Era of Globalization.' *Economic and Social Research Institute Journal*, 3 (2): 58–65.

Paul, T. 2013. 'Interrogating the Global/Local Interface: Workplace Interactions in the New Economic Spaces of Kolkata.' *Gender, Technology and Development*, 17 (3): 337–359.

Phadke, S. 2005. 'Unfriendly Bodies, Hostile Cities: Reflections on Loitering and Gendered Public Space.' *Indian Journal of Gender Studies*, 12 (1): 41–62.

Preeti and Rajeshwari. 2018. 'Violence Against Women in Haryana: Levels and Correlates.' *Punjab Geographer*, 14: 52–64.

Rajeshwari. 1996. 'Gender Bias in Utilization of Health Care Facilities in Rural Haryana.' *Economic and Political Weekly*, 31 (8): 489–494.

Rajeshwari. 2010. 'Levels and Determinants of Infant and Child Health and Nutrition in Rural Haryana, India.' *Malaysian Journal of Tropical Geography*, 41: 29–41.

Rajeshwari. 2011. 'Spatial Pattern of Child Nutrition in Rural Haryana: A Socio-Economic Analysis.' *Annals of the National Association of Geographers in India*, 33 (2): 56–69.

Raju, S. 2005. 'Limited Options – Rethinking Women's Empowerment 'Projects' in Development Discourses: A Case from Rural India.' *Gender Technology and Development*, 9: 253–271.

Raju, S. 2006. 'Contextualising Gender Empowerment at the Grassroots: A Tale of Two Policy Initiatives.' *GeoJournal*, 65 (4): 287–300.

Raju, S., and K. Lahiri-Dutt. 2011. *Doing Gender, Doing Geography: Emerging Research in India*, edited by Saraswati Raju and Kuntala Lahriri-Dutt. London: Routledge.

Raju, S. 2013a. 'The Material and the Symbolic: The Intersectionalities of Home-Based Work in India.' *Economic and Political Weekly*, 48 (1): 60–68.

Raju, S. 2013b. 'Women in India's New Generation Jobs.' *Economic and Political Weekly*, 41 (26): 16–18.

Raju, S., and D. Bagchi. 1993. *Women and Work in South Asia: Regional Patterns and Perspectives*. London and New York: Routledge.

Raju, S., and K. Lahiri-Dutt. 2018. *Doing Gender, Doing Geography Emerging Research in India*. New Delhi, London: Routledge.

Raju, S., P. J. Atkins, N. Kumar, and J. G. Townsend. 1999. *Atlas of Women and Men in India*. New Delhi: Kali for Women.

Raju, S., and T. Paul. 2019. 'Public Space and Place: 'New' Middle Class Women and the Gendered Interplay in Urban India.' In M. Bhatia (ed.) *Locating Gender in the New Middle Class in India*, pp. 99–128. Shimla: Indian Institute of Advanced Studies.

Ravera, F., B. Martin-Lopez, U. Pascual, and A. Drucker. 2016. 'The Diversity of Gendered Adaptation Strategies to Climate Change of Indian Farmers: A Feminist Intersectional Approach.' *Ambio*, 335–351.

Samanta, G. 2012. 'Unfolding the Debates Around Women's Empowerment in Development Discourse.' In K. K. Bag (ed.) *Changing Society, Culture and Its Impact on People*, pp. 73–82. Kolkata: Rupasi Bangla Publications.

Seemanthini, N. 2001. *Gender and Space: Femininity, Sexualization, and the Female Body*. New Delhi: Sage.

Sil, P. 2018. 'Creating New 'Places': Women and Livelihood in the Globalising Town of Burdwan, West Bengal.' In Saraswati Raju and Kuntala Lahiri-Dutt (eds) *Doing Gender Doing Geography*, pp. 108–128. New York, New Delhi: Routledge.

Sultana, F. 2018a. 'Water Justice: Why It's Important and How to Achieve It.' *Water International*, 43 (4): 483–493.

Sultana, F. 2018b. 'Gender and Water in a Changing Climate: Challenges and Prospects.' In Christiane Fröhlich, Giovanna Gioli, Francesca Greco, and Roger Cremades (eds) *Water Security Across the Gender Divide*, pp. 17–33. The Netherlands: Springer.

Tschakert, P., and M. Machado. 2012. 'Gender Justice and Rights in Climate Change Adaptation: Opportunities and Pitfalls.' *Ethics and Social Welfare*, 6 (3): 275–289.

Withers, C. W. J. 2009. 'Place and the "Spatial Turn" in Geography and in History.' *Journal of the History of Ideas*, 70 (4): 637–658.

14

BEYOND THE POSTCOLONIAL

Speculations on the Indian Contemporary

Yasmeen Arif

Robert J. Schiller's (2019) *Narrative Economics* is a compelling work on how events and 'stories' and their contagion can have an impact on how economics 'thinks' beyond statistics – a significant recalibration of a very objective social science. The more narratively oriented sociology and social anthropology, my disciplines of emphases in this chapter, have also long since moved into the more conventionally 'objective' realms of study, namely science and technology, life sciences, climate sciences, and so on, far beyond the human social that these disciplines were meant to circumscribe. In this milieu, the question of scientific method in sociology/social anthropology is not one that can be answered in that range of responses that draw on the binary divide between scientificity and subjectivity. In other words, the broad strokes that separate positivism, empiricism, objectivity, or arguments of reliability and validity (or the pursuit of 'truth', for that matter) to on one side and keep on the other, the intricacies of narrative structure, subjectivity, authenticity, plurality, representation, languages, and modes of writing are fading.

In this short reflection, I intend to propose a temporal frame – the 'contemporary', as elaborated below, to outline a set of responses to the question of 'scientific method' in the social sciences, by drawing attention to two notions. Firstly, I suggest a sensitivity to the crafting of an object of enquiry in the contemporary and, secondly, how that might influence response from methodological techniques and epistemological attitudes. The following sections are three parts of a reflection, each fulfilling the roles of proposal, context, and illustration, respectively. Overall, the discussion is written as a speculative essay rather than as a review of existing debates or of existing genealogies.

Proposal: Framing the Contemporary

I propose the contemporary as a spatial–temporal–material juncture of simultaneity, a recognition of material interconnectedness where research

DOI: 10.4324/9781003298908-21

concerns in sociology/social anthropology emerge in the assemblage of the local and the planetary as well as in the convergence of the past–present–future. In the current world horizon, issues of pressing concern, illustratively – climate, health, security, citizenship, identity, refugees, technology – are inevitably embedded in such assemblages. I do not allege unprecedented 'newness' in such assemblages, rather I underline the obvious epistemological tunnel vision that is evident when certain regions, peoples, time, or matter are not allowed to connect and resonate across geopolitical, ideological, or epistemological compartments.[1] Thus, the aspiration to understand a 'contemporary' positioning is to acknowledge that, in exploring the social, any object of enquiry is indeed shaped by such interconnections and resonances.[2] An echo may perhaps be heard here by those following the 'ontological turn' (Holbraad et al. 2017), or 'non-representational theory' (Thrift 2007), in social anthropology or variations of the Actor Network Theorizing made popular by Latour (2005). Others might find this no different from the political economy or world systems approach that saw a world interconnected by the stages of capitalism. While these theoretical reflections might be a matter of choice, my attempt in parenthesising the contemporary is to lead towards a methodological response that might be appropriate to it.

In this notion of the contemporary, a preceding position is Rabinow's (2008: 2), where he offers a definition of sorts, 'The contemporary is a moving ratio of modernity, moving through the recent past and near future in a (nonlinear) space that gauges modernity as an ethos already becoming historical'. An example clarifies his position better (2008: 3):

> [I]f one no longer assumes that the new is what is dominant ... and that the old is somewhat residual. Then the question of how older and newer elements are given form and worked together, either well or poorly, *becomes significant site of enquiry*.
>
> For example, the fact that the human genome has been mapped, and population differences at the molecular level identified, does not mean that older understandings of race disappear in the light of this new knowledge. But neither does it not mean that all of the older understandings of what constitutes difference undergo a total transformation (my emphasis in italics).

Rabinow makes an argument that suggests the reassembling of a civilisational temporal arc that underlines the knowing of biosocial life, where facts established by 'scientific' method collides with 'culturally' held meanings of race and difference. Arguing against any evaluations between 'new' and 'old', or respectively, 'better and progressive' as against 'bad and redundant', he thinks about the contemporary as an epistemological and methodological combination of such time-qualifiers on approaching a sociological/

anthropological 'site of inquiry'.[3] If time is his emphasis, then (there is, of course, much more in his argument), in addition, I would suggest a juxtaposition of space as well, as explained below.

The spatial and scalar strand that I pursue as a constituent of the contemporary is best expressed and reiterated in that ubiquitous problem in the social sciences – the scalar relationship of local particularity to universal theory. This relationship has been about the inherent hegemony of universals (read theory) over the empirical particular (read difference), and crucially, about how method and epistemology intervenes in this relationship. Another anthropologist who has written about the spatial prospect of contemporaneity in this particular–universal relationship is Marc Augé (1994) and his observations do point, in some ways, towards the direction I take here. In the preface to his book *An Anthropology for Contemporaneous Worlds*, he says (1994: x):

> The title of this volume combines a singular noun – anthropology – with a plural one: contemporaneous worlds. It is meant to register the double-movement universalization and particularization that is simultaneously affecting the entire planet. Social anthropology has always taken into account the context of the groups and phenomena it studied. Today, while multiplicity is being maintained or, more exactly renewed, that context has become for all cases, planetary.

His inflection seems to seek a kind of homology between singular planetary anthropology and multiple contexts on one side and the universal and the particular on the other. In a follow-up to the above, he says the following in the 'Afterword' (1994: 125–126):

> I have sought to show that in the unity and diversity of present contemporaneity, anthropology is not only possible but necessary. Necessary, because the question of social meaning, whether explicitly posed or implicitly present, is everywhere.
>
> an African-prophet healer, a group of architects working together on a development project, or a medical team trying to figure out how to intervene in this or that social or cultural milieu all constitute realities of the same nature. Adapting to changes in scale does not mean ceasing to privilege observations of small units, but rather taking into account the worlds that cross through them, overflow them and in so doing, continuously constitute and reconstitute them.

The important point of scale, parity, and interconnectedness that comes through is vital in the sense of contemporaneity that I propose – however, there is a twist here that I mention below and will underline rather heavily.

While these multiple realities and particulars are 'equal', and social meaning is indeed ubiquitous, the potential of articulating them as knowledges is dependent entirely on epistemic privileges that are embedded in methodological practices – which, in turn, are deeply embedded in spatial imaginations. Who gets to make meaning and from where becomes the important question, rather than whether that meaning is constructed 'scientifically' or not, whatever consensus or critique we might have on what is scientific.[4] In most instances, the inequity of privilege emanates from a geopolitical cartography, an acutely spatial knowledge politics. The critique against it appears in the form of a geopolitics of epistemology where 'the global south' is placed in opposition against 'the global north'; or, in another way, 'postcolonial' epistemological positions are asserted. In S/SA, a follow-up to this, in somewhat recent times, has been the political claiming of epistemological and methodological reparation through the decolonising movement.[5] Established as a call against racialised, colonial, masculinist (to which I add, Brahmanical) monopolies of knowledge production, in both material and intellectual practice, the decolonising call also underlines methodological perspectives, which must recognise and make valid the experiences, the voices, the world views of those routinely subjugated or invisibilised in S/SA knowledge genealogies.

An older version of these debates in classical SA was the alleged irrationality of the primitive 'others' versus the rational and reasoned discourse of the modern 'self'. A gendered version of this, of course, continued till recently in SA, the masculine author who spoke with scientific objectivity about the feminised irrational, cultural object. Critically important in these debates is also the hierarchy that is assumed within various social sciences, leave alone the continuous one between the 'hard' and 'soft' sciences. The challenge in this, within SA, is a double articulation – one, finding ways to balance out the epistemic inequities that are inherent in geopolitical knowledge production practices. Second, and this is a matter of my specific emphasis here, find methodologies for conceptual innovations that are adequate for global connections/resonances as well as local relevance. The following section outlines some institutional history and disciplinary precedents in India, hinting at the methodological or epistemological positioning of an 'Indian stance', if any, and what that might mean for a possible Indian contemporary.

Context: Proposing an Indian Contemporary

Two trajectories seem to dominate the historiography of S/SA in India, one of which goes to the heart of a methodological divide between sociology and social anthropology. A case in point is the MSc programme of anthropology at the University of Delhi as contrasted to the MA programme of sociology housed in the Delhi School of Economics at the same university – the nomenclature is pre-emptive of what the course contents are intended as. Space constraints limit a detailed discussion; however, a quick purview of the current course syllabi of the MSc course will show that social anthropology is one part of the syllabi, where the other part is largely under the rubric of physical anthropology, which includes project and practical work on topics like paleoanthropology, molecular anthropology, dermatoglyphics, biochemical genetics, somatometry, somatoscopy, serology, osteology, osteometry, craniometry, and so on. Social anthropology in these syllabi include courses on social theory and cover topics like that of the urban, the environment, tribes, kinship, marriage, religion, symbolism, and so forth. The methods and techniques accompanying these courses, commensurately, are divided between scientific, quantitative techniques and qualitative, ethnographic methodology. Overall, this kind of anthropology programme is reminiscent of the conventional anthropological programmes in North America where a similar programme will typically have four divisions: physical, linguistic, archaeological, and cultural or social anthropology. The MA programme in sociology, on the other hand, at Delhi University is more an extension of what is called social anthropology in the classification mentioned above – more on that shortly.

A small detour here: histories, 'originary' moments, and intellectual traditions of social anthropology and sociology in India and their ongoing trajectories have found narration as personality based and institutional accounts of how the disciplines have developed at various points in time.[6] A more generic commentary is Shiv Visvanathan's (2006) contribution in a landmark volume by Escobar and Rebeiro, *World Anthropologies*, and it is a useful frame with which to situate the discipline in India. Visvanathan's narrative identifies a set of almost chronologically contiguous phases, 'shifting scenarios', as he calls them, ones that articulate the many anthropological imaginations about India as well as the various institutional formulations they took. In this sense, his essay brings into relief the concerns of the many epistemic phases. Briefly, his starting point is the double-sided colonialist view – those who considered India to be 'a site to be surveyed and ruled', but also saw it as the venue for alternate epistemologies. While some phases privileged the civilisational lens – for instance, how does India mediate Western sciences and knowledges – others found the nationalist movement another lens with which to understand India.

Post-independence India saw the emergence of self-conscious intellectuals who struggled with the forces of the day – modernity, development, progress, and the Marxist critique in the unique theatre of tradition and modernity that India staged. The research agenda in fieldwork was focused on 'objects' like the 'Indian Village' as perhaps the real mirror of a microcosmic society that could reveal what authentic India was about. Straddling this was the massive influence of Louis Dumont and his driving conviction that India was to be seen as a theoretical whole (largely through an exposition of the Varna or Caste system of hierarchy) capable of standing as an 'other' to the West.[7] The 'new' discovery of India from within and outside (in varying orientations and agendas – amongst which was modernisation, progress, and scientific rationality) sustained a sociological imagination and excitement about internal hopes and desires as well as external curiosities and expectations of a newly independent subcontinent.

However, these eager explorations were to be rudely jolted out of complacent contentment, as Visvanathan (2006) notes – when Indira Gandhi imposed the emergency in 1975. The rude jolting forced the gaze away from village India, forcing it towards more disruptive locales and events. Clearly, the emergency marked a turning point in the historical records of independent India, changing fundamentally the ways in which social science or an anthropological orientation could document the country. The second rupturing moment, I would add, is that of the Sikh Carnage of 1984 following the assassination of Prime Minister Indira Gandhi. The language of violence in SA was a putative discourse through which social tragedies were beginning to unfold.[8] After the emergency and the 1984 event, in the last three decades or so, 'Indian' S/SA has witnessed a discursive expansion of issues from state, democracy and citizenship, communal violence and minority rights, urbanisms and developmentalisms to media, art, science, and technology studies – with a lot more interspersed.[9] The unquestioned credo was that S/SA in India is definitively about ethnographic fieldwork, and secondly, the point of the celebration was that India is no longer a field site for foreign researchers alone, Indian scholars have themselves been able to garner a vast body of empirical work on India. The analytical toolkit that accompanied all these developments reflected global theoretical environments from structural functionalism, political economy perspectives, postcolonial critiques, to post-structuralisms.

The latter half of the above historical sketch, arguably, develops significantly in the elite department established in the University of Delhi, in 1959, at the Delhi School of Economics (much in the lines of the London School), which I use as an exemplar here. One of the founding scholars of the department was M.N. Srinivas whose vision was to develop a unique disciplinary orientation that would encompass a blend of sociological and anthropological traits and rest on the double foundations of comparative work and field research.[10] The sociological arm, mirroring its North American counterpart, was to be about the study of one's own society, enriched by comparison, whereas the social–cultural anthropological arm was to be about embedded fieldwork. This was also the time when 'South Asia' as a label for an area studies programme came to be established across a spectrum of stellar universities in the United States. The sketch of the two departments at Delhi University does highlight an epistemological attitude – the study of human society, or the 'social' as such, was cohered in a classificatory system of two parts – one that divided the 'physical' objective, scientific aspects of a physical anthropology from the more 'social–cultural' subjective, ethno-graphic, qualitatively accessed anthropology.

In winding up this brief summary of the Indian context, a segue to what the Indian contemporary is will be the question of how an exploration of the Indian social, emerging from the historical narrative outlined above, might be placed epistemologically in the globally interconnected world. Secondly, what could contribute to conceptual innovations in ways that give mean-ing to that positioning as strength, rather than as articulations of colonial subjugation alone? The goal, succinctly, is to find a 'method' that explic-itly recognises a decolonising era when the exultant stance of multiplicity and heterogeneity against the ravages of universalising, colonising theory is somewhat past its prime. This is where I propose a movement beyond the postcolonial, where there is the potential of opening the defined limits of the postcolonial epistemological stance as an assertion of locality alone, or particularity, or authenticity measured in diversity alone. In addition, I also urge that the epistemic right of knowing and writing oneself against colonis-ing and hegemonic ways of knowing should aim more towards autonomy and less about geopolitical difference, opposition, or *re*-action. In this stance beyond the postcolonial, focusing on India through this lens is far less an exercise in finding Indian specificity – that, I expect, is an aporetic question. I would not like to arrive upon a sense of 'Indian' sociology or social anthro-pology – the variety of social meaning that this exercise might produce is going to be difficult, arguably, to fit into an existing category called 'India' – whether that is a geopolitical, historical, discursive or 'imagined' category.

The point that I reiterate again is that the question is about what con-tours a research objector query has in the contemporary and how might method respond to it. Notwithstanding the illustrative disciplinary division of accessing the social at the University of Delhi between physical anthro-pology and social anthropology, pedagogy in S/SA at large now encourages a 'mixed methods' approach combining both qualitative and quantitative toolkits are more the order of the current day.[11] The established mode of immersive, 'ethnographic' work is also now an amalgamation of various perspectives and practices as well as the focus of very complex criticism. The mixed-methods approach is considered to have been the way out of the paradigm wars between the 'qualitative and the 'quantitative' – it was ushered in as a way of ensuring the best possible – meaning doubly valid – research results in social science queries.[12] Both were meant to rectify the 'representative' gaps left by the other and hence the best way forward, methodologically.

Condensing a wide array of perspectives that could be aligned with the mixed methods approach, the methodological spectrum that can be mobi-lised to address an object of enquiry can range, currently, on the one hand, from the avowedly subjective realms (the qualitative) of 'autoethnogra-phy', where the experiencing singular subject is the producer of data and

302

social knowledge. On the other hand, the enquiry can also accommodate responses that involve the mining of large data sets with statistical tools of great finesse (the quantitative).[13] The simple but serious point seems to be as follows: what is the nature of the social phenomena that is sought to be addressed, investigated, and represented, which can lead to the production, so to speak, of reliable, valid, and, more importantly, *relevant* sociological/ social anthropological knowledge. Clearly, following the twists and turns of representation of the qualitative and the quantitative, the underscored point is this: *method* is required to fundamentally acknowledge and respond to the changing social environment in which questions are formulated. And added to that responding, there must also be a clear methodological awareness as well as epistemological calibration of historical continuities and ruptures in that social object of enquiry.

However, the potential of epistemic and methodological calibration in India is a more nuanced problem of politics, location, and privilege. An illustration can be made to nudge along this argument – the continuing conditions of violence, humiliation, and subjugation in Dalit experience. This example can bring the focus back to the problem of how epistemic inequities are connected to the methodological process of representation and the possibilities of knowing. Dalit articulations in the production of social knowledge are such claims for epistemological positionings that need to attain validity, inclusion, and representation. The accompanying methodological claims that respond to that question may well be auto-ethnography or biographical literature, which rebels at any 'universalistic', homogenising sociological and social anthropological theory that explains the caste system, but barely acknowledges the epistemic silences. Jaware's (2019) point of anthropological and sociological complicity in Dalit subjugation is a powerful statement here when he suggests that studying the caste system 'objectively' – say, purity and pollution practices, hardly addresses the brutality of that reality. On the one hand, excavating the historical record of subjugation and silencing is an ethical motivation that squarely challenges continuing epistemological inequities. And the 'evidence' for that record is the question of both methodological and epistemological recalibration where the production of knowledge through, say, 'auto-ethnography' is aimed at making explicit that lacuna. On the other hand, there is also an urgent need to excavate what allows such oppressive knowledge production practices to flourish in a mature post-colony, or have governance measures of representation been socially and materially effective. In interpreting both, the quantitative data that could be generated to explore the above queries will remain a necessity of sorts. Both combined will endorse an epistemological intention that speaks of a temporal imagination, which reaches into archives, faces the present, and speculates the future.

At the same time, in the contemporary that is under purview here, what spatial reach do Dalit experiences have, as an object of enquiry? Does an epistemological parity show with the 'Black Lives Matter' movement, for instance? Local voices have made that connection evident in social media platforms and in some academic or popular writing – Does this kind of emerging resonance demand a methodological approach that can work across national boundaries in understanding the claim of Dalit voices? The question is not about comparison alone – that is a well-formulated methodological position that has its own points of concern. The question is how does the contemporary reveal itself in these networked connections – material, intellectual, and historical – and how much do those perspectives help in understanding 'local' ecologies as much as planetary ones. Finally, this move is a methodological movement that enables renewed connection of local specificity with global concept. Perhaps, my plea is that the epistemological urgency of the contemporary is indeed to find global parities in social experience, where solutions might as well be local, but a theoretical commonality is a serious requirement.[14]

Another example can be made of potential futures without 'continuities' with the historical past, for instance, social science questions that turn towards digital life which likely needs a possible combination of data sets as well as virtual ethnographies that are not necessarily bound to any geographical location in the conventional way. In such an illustration, the exploration of digital labour in India, for example, will require an eye on globally networked technologies and another eye on what creates local technological conditions, inequality being an undoubted emphasis. Once again, the connection between planetary practices and local specificities emerge and once that connection is unravelled, it will become important to look at what role the locally 'different' specificity might offer to global theory and practice.

In the last section below, I offer a sketch of my current work as an illustration, which reflects the material–spatial–temporal assemblage I have argued for. And, in some resonance with the animating themes of the volume, they bring the question of science and the social in close conversation, especially in the contemporary. The illustration speaks with the arguments made so far – showing how the crafting of an object of enquiry might bring interdisciplinary and theoretical genealogies into conversations that reflect epistemological concerns at large in the relationship between science and the social. In addition, by posing a specificity question from 'India' in that larger theoretical universe, global connectivity as much as local specificity reveals the importance of the concern with contemporaneity I have gestured towards, especially as seen from where I write.

Illustration: Life and the Emerging Social

Summing up my arguments so far, the study of human social life seems to be about a double-sided challenge – firstly, the adequate framing of an object of enquiry that is attendant to an empirical excavation of the social in situ (cultural–social embeddedness). Secondly, its abstraction into a social analytic that can be theoretical in so much that it can have potential global resonance and equitable expression. Both of the above, together, need methodology and epistemology that is reliable and relevant. From empirical location to analytics, the relationship between a research problem's shape on the ground and its explanation in terms of the global and what that relationship is at any moment in time is also the way in which that object of enquiry finds its *contemporary* contours. Following from some of my earlier work, I now pose the social sensing of life as a rubric of enquiry, among other concerns, to explore biopolitical questions of life, law, identity, and the social. Writing this chapter at the time of an unprecedented event, the COVID-19 pandemic, it is impossible not to be aware of the challenges that the combined questions of the social and the biological pose – the biosocial, as it were. In addition and not in the least, it is also the apparent abrupt consciousness of a global contemporary with a peculiar relationship between the local and the global. How do we begin facing that challenge from India? The following outlines the methodological and epistemological questions that pre-empt this challenge.

That the biological and the sociological run together in understanding human existence has been an irrefutable awareness, but not always an acknowledgement. In that history of an interdisciplinary relationship, the turn of phrase, biosocial, is not new. Taking a short detour into a very complex historical genealogy, I take recourse to a few social science interventions, mainly from sociology that have left a mark on the 'dialogue' between the social and the biological. As much of the syllabi in sociology/social anthropology that we continue to teach in India or elsewhere will attest our ubiquitous 'founding fathers', for example Emile Durkheim or Max Weber, fall amongst those who worked with a notion of establishing a distinct 'science' of society, measuring their independence from the prevailing simplistic biological modelling of society as 'organism' but worked with what Maurizio Meloni et al. states as (2016a: 9), 'a proliferation of conceptual hybrids and the intense – but unnoticed, or neglected – transfer of knowledge' between science and the social. As Meloni explains, historically, the illusory opposition between life sciences and sociology was also because of the way in which either science had progressed.[15]

To come to point, when the life sciences came to be encapsulated by the gene or genetics, a literal substantial separation of what the subject matter of biology will be and that of sociology must be came to pass. As Meloni (2016b) explains, the gene as the pathway of biological 'hard heredity', impervious to change, sealed itself off from the genealogies of sociocultural, malleable, and soft 'heritage' that the social could excavate and explore. That in itself was a wedge driven between two epistemological perspectives conventionally attached to either the social (soft) or the scientific (hard), on two sides of that mirror of the biosocial. However, into the contemporary, in what Meloni et al. (2016a) call a 'social turn' in biology, there seems to be some noticeable proliferation and perhaps excitement even about the kind of effects the social might have on these hitherto closed chambers of bioscientific matter and knowledge production.

Meloni and his colleagues write about the ongoing changes in the field of genetics and epigenetics and I quote their work in some detail as it pinpoints the kind of fascination I would share. Meloni et al write (2016a: 10):

> This new conception of the genome is increasingly represented today as a 'vast reactive system'..., an 'exquisitely sensitive reaction (or response) mechanism ... for regulating the production of specific proteins in response to the constantly changing signals it receives from its environment'
>
> However, not only is the human epigenome (the set of epigenetic changes in an organism) seen today as modulated by social experiences, but the same genome, which was supposed to be immune from this level of contingency, is increasingly understood as falling *'within the parameters of the human life span'*... This is a major shift from the fixed view 'that came into being through the massive sequencing efforts of the 1990s and 2000 in which genomes were understood as 'the same in every cell of the body for all of that body's life' (2016a: 153).

These sorts of discoveries make the biological–social boundary far more permeable than was reckoned with, but once again, these are concessions that are far more on the side of the sociologists rather than those of the life scientists. This permeability between the biological and the social leads to the literature in anthropology, where empirical and ethnographic work has long challenged the nature–culture divide that the sociological literature above has suggested and subsequently contested. The point I underline here is the necessary interaction between the qualitative methods of soft science approaches and the quantitative reliability of the hard science practices in posing these questions and finding responses to them. Nonetheless, they both imply a theoretical universal, theoretical reach with more or less global relevance.

Privileging and selecting the work and authors that support my line of argument, I turn to the ideas that Ingold and Palsson (2013) discuss in their formulation of what they call 'bio-social becomings'. Resolutely Deleuzian in their approach, they write in the front page of their volume that human life is a 'becoming' that is 'neither given in the nature of our species nor acquired through culture, but forged in the process of life itself'. Their critique is directed to the fallacies inherent in a Neo-Darwinian model of explaining 'culture' and its transmission, where they say that the

Neo-Darwinist position follows the argument summarised below (Ingold and Palsson 2013: 5):

> The idea is to come up with a model of observed behavior, a 'culture type' (strictly analogous to the 'genotype' of biology), that is entirely context-independent. It is then supposed that this model is pre-installed inside the heads of individual carriers whence it is alleged to generate the described outcomes under the particular environmental or contextual conditions they happen to encounter. Thus in effect, is culture 'ultimately' explained by culture?

This rather alarming and fallacious way of delimiting culture types is an illustration of how a biological perspective, insularly limited to its own epistemological truths, attempts to co-opt and explain the social and cultural. Notwithstanding, either perspective of the sociological or the biological articulates a fundamental set of questions – the evolution of human life, how the individual relates to a population, what are the modalities of reproduction in either paradigm (like the heritage and heredity distinction above), and whether it is about social or genetic 'traits'. The singularly important idea that Ingold and Palsson propose is this (Ingold and Palsson 2013: 6): 'Evolution, in our view, does not lie in the mutation, recombination, replication and selection of transmissible traits. It is rather a life process. And at the heart of this process is ontogenesis'.

I cannot detail here the nuance this proposition deserves; however, what Ingold and Palsson recommend is a thorough refurbishing of our divided knowledge systems and understand that *the biological and the social are the same*. Dramatic as this pronouncement might sound, it captures the point about an ontogenetic orientation, where both the social and the biological organisms develop together, where neither is complete in their ability to explain the trajectory of evolution, and that this trajectory is infinite. In that sense, the notion that is more befitting here is *life* as an ontology of becoming. This holds together with the approach that neither is the gene a stable entity nor is culture (which is well-known), that the individual is not a stable carrier, nor is the population. What remains constant though is the idea that both do travel together in this thing called life. Once again, I use their words to underline the significance of this development (Ingold and Palsson 2013: 9):

> That life unfolds as a tapestry of mutually conditioning relations maybe summed up in a single word, social. All life, in this sense, is social. Yet all life, too is biological in the sense that it entails processes of organic growth and decomposition, metabolism and respiration, brought about through fluxes and exchanges of materials across the membranous surfaces of its emergent forms. It follows

that every trajectory of becoming issues forth within a field that is intrinsically social and biological, in short bio-social.

Reading this perspective into the discussion so far, two notions find shape. One, an enhanced dynamism and fluidity in hitherto unacknowledged ways in given 'natural' matter, for example, the gene motif of biological deliberations, and, two, the crucial invocation of the question of life as an expanding arc of the biological *and* the social *together*, rather than as a product of a possible exchange over the 'boundaries' between them. If the above is a stated human condition where the social and the biological are co-constituents of human life, then that possibility is universal relevant on a planetary scale. This small sketch indicates the epistemological variety that can be summoned with both hard and soft science methodologies so as to propose, potentially, universally relevant conceptualisations in the realms of the biosocial.

What might a view from an 'Indian specificity' look like in this vector of the biosocial, assuming a vantage position in the current COVID-19 contemporaneity? Without doubt, the material factors (still planetary or universal in concept) of population, density, class, economy, governmental policy, and medical care/capacity will fall in line. And further, without doubt, they will also be the addition of religion and caste as the infamous Indian contenders to specificity. I will leave the substantial elements of that argument out of this discussion to make space for the argument of method and epistemology.[16] The questions that could be posed here are as follows: What might be a way of capturing this movement between the particular of Indian caste and religion and the universal of the virus – what social and political practices will mediate that relationship? Or, how can the interconnections within the assemblage of medical care, economies of exclusion and inclusion, spatialised governmentalities, varying densities of cities, and the variations of social distancing be measured? The formulation of these sorts of 'contemporary queries' will indeed be the assembly of method and epistemology that I argue for here.

Comaroff and Comaroff's (2003) notion of 'anthropology on an awkward scale' is an example of the kind of assembly I suggest. Speaking both of ethnographic methodology and epistemic implication, they discuss this possible anthropological approach in terms of their research on the rise of an 'occult economy' (which implies practices and beliefs that connect magical means and mysterious techniques to the materialisation of wealth) in South Africa. Investigating the peculiar appearance of 'zombies' in Mafeking, they make an interpretive suggestion that the figure of the zombie, in effect, is a peculiar product of the interstices of neoliberal capitalism and vernacular ways of refracting multifaceted experiences of globalisation, poverty, alienation, and so forth. It is a product that does not find interpretive fullness in the ethnographic limitations of the locality, say, in relations to sorcery and witchcraft, but rather in a social imaginary that is surmised from an 'awkward' ethnography that starts with something found *in situ* but whose explanation marks the movement from the local to the supra-local, the concrete to the conceptual. Keeping their substantive arguments aside for the moment, and keeping the focus on the methodological aspect, I quote (Comaroff and Comaroff 2003: 151):

> By what ethnographic means does one capture the commodification of human beings in part or in whole, the occult economy of which it is part, the material and moral conditions that animate such an economy, the new religious and social movements it spawns, the modes of producing wealth which it privileges, and so on? Inherently, awkward of scale, none of these phenomena are easily captured by the ethnographer's lens. Should each of them nonetheless be interrogated purely in their own particularity, their own locality? Or should we try to recognize where, in the particularity of the local, lurk social forces of larger scale, forces whose sociology demands attention if we are to make sense of the worlds we study without parochializing and, worse yet, exoticizing them?

In many ways, the sense of the contemporary is perhaps the common condition which implies that localised ways of living or, in other words, heterogeneous and multiple contexts are first connected to larger and expanding material and discursive universes. Second, they are just as well intra-connected within local, cultural, or institutional practices. For the pursuit of anthropology, this ontology of global connectivity and the local interface is also about the likely advent of new forms of socialities, ethics, and politics, economies and practices that need apprehension through appropriate ethnographic method and fieldwork. If the COVID-19 milieu is to continue as an example, the unfolding of these relations will be in vectors of the evolving and the emergent.

A methodological implication, thus, of mapping such global connectivities and local interfaces is in what Maurer (2005: 2) underlines as a kind of ontological connections and calls them 'autodocumentations', which is akin to my notion of emergence in anthropological ways of knowing.

> As autodocumentary, these fields always oscillate between the particular and the general ("this patient" vs. "medicine", "this case" vs. "the law" ... "this news story" vs. "the media" ...) and they make their claims to knowledge on the basis of this oscillation.
>
> These characteristics – bleeding across the frame, hybridity, autodocumentation or reflexivity and the continual shift in perspective between general and particular to generate knowledge – are also, of course hallmarks of anthropology.

For my arguments here, Maurer suggests three interrelated parameters – firstly, the social anthropological imagination, when it apprehends this contemporary sense of emergent forms, finds its articulation in the simultaneous drawing up of descriptive, methodological, and analytical scripts that are continually and processually making the contours of the object (the complex hybrids) itself clearer (they are 'autodocumentary'). Secondly, these are processes that illustrate the negotiations between universals and particulars, or even empirical specificities and theoretical generalisations. Thirdly, these negotiations are expressive of the base anthropological reflexivity of social cultural study ('our' societies and 'theirs', the human condition).

In this schematic of methodological devices and modalities above, there seems to be a ready template for a critical anthropological method of and in the contemporary. With due critical reflection then, firstly, recognition of complex hybrids has been a known social science perspective, especially in a post-structuralist or postmodernist stance – the added inflection here is the continually shaping objects of enquiry – ones that underscore trans-local connective genres, on the one hand, and particular–general interfaces on the other. Secondly, and significantly, this template is about locating epistemological positions that seemingly remain embedded in regimes of locational hierarchy that continue to be disavowed – the questions of gender, caste, and religion as markers of these positions come to the fore in India – alongside the already accepted points of contention in global geopolitical hierarchies. In the light of such a critical milieu, what sociological/social anthropological work can be done with an empirical connect between, say, the pandemic world experiences at large and on the other plane – the urban political ecologies of Delhi, the returning migrant labour in rural India, the spatial densities of Dharavi in Mumbai? As Maurer says above, the necessary epistemological autodocumentations will be the oscillations between the fields of the particular and the general. To this, I add the locational, often hierarchical and subjugated, positioning of those epistemic positions from India, not as entrenched or insular ways of 'knowing' the pandemic but rather as important contributions to the global knowing of the pandemic.

To conclude then, the problem of epistemology and knowledge production seems to have been trapped within the two inadequate stances of the last decades of the 20th century. In Canclini's (2006: 296–297) words:

[O]n the one hand, the entrenchment of certain African, Asian or Latin American thinkers with 'their own ways' of producing knowledge and developing culture; on the other hand, postmodern narratives – particularly influential in metropolitan anthropology and cultural studies – which carried to an extreme the praise of difference and the positioning of the autonomy of the forms of knowledge of each ethnic group, gender, country, or subaltern group, as a supreme value.

The inadequacy does not, of course, lie in these knowledges themselves, but rather in their inability to transcend their insularities and their relativisms to comprehend a global order that links each of these knowledges, each of their positionings, or their vulnerabilities in an incessantly interactive web. The challenge then is not just to acknowledge the presence of multiplicities, or rather multiplicities as fragments but to fathom how these multiplicities/fragments are interlinked and interdependent in a current world across complexities of geography and history. And it is again this interdependent web that dismisses, devastates, challenges, and delegitimises ways of being and knowing, making some more vulnerable than others. Staying with the much too contemporary example, even under the influence of an allegedly homogenising virus, the power of particulars and the critical need for their articulation are concerns that the current pandemic has made all too visible.

I need to emphasise again the basic blending of these epistemological claims with the mixed method approach that necessarily requires a blending of conventional bioscience with social realities.[17] Clearly, a sociological or anthropological attempt that can tackle this horizon will require a combined crafting of conceptual innovation and empirical rigour. It has to avoid the obvious pitfalls of totalising discourse or fragmentary ethnocentrisms. In my illustration of the notions of the biosocial and their connections to the current pandemic, I have highlighted the set of concatenated features that lead from disciplinary relationships to local empirical motifs embedded in global developments. At one level, it traces the wider discursive, theoretical implications of the social, the biomedical, and the governmental. Each will have gaps that the other can, through methodological intervention, make rigorous. On another level, it locates and connects empirical particularities and vernacular specificities through which and towards which the global channels its negotiations. Links between the observed particularities with greater world systems, historical developments, and conceptualisations in theory become evident much in the way that my earlier references to developments in the biosocial discourse. That indeed is the importance of crafting a research object of enquiry which takes into account the multiple perspectives and 'data' required to find an adequate response or answer.

What ways of knowing, what epistemic stances, what methodological tools can comprehend this challenge and find ways, simultaneously pliant and robust, which can comprehend multiplicities and acknowledge fragments, yet comprehend their static or kinetic place in an interactive, often abstract web. In another way, what bridges of understanding can make vivid the connection between multiple saturated times and places, multitudes of groups, world views, ways of living, and their unequal relations to a planetary web. I expect a response to these queries would lay out the pathways to understanding the relationship between method and social sciences in the contemporary.

Notes

1 This kind of global interconnectedness which goes unacknowledged has been well discussed in historiography. Historians I will privilege on this are Subrahmanyam (1997) and Morss (2009).

2 See Arif (2014) for some discussion on this.

3 There is a selective privileging here of Rabinow amongst others. As I illustrate later, I am focusing on the biosocial realm of sociological research so as to drive home the methodological orientation of those research areas that blur the line between 'objective' and 'subjective' methods.

4 I do not want to comment on the epistemic privileges and hierarchies that abound in the hard sciences or in the scientific community as such – there is discussion on that, amongst which, Bruno Latour's is well-known from the anthropological perspective.

5 See, for example, Harrison (2010) and Mignolo (2009).

6 The much too brief sketch of the Indian context summarised here in lieu of any history is a description aimed at supporting the points that I argue for here. For the biographical and institutional aspects, see, for example, an older essay by Srinivas and Panini (1973) and more recently, Uberoi et al. (2007) or Deshpande (2018).

7 Caste is too large an archive in S/SA in India to summarise here. Suffice to say that there is enough critique of Dumont's position. In addition, there have been actual misguided denials of casteism (see Visvanathan [2001] for a perspective) in sociological discourse, as much as continuing research and debate – another nuance of the Indian contemporary.

8 See Veena Das (1985).

9 A recent volume by Sanjay Srivastava, Yasmeen Arif, and Janaki Abraham et al. (2018) is a good survey of S/SA in India in the recent decades.

10 See Andre Beteille (1993), "Sociology and Anthropology: Their Relationship in One Person's Career". *Contributions to Indian Sociology* (n.s.) 27(2): 291–304.

11 Both quantitative and qualitative methodologies are taught in the MA programme in sociology; however, a detailed quantitative curricula has been a recent development. At any rate, students are expected to understand the necessity of both in developing their methodological approaches.

12 A good example, among a vast collection, is Clark and Creswell (2008).

13 The distinction between sociology and social anthropology in the North American framework followed a classical distinction between statistical work in the former and ethnographic work in the latter – respective syllabi in each included appropriate instruction. However, it is the blending of that is mostly evident for a long time now, not just in the United States, but worldwide. No quick summary of this blending can be made – multi-sited ethnography, emic or etic views, archives, and historical sociology alongside fine-tuned statistical tools can potentially animate a research project.

14 Wilkerson's (2020) recent prize-winning *Caste: The Lies that Divide Us* may be an example of conceptual fluidity that helps connect, rather than divide on the grounds of specificity, authenticity, and difference. This example is by no means a review statement of the book. See Arif (2014) for some discussion on this kind of argument.

15 Meloni et al. (for example, 2016a) and his colleagues writes a clear and succinctly illustrative history of this relationship. I convey here a small part of this.

16 It is much too early to deliver any firm analytics on the "Covid Social"; however, a reflection on this is in my 28th April blog in Diacritics – https://www.diacriticsjournal.com/life-and-the-interim/

17 My endorsing of the mixed methods approach for my project does not exclude 'single' method approaches at all – rather, method, at any point, should best be a reflection of the research object at hand.

References

Arif, Yasmeen. 2014. 'The Audacity of Method (Special Article).' *Economic and Political Weekly*, L (1): 53–61.

Augé, M. 1994. *An Anthropology for Contemporaneous Worlds*. Stanford: Stanford University Press.

Beteille, A. 1993. 'Sociology and Anthropology: Their Relationship in One Person's Career.' *Contributions to Indian Sociology*, 27 (2): 291–304.

Buck-Morss, Susan. 2009. *Hegel, Haiti and Universal History*. New York: Zone Books.

Canclini, Nestor Garcia. 2006. 'Totalizations/Detotalizations.' *Theory, Culture and Society*, 24 (7–8): 296–301.

Clark, V., and Creswell, J. 2008. *The Mixed Methods Reader*. Los Angeles: Sage.

Comaroff, J., and Comaroff, J. 2003. 'Ethnography on an Awkward Scale: Postcolonial Anthropology and the Violence of Abstraction.' *Ethnography*, 4 (2): 147–179.

Das, V. 1985. 'Anthropological Knowledge and Collective Violence: The Riots in Delhi, November 1984.' *Anthropology Today*, 1 (3): 4–6.

Deshpande, S. 2018. 'Anthropology in India.' In Hillary Callan (ed.) *The International Encyclopedia of Anthropology*. Hoboken, NJ: Wiley Blackwell.

Harrison, Faye V. 2010. *Decolonizing Anthropology: Moving Further Toward an Anthropology for Liberation*. Arlington, VA: Association of Black Anthropologists, American Anthropological Association.

Holbraad, Martin, and Pederson, Morten A. 2017. *The Ontological Turn: An Anthropological Exposition*. Cambridge: Cambridge University Press.

Ingold, T., & Palsson, G. (eds). 2013. *Biosocial Becomings: Integrating Social and Biological Anthropology*. Cambridge: Cambridge University Press.

Jaware, Aniket. 2019. *Practicing Caste: On Touching and Not Touching*. New Delhi: Orient Blackswan.

Latour, Bruno. 2005. *Reassembling the Social: An Introduction to Actor Network Theory*. New York: Oxford University Press.

Maurer, B. 2005. 'Introduction to "Ethnographic Emergences."' *American Anthropologist*, 107 (1): 1–4.

Meloni, Maurizio. 2016b. 'From Boundary-Work to Boundary Object: How Biology Left and Re-Entered the Social Sciences.' *The Sociological Review Monographs*, 64 (1): 61–78.

Meloni, Maurizio, Williams, Simon, and Martin, Paul. 2016a. 'The Biosocial: Sociological Themes and Issues. Editorial Introduction.' *The Sociological Review Monographs*, 64 (1): 7–25.

Mignolo, Walter D. 2009. 'Epistemic disobedience, independent thought and decolonial freedom.' *Theory,Culture and Society*, 26(7–8): 159–81.

Rabinow, Paul. 2008. *Marking Time: On the Anthropology of the Contemporary*. Princeton: Princeton University Press.

Schiller, Robert J. 2019. *Narrative Economics. How Stories Go Viral and Drive Major Economic Events*. Princeton: Princeton University Press.

Srinivas, M. N., and Panini, M. N. 1973. 'The Development of Sociology and Social Anthropology in India.' *Sociological Bulletin*, 22 (2): 179–215.

Srivastava, Sanjay, Arif, Yasmeen, and Abraham, Janaki (eds). 2018. *Critical Themes in Indian Sociology*. New Delhi: Sage.

Subrahmanyam, Sanjay. 1997. 'Connected Histories: Notes Towards a Reconfiguration of Early Modern Eurasia.' *Modern Asian Studies*, 31 (3): 735–762.

Thrift, Nigel. 2007. *Non-Representational Theory: Space, Politics, Affect*. New York: Routledge.

Uberoi, P., Nandini, S., and Satish, D. (eds). 2007. *Anthropology in the East: Founders of Indian Sociology and Anthropology*. New Delhi: Permanent Black.

Visvanathan, S. 2001. 'The Race for Caste. Prolegomena to the Durban Conference.' *Economic and Political Weekly*, 36 (330): 3123–3127.

Visvanathan, S. 2006. 'Official Hegemony and Contesting Pluralisms.' In Gustavo Lins Ribeiro and Arturo Escobar (eds) *World Anthropologies: Disciplinary Transformations Within Systems of Power*, pp. 239–258. London: Routledge.

Wilkerson, I.2020. *Caste: The Lies That Divide Us*. New York: Random House.

15

TOWARDS NEW ECOLOGIES OF METHOD

A Speculative Afterword

Sasheej Hegde

> There is not a single philosophical method, though there are
> indeed methods, different therapies as it were.
> (Wittgenstein [1953] 2009: 57e, #133d)

The difficulty in speaking cogently about a volume such as this one, which traverses the question of 'method' *within* different disciplines and across three domains of enquiry (namely, humanities, natural sciences, and social sciences), is bound to be a sharp and decisive one. This is further accentuated by the fact that each of the contributors is bringing a distinct set of questions to bear upon the orders of their elucidation. Without doubt, I have no intentions of working diligently through the spaces specific to each contribution – I have but waded summarily through them – nor will I be venturing to work off any specific formulation contained therein. Our editors' skilful order of narration (as epitomised by their expansive and lucid introduction) and the sequencing of the volume (as yielded through its partitioning) does set the terms for my reflection, even though (as part of my speculative thrust) I strive to break free of the constraints. I certainly try to think actively with the agenda – or agendum – underscoring the volume, as it modulates its focus on 'scientific method' in the context of specific disciplines, whether it be linguistics, literary studies, and philosophy (in the context of the humanities); mathematics, biology/genetics, chemistry and physics (across the natural sciences); and critical political studies, anthropology, economics and women's studies, social theory and historical epistemology, psychology and critical sociology (broadly, the social sciences per se).

To be sure, the editors' do not venture any unifying thrust – save some trenchant remarks against objectifications of 'scientific method' and/or against 'scientism' per se, as also against what is termed 'scientification' – while retaining the singularity of the focus on 'scientific method' *within* specific disciplines. A key term, 'silos', partakes of their movement across the

DOI: 10.4324/9781003298908-22

volume and the way in which the various contributions get to be structured. Much as I would have liked to dwell on this key term – both etymologically and pragmatically as a kind of device for structuring a conversation between and across disciplines – I shall desist from that move, and concentrate on the following orders of appraisal: one having to do with the 'disciplinary' locus of the volume; two, with the movement of contextualisation, as staged across internally by the essays and gestured at in the editorial introduction; and the, third, implicating more directly the question of method and 'scientificity' per se in a context where the lines to be drawn between the humanities, natural sciences, and social sciences are blurred. Indeed, in this last order of invocation particularly, I am pushing forward the agenda of my own research, even as I partake of the ambitions underscoring the current volume and the enthusiasms of its editors and contributors (all Indian scholars, incidentally, and situated in Indian institutions). The owl of Minerva may well return home at dusk, but can we be thinking through the dawn as well (or even mid-day, for that matter)?

In Flight: The 'Disciplinary' Locus and Beyond

Taking a measure of the limits of extant formulations within a 'discipline', as also the apparatus of distinctions germane to it, could yield the conditions for reinterpreting the developments internal to the discipline in question.[1] This, clearly, has been the guiding thought informing the volume and its contributors. And, yet, it seems to me that the problem (or, for that matter, any axis of critical discourse) cannot be reduced to that which makes it possible. There is the further question – in a word, what would be the condition of this condition? – which would have to be addressed. This is all the more since the imperative to posit new contexts, and to reconfigure our access to established contexts, has framed the manner in which the editors and our contributors have sought to traverse their respective disciplines and oversee its translation across spheres of method. Even more insistently, perhaps, we would have to take up for scrutiny a question that has formed the basis of meta-theoretical reflection across the disciplines, as it were, namely, the issue of cross-disciplinarity or even 'post-disciplinarity'. The editors certainly seem to gesture at this, when they observe that 'The essays in the volume attempt to find, and tell, the stories of method in their own disciplines, *stepping in and out of the disciplinary gates*' (emphasis added) – although the weight of the point is negated somewhat by their appended footnote to the effect that 'Exploring the method question in interdisciplinary domains would be another volume'. We will let that latter point slip by and concentrate on the prior admission.

Doubtless, as inherited academic forms of study and research methodology become more fluid and multiple, there is always an increased pressure to direct and constitute a more encompassing logic of (and for) enquiry.

Strangely enough, the form and status of that kind of overarching 'logic' has never been thoroughly examined, whether it be in terms of remonstrations against disciplinarity per se or adducing to 'cross/post-disciplinarity'. Indeed, to stick to our volume, the 'disciplinary narrations' effected here are not each adducing to a new kind of totality that would require another authoritative normative summation; nor is it, for that matter, a question of simply describing and cataloguing a diverse range of practices and methods within particular disciplines. Rather, I would like to think that, in urging 'to find, and tell, the stories of method in their own disciplines, *stepping in and out of the disciplinary gates*', the 'narrations' effected demonstrate something crucial: namely the point whether a purely disciplinary capacity (that is to say, a grounding in one's own discipline) could envisage alternative perspectives which by definition a disciplinary capacity cannot seemingly occupy.

I realise the point as just stated is somewhat vague, perhaps even whimsical, and I do not have the time or the space to clarify it.[2] I can stress, yet, that my own disciplinary locus has been sociology, although as one continually traverses the discipline and re-engages with its corpus, it has become even more impossible to define what a 'sociological' orientation consists in. Again, this confoundedness – or indeterminacy, if you will – cannot be limited to the contemporary situation which the discipline and its practitioners find themselves; in fact, it has been part and parcel of its trajectory through its founding moments in various national contexts. In transhistorical terms too, it is quite difficult to determine just what is internal to the discipline and what is external to it – often with its central figures drawing from within 'disciplinary gates' and from outside them as well.[3]

At any rate, as a thinking practitioner of the very craft of scientific enquiry across domains, one is quite exercised by a tendency to see 'interpretation' as a condition of *any* (or all) judgement about disciplines and their defining problems (or problematics). Yet, it seems to me that it is the very generality of this thesis about interpretation which makes it very suspect across the humanities and social sciences, including the natural sciences. Of course, I do not want to criticise this or that theory of interpretation, so much as to raise a doubt about the tendency to theorise 'judgement' as interpretation *tout court*.[4] Basically, we must confront the inner coherence (or otherwise) of this possibility – in fact, endemic to the humanities and social sciences howsoever conceived – as indeed of the wider point being urged here about 'disciplinarity' and disciplinary capacity. We must examine more intently what this would yield about both the genealogy of disciplines and their epistemology, and, not the least, the question of the contextualisation (at once, historical and epistemological) that could be productive of enquiry across 'mappings' of scientific method.

I will move now quickly into my second axis of appraisal, bearing particularly on this question of contextualisation. Hopefully, this will further ground our contentions about the disciplinary locus and beyond as sounded in the previous section.

Centres of Variation: Contextualisation and 'Practice Turn'

I would like to think that what the volume and its chapters are attempting, strictly, is forging a discussion about 'scientific method' across two levels (or, better still, centres of variation) – what we might call the *methodology of a discipline*' (or a specific disciplinary track within a discipline) which may also incorporate a philosophically interesting sense of 'method' as a form of choice (say) between competing theories, and a *meta-methodology*' (in terms of the philosophy of science that justifies the moves made within a discipline).[5] No one can doubt that a discipline – even one staged as an 'unstable compound' – is 'methodological' in the sense that it employs various procedures and heuristics as part of its research and/or investigation; it could also combine this with a focus on competing theories, while also expressing a preference for one over the other. The 'meta-methodological' level can accrue, as specified, in the context of assessing particular moves, asking frontally as to whether the 'methods' of a discipline are any good (as also about what is further required, or even possible, within a discipline). I wish I could illustrate this more concretely with the individual chapters in mind, although it must be conceded that the levels of contextualisation are never articulated as such.[6] All the same – and this is important – even as these two levels are straddled (with the complications, as disclosed), there is a further staging register that I discern across the space of the volume and its contributions, which needs to be addressed. Indeed, quite unlike the levels of contextualisation specified, this staging register is very much available and writ large across the work as a whole. Let me try and concretise this register, following which I will make a further critical comment bordering on the question of contextualisation and the genealogical study of disciplines and their method.

For my purposes, I see this staging register as incorporating three elements, each of which is disclosed in the epigraphs lacing the editors' introductory chapter. I will not reproduce the full citations here, but will work off the key phrases in the three quotes: one, 'the conventional and artifactual status of our forms of knowing' (from Steven Shapin and Simon Schaffer); two, 'European monopoly of the scientific method' (J.P.S. Uberoi); and, three, 'But at the same time, stereotypes are never idle' (Evelyn Fox Keller). Of course, I am not going to examine the specific contextuality attaching to these remarks, embed them in the work of the scholars named; they yet together, in their distinct emphases and turns, constitute broadly the staging register of our volume. More concretely, they encapsulate a commitment to address forms of scientific practice not in the abstract but in practice. As such, they accentuate what has been rendered in the wider literature devoted

318

to the philosophy, history, and social studies of science as the 'practice turn' (see Soler et al. 2014, for a detailed exploration of this trend).[7] I am also willing to concede a tension between the centres of variation identified (the 'methodology' and 'meta-methodology' distinction urged above) and the staging register as just epitomised. But importantly, far from undermining each other, the tension only accentuates the question of contextualisation.

In fact, I can push this last claim even further, for while the 'practice turn' is quite right to affirm that many philosophers are not concerned with what scientists actually do (rather working off scientists' rational reconstructions of what they actually did) and even approach science as interested in truth, theory, statements, or propositions about the world (often underestimating experimentation, intervention, and the technical transformation of the world), the crux is in determining the 'actual' practice and the challenge of theorising and of representing the world. I cannot quite see how this crux can be addressed without some recourse to the centres of variation as urged above. At any rate, this is a heavy question incorporating the dimensions of both history and epistemology as part of a composite exercise – without necessarily translating always into the continental exercise of 'historical epistemology' (for the latter, see Rheinberger 2010, 2013; Pena-Guzman 2020).

To return, then, to our staging register as epitomised by the three orders of remarks as highlighted above, there is more than a suggestion about the 'practice turn' underwriting them. What is being attempted is a 're-enactment' of that very turn in a renewed exercise of contextualisation, implicating at once specific registers of the history of disciplines within and across the humanities, natural sciences, and social sciences. The emphases writ large over the space of the individual contributions, each highlighting in their own specific ways 'the conventional and artifactual status of our forms of knowing' – as indeed the point that any (or all) 'stereotypes are never idle' in the practice of disciplines – and contesting the wider 'European monopoly of the scientific method' are all indicative of a politics of knowledge and location which grounds the overall work of the volume and its editorial anchoring/leveraging.[8] Obviously, such a contextualising manoeuvre is not without its perplexities – and more astute commentators would certainly pick on them – but I want to sound something else out which is in keeping with the thrust of my own order of work (as represented in Hegde 2014b).

All too often, we critical practitioners of the craft of 'method' have tried to make the contemporary organisation of academic knowledge answer to every circumstance, both internally and externally, in the process bypassing the methods (and meta-methods) internal to disciplines. Of course, we must all eschew a strong nominalism about the knowledge yielded by practices internal to disciplines, which also means coming to terms with what has been called – within the 'practice turn' as a whole and, arguably, also quite independent of it – 'constructivism'.[9] Indeed, where one would be predisposed to reify disciplines, to treat them as fixed and given, 'constructivism' seeks precisely to complicate the axis of such a framing. By introducing further agendas that themselves encounter previously introduced agendas – and thus 'over-determined' spheres of practice – constructivist strands often note with mock surprise that, within the evolving structures of disciplines, frameworks and commitments appear as at once forming and deforming of scientific practice. Consequently, I am always interested to push 'constructivism' less in the direction of an evaluation of the politics of knowledge than in the direction of the appropriate epistemological protocols that can (or should) govern enquiry. Allow me to conclude this line of prognosis by quoting more fully from my own previous work:

> The task cut out for criticism, either epistemological or sociopolitical, is to determine whether the 'politics of knowledge' theme is less a description of how the world is and more an image in which the world is being made. Indeed, the uneasy amalgam of constructivist language and essentialist argumentation – as also the peculiar positioning of academics as both analysts and protagonists of identity politics – which one find across the spaces of intellectual practice today is hardly a way out of these quandaries.[10]
>
> (Hegde 2014b: 72–73)

All the same, quite repeatedly – and programmatically, as well – I realise that our attempts to be 'realists' can fail, so either we need more realism about realism (that is, a more 'realistic' conception of what realism amounts to) or we need to acknowledge that realism is still too hard for us across the disciplines.[11] Either way the question is a complex one; and, in this light, let me shift to my final order of comment, implicating more directly the question of method and 'scientificity' per se in a milieu where the lines to be drawn between the humanities, natural sciences, and social sciences are blurred (one approximation, at my end, is Hegde 2018).

'New' Ecologies of Method: Final Flourish

I would like to think that my subtitle here – echoing a part of my main title – names a fundamental idea that the Weimar theorist Walter Benjamin draws out in the preface of his *The Origin of German Tragic Drama* (Benjamin [1977] 1985: 27–56).[12] The preface, as readers will recall of that challenging work, was cryptically titled 'Epistemo-Critical Prologue', where Benjamin sought to lay out the ground rules of his method, principally the idea that the practice of critique (in his case) must be attentive to both the work of art and its critical afterlife. As such, he sought to address the 'immanent tendency' of the work, which for him was the reflection that lies at its basis and is implanted in its form; and, in this wake, held out to the possibility of a

320

different practice of criticism, one which is both 'estimative' (not quite only of the single work in question but also of its relation to all other works) and 'inventive' (of the very idea of art). Benjamin was very clear that critique must disturb the identity of a work, but it must do so without transforming its immanent possibilities neither into necessities nor by dissolving them into contingencies. This latter idea does inform a lot of my procedures of enumeration here, in the context of the foregoing pages. But there is a passage in Benjamin's prologue that I particularly want to draw attention (and I shall cite it in full):

> But in the last analysis a methodology must not be presented in negative terms, as something determined by the simple fear of inadequacy on a factual level, a set of warnings. It must rather proceed from convictions of a higher order than are provided by the point of view of scientific verism. Such verism must then, in its treatment of the individual problem, necessarily be confronted by the genuine questions of methodology which are ignored in its scientific credo. The solution of these problems will generally lead to the reformulation of the whole mode of questioning along the following lines: how is the question, 'What was it really like?' susceptible, not just of being scientifically answered, but of actually being put.
>
> (Benjamin [1977] 1985: 41–42)

This passage – Benjamin's credo, really[13] – could be used to traverse extant debates in the philosophy and methodology of the social sciences over the years, although many of those debates arguably are passe today. My concern, yet, is with something else here, which has to do with a discursive impasse that Benjamin (as cited above) is reminding us about, particularly in relation to questions of methodology (as indeed meta-methodology). Even as he is urging that there is more to methodology than simply collecting 'facts' or faltering at that level, he is also telling us that the central question must concern less 'scientific verism' – the answer to the question 'What was it really like?' – than the very possibility of posing that question (or, quite plainly: How can we be both truthful to 'facts' and raise questions about the facticity of the facts so revealed?). Indeed, in this light, how can we have it both ways in questions of methodology and meta-methodology? Can we have it both ways? I suppose there can be no easy answers to these questions, and my allusion to a discursive impasse is a reference to the difficulties involved in answering these (and such other) questions.[14]

All the same, it must be admitted that the current 'dogma' of method – where research comes to be seen as one of providing a background of practice that explains and grounds cognition (or cognitive capacity) as such – comes naturally to us because it coheres with the thought that the point of any research can only lie in some describable state of affairs that is sought to be brought about. That is to say, the actions, events, and circumstances that research broadly feeds off (or is about) can be seen as intelligible only if they are portrayed as attempts to work on the world. All research is, on this view, a mode of production guided by a determinate description of a possible world at which it aims, and which remains (or obtains as) 'true' even if the object at which the researcher aims is nothing other than the occurrence of the object itself.[15] Of course, as transposed to the zone we are here occupying, namely, 'new' ecologies of method, the point necessarily is about 'constitution', taking this term to designate the way in which given measures of truth are embodied in our understanding and, accordingly, structure our experience of object-domains. In this broad sense, we are all phenomenologists, and the division structuring many discussions of method (or even meta-methodology) – the division between 'positivism' and 'phenomenology' – is a false one in cognitively operational terms (even as the division may be instantiated in historical terms). As such, this volume's investment in the 'scientific method' *within* disciplines and *across* domains of scientific enquiry translates into a genuine conviction that we must continually return to the frames that structure and shape our methodological and substantive positions and in the process yield a more compelling vision of our disciplinary practices.

I wish to press for more, however, considering that the older division between the 'natural sciences' and the 'human sciences' are no longer tenable in their commensurate terms. In other words, even as it would appear possible to discuss the 'human' sciences without reference to the 'natural' sciences (as well as the converse), there are specifically two points that would mitigate against the notion that the domain of the sciences, *whether humanistic or otherwise*, could be devoid of (shall we say) 'relationality'. For one, increasingly in recent times, topics or issues considered to be of interest to the human sciences (such as morality, politics, memory, the nature of rationality, aesthetics and value, the domain of desires, emotions and affects, and so on) have come to engage a whole new array of scientific disciplines (as constituted by the cognitive and neurosciences, primatology, evolutionary biology, genetic sciences and biomedicine, computer science, and even behavioural economics). The domain of the sciences as a whole has tended to converge, which has meant that they have to be rearticulated through a more active consideration of the creaturely and relational element that the disciplines as compounded elements now both presuppose and imply. Secondly, and summarily, inasmuch as the sciences have tended to converge, new lines may yet have to be drawn in reconstituting the grounds of both scientific research and even disciplinarity per se.[16]

Allow me, in this charged and convergent context, to conclude with an explicit thought about the 'new' (or, better still, 'the conditions of the new'). I realise there are many ways of posing the question of 'new', even as the new, as a condition that did not exist before, must always, and necessarily, be approached as a 'moving ratio'.[17] However, as Gilles Deleuze had sharply explained (in deference to Whitehead), 'the abstract does not explain, but must itself be explained; and *the aim is not to rediscover the eternal or the universal, but to find the conditions under which something new is produced (creativeness)*' (Deleuze [1987] 2007: vii, emphasis added). While, clearly, Deleuze was outlining an imperative for contemporary thought incorporating more actively both science and metaphysics (see Smith 2007, for absorbing discussion, placing in perspective the elements of Deleuze's engagement with the 'new'), I would be more inclined to push the direction of this thought into our ways of negotiating the work of disciplines, even striking a critical note about the standards of reflexivity being brought to bear on that undertaking.[18] Undoubtedly, as writ large across the space of this Afterword, our interest with the work of disciplines – as indeed their broader rubrics that we designate as 'humanities', 'natural sciences', and 'social sciences' – is less as the site where strains of given practices of knowledge have sought to query their foundations than as the theatre in which the structure of knowledge about a given domain and its relation to the epistemic practices configuring it can be staged as questions. Can we, then, have the 'new' ecologies of method that each and all of us – this volume, its editors and contributors (including myself), as also disciplinary practitioners and readers at large – have always been striving for but never quite approximating into primary determinations of the 'scientific method'? To recall our primary overture: the owl of Minerva may well return home at dusk, but can we be thinking through the dawn as well (or even mid-day, for that matter)! It has been an honour to be given the opportunity of this Afterword.

Coda

Surely, it would not have escaped anyone's attention that our approach to disciplines, and the frameworks of thought sounded out as bearing on them, is a totalising one. Such an operation is not without its speculative streaks and other associated perplexities. If theoretical abstraction is not something given in immediate experience, then it is pertinent to worry about the potential confusion of the concept – in our instance here, 'scientific method' – with the process that it designates, as also the possibility of taking the abstract 'representation' for the method as practised. In the long run, I guess, there is no warding off a thought and its representation from its idealistic recuperations. But surely intellectual work in the present demands that we strive to forestall the dangers of both conceptual reification and ethnocentricity. My acknowledgements follow.

Acknowledgements

I am thankful to the volume's editors, Gita Chadha and Renny Thomas, for the opportunity of staging this 'Afterword'. They wanted a 'Foreword', but one long and open conversation with them, as indeed my perusal through the volume, set in motion this speculative flourish. Hopefully, this exercise enhances the promise and scope of the volume. The usual disclaimers apply.

Notes

1   There is, of course, the question of the 'object' which, as I take it and as the question itself signifies, can and must constantly shift with the orders of study and reflection internal to a discipline. Doubtless, this can also be a historical order of appraisal. In my own space of work – as epitomised in Hegde (2014b) – I have tried to forward an idea of disciplines as 'unstable compounds', drawing on the cultural historian Collini (2001) for the purpose.

2   For an order of elaboration, the reader is directed to Hegde (2018), although Hegde (2014b) also translates into the point (a principle, really, that has formed one of the bases of my engagements with the structure and dynamics of disciplines). But note also the clarification that follows in my text above.

3   In strict terms, this is not quite unique to sociology per se; it can be read across the various disciplines that constitute the humanities and the social sciences, whether in India or elsewhere. I would like to think that this is quite true of the various sciences as well, across even the diverse subfields and specialisations that have come to constitute the spaces of scientific knowledge overall. The respected historian of science, Steven Shapin, has an absorbing essay (Shapin 1992), which I keep going back to often.

4   Again, this is not a claim to 'authority' or mastery, but see my Hegde (2021; as also 2014a).The brevity of my formulations here obviously requires this further referencing, I am afraid I will be calling attention to my 2014a piece later too.

5   I realise the distinction, as represented herein, is somewhat elusive, and even arbitrarily formulated. Singularly, as rendered, the *meta-methodological* level suggests a higher-level contextualisation than the one on offer at the level of the *methodology of a discipline*', which can induce a further problem about how that discussion is to be staged. For one, if the meta-methodology of a discipline (or a track within the discipline) employs the structure and type of reasoning as the methodology it is meant to justify, one is obviously caught in a circle. Alternatively, if one's meta-methodology is quite set apart and distinct, then it itself will demand its own justification and a regress can ensue. See also Nola and Sankey (2007), who use this framing to anchor a fairly comprehensive discussion about 'scientific method'. I am not entirely sure, however, about whether they are open to all the complexities and attendant perplexities of their framing. It is in this spirit that I proceed ahead in my text above.

6   Arguably, there is an unevenness across the space of the volume, with the part constituted as 'Natural sciences' going somewhat easy on the 'meta-methodological' level and the parts represented as 'Humanities' and 'Social Sciences' going heavier on this level. Having said so, I realise that such a labelling is contestable from within the contributions themselves, that indeed the chapters are each developing this general structure of contextualisation in their own specific ways. The distinction, carefully articulated and more importantly demonstrated across articulations of 'scientific method' in the various disciplines, could still be helpful in clarifying the specific 'problems' that animate the practices of disciplines. The choices operable at the 'methodological' level need not obtain at the meta-methodological level, even as the latter may yet be indecisive at the

adjudicatory level. All the same, the axis of this complication is best addressed contextually on a case-by-case basis. See also Dicken (2018).

7 As with all 'turns', it is difficult to name the moment when the 'turn' in question – in our instance, the turn to practice in the study of science – took place (as also what came before to warrant the 'turn'). Personally though, I am agnostic about all proclamations of 'turn', even as, in the spirit urged by our prefatory remarks, I am quite willing to take note of them. The Soler et al. (2014) volume cited above broadly suggests that the practice turn can be traced back to Thomas Kuhn, although importantly Hacking (1983) is often presented as the work that led to the 'practice turn'. We need not mediate these claims here, although it is important to note that most philosophers of science today have been influenced by the general post-Kuhnian commitment to the history of science and a focus on the special sciences.

8 In an open conversation with the editors', both of them highlighted the 'positionality' underlying the volume, drawing as much from feminism as from post-colonial and critical theory, and underscoring the wider politics of knowledge in which their own individual work is heavily invested. The critical order of my remarks that follow in the text is certainly meant to dissolve the orders of this commitment, although it does complicate the exercise.

9 To be sure, in the context of the 'practice turn' in the philosophy, history, and social studies of science, the label 'constructivism' broadly designates an orientation towards practice-oriented explanations, where the explanandum is, as it always was, the agreement between theory and reality (see, particularly, the Editors' introduction to Soler et al. (2014), esp. pp.1–44; as also the contribution of Lynch (2014) in the same volume). There is, yet, an ambiguity here, for even as it is noted that our familiar notions of 'representation', 'explanation', 'lawfulness', and the like need to be reconceived in light of 'practice', it is still unclear whether the effort is to 'fit' these notions into a conception of science as the pursuit of theory and reality, or whether the effort also is to reconceive the very idea of scientific knowledge production? Indeed, this predicament is reflected in the claims underscoring the turns of phrase highlighted from Shapin and Schaffer and Keller, respectively, especially when one attempts to run them together as part of the 'practice turn'. But, of course, there is more to the work of these scholars than what I am deducing here, and strictly they are not reducible to the 'practice turn'. Our point can still stand, though.

10 Explicitly, in the context of this volume and its probing, one may extend this point to the epistemological undertakings of 'standpoint theory' as well. For more on the point, see Wolfe (2000: 167–175) among others. Such an order of appraisal also implicates the oddly central status of the figure of the observer in theories of this sort; the current orders of my teaching and research tries to engage this ground more fully and programmatically. Note also the observation that follows in our text above.

11 But, of course, for a different line of prognosis along these lines – without necessarily implicating the ways as just urged – see Sankey (2017).

12 I realise that there is a more reliable and authoritative translation now available for this work, but for our purposes here, I continue with the 1985 Verso edition.

13 For, as Benjamin has further observed, with specific reference to the coherence provided by the realm of knowledge: 'Knowledge is possession. Its very object is determined by the fact that it must be taken possession of – even in a transcendental sense – in the consciousness. The quality of possession remains. For the thing possessed, representation is secondary; it does not have prior existence as something representing itself ... For knowledge, method is a way of acquiring its object – even by creating it in the consciousness' (Benjamin [1977] 1985: 29–30).

The gnomic content of these lines notwithstanding, they instantiate a form of 'constructivism', which consists, not in denying that reality, an external world of reference, is obtaining and/or that 'reality', is subservient to the framework which organises it; rather, that in the effort to define a method adequate to its object, the object is being taken 'possession' of from the point of view of the way in which it can be known rather than the ways in which it exists. The cognition which occurs here is of an order we can term 'epistemological', albeit constrained by a tendency internal to modern intellectual culture – namely, the tendency to think out the question of what something *is* in terms of how it is *known*. I am afraid I will have to leave this here, even as it adds a dimension to our point about 'realism' concluding the second section above.

14 Be that as it may, it must still be emphasised that the centres of variation as disclosed through the levels of the 'methodology of a discipline' and its 'meta-methodology' can be constituted as much by an insistence on the primacy of practice as by an attention to the 'effects' of truth as it were. One may, in the latter light specifically, speak of 'practices of truth', but my mind is somewhat fuzzy here. Surely, the intention cannot be to institute a new binary between claims to the primacy of practice and claims to the primacy of truth. See also the redoubtable Williams (2002), which opens with the problem of 'truthfulness and truth' (or, rather, the 'tension between the pursuit of truthfulness and the doubt that there is (really) any truth to be found' [ibid.: 2]). Note also the submissions that follow in my text above.

15 Indeed, this may well mark a 'performative' appreciation of Benjamin's credo – of the kind which says, with specific reference to research methods in the social sciences, that the latter do not simply describe the world as it is, that they also 'enact' it (see, among others, Law and Urry 2004; as also Denzin 2001).I have engaged this ground more fully in Hegde (2014a), and therefore will not elaborate (although some of my overtures in this subsection do bear the brunt of this piece – incidentally, I am particularly fond of this piece!).The problems get complicated because the 'performative' stance is not limited only to the social sciences per se, but straddles very many domains of scientific enquiry across the disciplines. For some compelling thoughts that bear on this last point, see Kukla (2008).

16 The caveat 'whether humanistic or otherwise' figuring in the early part of this paragraph is important. I have stressed on the role of this caveat elsewhere – see Hegde (2018) – which also redesigns the 'two cultures' debate fostered by C.P. Snow as naming a phenomenon that has to do with contradictory dispositions internal to the practice of enquiry (whether humanistic or otherwise). The relevant references can all be sourced therein; likewise for our second admission summarily postulated here, although see also Hacking (1995). All this, of course, is not to imply that the distinctiveness of the humanities/social sciences cannot be rearticulated, which in fact my 2018 piece tries to do reflectively and constructively.

17 I am drawing the phrase 'moving ratio' from Paul Rabinow (2008), who uses it to designate the 'contemporary' as something that can be both equated with and differentiated from the 'modern'. As he formulates, 'Just as one can take up the "modern" as an ethos and not a period, one can take it up as a moving ratio', while going to maintain: '*The contemporary is a moving ratio of modernity, moving through the recent past and near future in a (nonlinear) space that gauges modernity as an ethos already becoming historical*' (Rabinow 2008: 2, as in original). I am not entirely sure whether current strands of critical theorising in India – as epitomised in our volume, as also Banerjee et al. (2016) – have approximated to this absorbing formulation.

18 One is certainly not implying that the current arrangements of disciplinarity do not leave a lot to be desired. But this is a project on which me must embark with great care. My 'Concluding Postscript' to Hegde (2014b) has some historical thoughts bearing on this, although it is also implicit to the foregoing order of my remarks here. In more recent times, though, I have found it more meaningful and productive to graft this on to the nomenclature of 'exact' and 'conjectural' sciences. Doubtless, this nomenclature has a shady and shifting dynamics in the 20th-century thought as a whole, and requires to be augmented with the fuller recognition of the cascading effects of computing technologies in our own times. I am afraid I cannot disclose more for reasons of space, but those interested could look at, among others, Elmer (2000) and Stengers (2005). Note, yet, the further claim about the 'new' that we go on to make in our text above.

References

Banerjee, Prathama, Nigam, Aditya, and Pandey, Rakesh. 2016. 'The Work of Theory: Thinking Across Traditions.' *Economic and Political Weekly*, Vol. 51 (No. 37), pp. 42–50.

Benjamin, Walter. [1977]1985. *The Origin of German Tragic Drama*. London: Verso.

Collini, Stefan. 2001. 'Postscript: Disciplines, Canons, and Publics: The History of 'The History of Political Thought' in Comparative Perspective.' In D. Castiglione and Iain Hampsher-Monk (eds) *The History of Political Thought in National Context*. Cambridge: Cambridge University Press, pp. 280–302.

Deleuze, Gilles. [1987]2007. 'Preface to the English Language Edition.' In Gilles Deleuze and Claire Parnet (eds) *Dialogues II*. New York: Columbia University Press, Revised Edition, pp. vii–x.

Denzin, Norman K. 2001. 'The Reflexive Interview and Performative Social Science.' *Qualitative Research*, Vol. 1 (No. 1), pp. 23–46.

Dicken, Paul. 2018. *Getting Science Wrong: Why the Philosophy of Science Matters*. London: Bloomsbury.

Elmer, Jonathan. 2000. 'Blinded Me With Science: Motifs of Observation and Temporality in Lacan and Luhmann.' In William Rasch and Cary Wolfe (eds) *Observing Complexity: Systems Theory and Postmodernity*. Minneapolis, MN: University of Minnesota Press, pp. 215–246.

Hacking, Ian. 1983. *Representing and Intervening: Introductory Topics in the Philosophy of Natural Science*. Cambridge: Cambridge University Press.

Hacking, Ian. 1995. 'The Looping Effect of Human Kinds.' In Dan Sperber, David Premack, and Ann James Premack (eds) *Causal Cognition: A Multidisciplinary Debate*. Oxford: Clarendon Press, pp. 351–383.

Hegde, Sasheej. 2014a. 'A Measure of Truth: Proposals for a Method-Centred Research Pedagogy.' *Economic and Political Weekly*, Vol. 49 (No. 5), pp. 63–68.

Hegde, Sasheej. 2014b. *Recontextualizing Disciplines: Three Lectures on Method*. Rashtrapati Nivas, Shimla: Indian Institute of Advanced Study.

Hegde, Sasheej. 2018. 'What Humanities/Social Sciences Can Mean: Transmuting the 'Two Cultures' Idea.' In Mrinal Miri (ed.) *The Place of Humanities in Our Universities*. London: Routledge, pp. 176–188.

Hegde, Sasheej. 2021. 'Disciplinary Cross-Currents: Rebounding on the Heuristics of Inquiry.' *Sociological Bulletin*, Vol. 70 (No. 2), pp. 259–263.

Kukla, Rebecca. 2008. 'Naturalizing Objectivity.' *Perspectives on Science*, Vol. 16 (No. 3), pp. 285–302.

Law, John, and Urry, John. 2004. 'Enacting the Social.' *Economy and Society*, Vol. 33 (No. 3), pp. 390–410.

Lynch, Michael. 2014. 'From Normative to Descriptive and Back: Science and Technology Studies and the Practice Turn.' In Lena Soler, Sjoerd Zwart, Michael Lynch, and Vincent Israel-Jost (eds) *Science After the Practice Turn in the Philosophy, History, and Social Studies of Science*. New York: Routledge, pp. 93–113.

Nola, Robert, and Sankey, Howard. 2007. *Theories of Scientific Method: An Introduction*. Stockfield: Acumen Publishing.

Pena-Guzman, David M. 2020. 'French Historical Epistemology: Discourse, Concepts, and the Norms of Rationality.' *Studies in History and Philosophy of Science*, Vol. 79, pp. 68–76.

Rabinow, Paul. 2008. *Marking Time: On the Anthropology of the Contemporary*. Princeton, NJ: Princeton University Press.

Rheinberger, Hans-Jorg. 2010. *On Historicizing Epistemology: An Essay*. Stanford, CA: Stanford University Press.

Rheinberger, Hans-Jorg. 2013. 'My Road to History of Science.' *Science in Context*, Vol. 26 (No. 4), pp. 639–648.

Sankey, Howard. 2017. 'Realism, Progress and the Historical Turn.' *Foundations of Science*, Vol. 22 (No. 1), pp. 201–214.

Shapin, Steven. 1992. 'Discipline and Bounding: The History and Sociology of Science as Seen Through the Externalism-Internalism Debate.' *History of Science*, Vol. 30 (No. 4), pp. 333–369.

Soler, Lena, Zwart, Sjoerd, Lynch, Michael, and Israel-Jost, Vincent (eds). 2014. *Science After the Practice Turn in the Philosophy, History, and Social Studies of Science*. New York: Routledge.

Smith, Daniel W. 2007. 'The Conditions of the New.' *Deleuze Studies*, Vol. 1 (No. 1), pp. 1–22.

Stengers, Isabelle. 2005. 'Introductory Notes on an Ecology of Practices.' *Cultural Studies Review*, Vol. 11 (No. 1), pp. 181–196.

Williams, Bernard. 2002. *Truth and Truthfulness: An Essay in Genealogy*. Princeton, NJ: Princeton University Press.

Wittgenstein, Ludwig. [1953]2009. *Philosophical Investigations*. Oxford: Wiley-Blackwell, 4th Edition.

Wolfe, Cary. 2000. 'In Search of Posthumanist Theory: The Second-Order Cybernetics of Maturana and Varela.' In William Rasch and Cary Wolfe (eds) *Observing Complexity: Systems Theory and Postmodernity*. Minneapolis, MN: University of Minnesota Press, pp. 163–197.

INDEX

Note: **Bold** page references indicate tables. Page numbers followed by "n" refer to notes.

For Product Safety Concerns and Information please contact our EU
representative GPSR@taylorandfrancis.com
Taylor & Francis Verlag GmbH, Kaufingerstraße 24, 80331 München, Germany

www.ingramcontent.com/pod-product-compliance
Lightning Source LLC
Chambersburg PA
CBHW052118230326
41598CB00080B/3831